Charles Kingsley

**The Boys and Girls Book of Science**

Charles Kingsley

**The Boys and Girls Book of Science**

ISBN/EAN: 9783337034580

Printed in Europe, USA, Canada, Australia, Japan

Cover: Foto ©berggeist007 / pixelio.de

More available books at **www.hansebooks.com**

# THE BOYS' AND GIRLS' BOOK OF SCIENCE.

JAMES WATT INVENTING THE STEAM-ENGINE.

WITH NUMEROUS ILLUSTRATIONS

STRAHAN AND COMPANY LIMITED
34 PATERNOSTER ROW LONDON
1881

## ADDRESS TO BOYS AND GIRLS.

MY DEAR BOYS AND GIRLS,—When I was your age there were no such children's books as there are now. Those which we had were few and dull, and the pictures in them ugly and mean, while you have your choice of books without number, clever, amusing, and pretty, as well as really instructive, on subjects which were only talked of fifty years ago by a few learned men, and very little understood even by them. So if mere reading of books would make wise men, you ought to grow up much wiser than us old fellows. But mere reading of wise books will not make you wise men; you must use for yourselves the tools with which books are made wise, and that is—your eyes and ears and common sense.

Now, among those very stupid old-fashioned boys' books was one which taught me that, and therefore I am more grateful to it than if it had been as full of wonderful pictures as all the natural history books you ever saw. Its name was "Evenings at Home," and in it was a story called "Eyes and No Eyes," a regular old-fashioned, prim, sententious story; and it began thus:—

"Well, Robert, where have you been walking this afternoon?" said Mr. Andrews to one of his pupils at the close of a holiday.

Oh, Robert had been to Broom Heath and round by Camp Mount, and home through the meadows. But it was very dull. He hardly saw a single person. He had much rather have gone by the turnpike road.

Presently in comes Master William, the other pupil, dressed, I suppose, as wretched boys used to be dressed forty years ago, in a frill collar and skeleton monkey-jacket, and tight trousers buttoned over it and hardly coming down to his ankles, and low shoes which always came off in sticky ground; and terribly dirty and wet he is; but he never, he says, had such a pleasant walk in his life, and he has brought home his handkerchief (for boys had no pockets in those days much bigger than keyholes) full of curiosities.

He has got a piece of mistletoe, and wants to know what it is; and he has seen a woodpecker and a wheat-ear, and gathered strange flowers on the heath, and hunted a peewit because he thought its wing was broken, till, of course, it led him into a bog, and very wet he got. But he did not mind it, because he fell in with an old man cutting turf, who told him all about turf-cutting and gave him a dead adder. And then he went up a hill and saw a grand prospect, and wanted to go again and make out the geography of the country from Cary's old county maps, which were the only maps in those days. And then, because the hill was called Camp Mount, he looked for a Roman camp, and found one; and then he went down to the river and saw twenty things more, and so on, and so on, till he had brought home curiosities enough and thoughts enough to last him a week.

Whereon Mr. Andrews, who seems to have been a very sensible old gentleman, tells him all about his curiosities. And then it comes out—if you will believe it—that Master William has been over the very same ground as Master Robert, who saw nothing at all!

Whereon Mr. Andrews says, wisely enough, in his solemn, old-fashioned way:—

"So it is. One man walks through the world with his eyes open, another with his eyes shut; and upon this difference depends all the superiority of knowledge which one man acquires over another. I have known sailors who had been in all the quarters of the world, and could tell you nothing but the signs of the tippling-houses and the price and quality of the liquor. On the other hand, Franklin could not cross the Channel without making observations useful to mankind. While many a vacant thoughtless youth is whirled through Europe without gaining a single idea worth crossing the street for, the observing eye and inquiring mind find matter of improvement and delight in every ramble. You then, William, continue to use your eyes. And you, Robert, learn that eyes were given to you to use."

So said Mr. Andrews; and so I say to you. Therefore I beg all good boys and girls among you to think over this story, and settle in their own minds whether they will be Eyes or No Eyes; whether they will, as they grow up, look, and see for themselves what happens; or whether they will let other people look for them, or pretend to look, and dupe them, and lead them about—the blind leading the blind, till both fall into the ditch.

I say "good boys and girls;" not merely clever boys and girls, or prudent boys and girls; because using your eyes or not using them is a question of doing right, or doing wrong. God has given you eyes, and it is your duty to God to use them. If your parents tried to teach you your lessons in the most agreeable way, by beautiful picture-books, would it not be ungracious, ungrateful, and altogether naughty and wrong, to shut your eyes to those pictures, and refuse to learn? And is it not altogether naughty and wrong to refuse to learn from your Father in heaven, the Great God who made all things, when He offers to teach you all day long by the most beautiful and most wonderful of all picture-books, which is, simply all things which you can see, and hear, and touch, from the suns and stars above your heads, to the mosses and insects at your feet? It is your duty to learn His lessons, and it is your interest likewise. God's Book, which is the Universe, and the reading of God's Book, which is Science, can do you nothing but good, and teach you nothing but truth and wisdom. God did not put this wondrous world about your young souls to tempt or to mislead them. If you ask Him for a fish, He will not give you a serpent. If you ask Him for bread, He will not give you a stone.

So use your eyes and your intellect, your senses and your brains, and learn what God is trying to teach you continually by them. I do not mean that you must stop there, and learn nothing more: anything but that. There are things which neither your senses nor your brains can tell you; and they are not only more glorious, but actually more true and more real, than many things which you can see or touch. But you must begin at the beginning in order to end at the end; and sow the seed if you wish to gather the fruit. God has ordained that you, and every child which comes into the world, should begin by learning something of the world about him by his senses and his brain; and the better you learn what they can teach you, the more fit will you be to learn what they cannot teach you. The more you try now to understand *things*, the more you will be able hereafter to understand men, and That which is above men. You begin to find out that truly Divine mystery, that you had a mother on earth, simply by lying soft and warm upon her bosom; and so (as our Lord told the Jews of old) it is by watching the common natural things around you, and considering the lilies of the field, how they grow, that you will begin at least to learn that far Diviner mystery—that you have a Father in heaven. And so you will be delivered (if you will) out of the tyranny of darkness, and distrust, and fear, into God's free kingdom of light, and faith, and love; and will be safe from the venom of that tree which is more deadly than the fabled

Upas of the East. Who planted that tree I know not, it was planted so long ago; but surely it was none of God's planting, neither of the Son of God: yet it grows in all lands, and in all climes, and sends its hidden suckers far and wide—even (unless we be watchful) into your hearts and mine. And its name is the Tree of Unreason, whose roots are conceit and ignorance, and its juices folly and death. It drops its venom into the finest brains, and makes them call sense nonsense, and nonsense sense; fact fiction, and fiction fact. It drops its venom into the tenderest hearts, alas! and makes them call wrong right, and right wrong; love cruelty, and cruelty love. Some say that the axe is laid to the root of it just now, and that it is already tottering to its fall; while others say that it is growing stronger than ever, and ready to spread its upas-shade over the whole earth. For my part, I know not, save that all shall be as God wills. The tree has been cut down already, again and again, and yet has always thrown out fresh shoots, and dropped fresh poison from its boughs. But this at least I know, that any little child who will use the faculties which God has given him, may find an antidote to all its poison in the meanest herb beneath his feet.

There—you do not understand me, my boys and girls; and the best prayer I can offer for you is, perhaps, that you should never need to understand me; but if that sore need should come, and that poison should begin to spread its mist over your brains and hearts, then you will be proof against it, just in proportion as you have used the eyes and the common sense which God has given you, and have considered the lilies of the field, how they grow.

CHARLES KINGSLEY.

# CONTENTS.

| | Page |
|---|---|
| ABOUT A FLY ... | 1 |
|    With 7 *Illustrations.* | |
| EARTHQUAKES ... | 14 |
|    With 1 *Illustration.* | |
| FRUIT ... | 26 |
|    With 3 *Illustrations.* | |
| WASPS AND PAPER-MAKING ... | 31 |
| BESSIE'S CALENDAR ... | 44 |
|    With 1 *Illustration.* | |
| SILK AND SILKWORMS ... | 48 |
|    With 1 *Illustration.* | |
| WHAT HAPPENS IN GARDENS ... | 59 |
|    With 1 *Illustration.* | |
| A LUMP OF COAL ... | 65 |
|    With 2 *Illustrations.* | |
| THE SPIDER AND ITS WEBS ... | 72 |
|    With 1 *Illustration.* | |
| VOLCANOES ... | 81 |
|    With 3 *Illustrations.* | |
| BEAVERS AND BUILDING ... | 91 |
|    With 1 *Illustration.* | |
| LIGHT ... | 105 |
|    With 3 *Illustrations.* | |
| AMONG THE BUTTERFLIES ... | 118 |
|    With 13 *Illustrations.* | |
| POND LIFE ... | 185 |
|    With 1 *Illustration.* | |
| INSECT PETS ... | 195 |
|    With 16 *Illustrations.* | |
| BEES AND BEEHIVES ... | 231 |
|    With 12 *Illustrations.* | |
| BUTTERCUPS AND DAISIES ... | 245 |

## CONTENTS.

| | Page |
|---|---|
| ANTS AND ANT-HILLS ... ... ... ... ... ... ... | 257 |
| *With 1 Illustration.* | |
| ROOKS AND THEIR RELATIONS ... ... ... ... ... ... | 270 |
| *With 1 Illustration.* | |
| THE LUNAR HALO ... ... ... ... ... ... ... ... | 281 |
| *With 1 Illustration.* | |
| COALS AND COLLIERS ... ... ... ... ... ... ... | 287 |
| *With 1 Illustration.* | |
| WILD FLOWERS ... ... ... ... ... ... ... | 295 |
| *With 1 Illustration.* | |
| BEES AND THEIR BUSINESS ... ... ... ... ... ... ... | 302 |
| *With 2 Illustrations.* | |
| STEAM AND THE STEAM-ENGINE ... ... ... ... ... ... | 309 |
| *With 1 Illustration.* | |
| OUR IRONCLADS ... ... ... ... ... ... ... | 319 |
| *With 1 Illustration.* | |
| BULBS ... ... ... ... ... ... ... ... ... | 328 |
| *With 2 Illustrations.* | |
| CATS, DOGS, AND OTHER HOUSEHOLD PETS ... ... ... | 333 |
| *With 1 Illustration.* | |
| SUMMER GRASS ... ... ... ... ... ... ... ... | 340 |
| *With 1 Illustration.* | |
| ABOUT A CATERPILLAR ... ... ... ... ... ... | 345 |
| *With 1 Illustration.* | |
| THE ACANTHUS ... ... ... ... ... ... | 351 |
| *With 1 Illustration.* | |
| THE MICROSCOPE ... ... ... ... ... ... ... ... | 353 |
| WATERY WASTES ... ... ... ... ... ... ... | 360 |
| *With 1 Illustration.* | |
| ANIMAL DEFENCES ... ... ... ... ... ... ... | 366 |
| *With 1 Illustration.* | |

## ABOUT A FLY.

*"Little things on little wings*
*Bear little souls to heaven."*

AND this is more than can be said of many of the greatest, rather let us say the largest of things; for greatness—true, genuine greatness—is not a thing of bulk alone, save as its elasticity can be alike compressed into the microscopic space of the quasi-invisible, or can fill a space which our minds can never take in. So also do we find the largest amount of utility in the smallest of creatures, while the hugest of beings leaves us in admiring doubt as to the amount of benefit conferred by it on the creation which it seems to crown if not to adorn.

But what shall we say of the Fly, poor little thing? What use is it of in that "economy of nature," whereof philosophy boasts the power of explaining the hidden workings and intricate machinery? To what extent does the buzzing creature, so ruthlessly poisoned, so suddenly dashed to the ground, preserve that "balance of being," which if once interrupted, as in the case of the eagles and the falcons, avenges itself; as by the multiplication of the wild pigeons, rats, and other devourers of the farmer's produce?

To crawl on the ceiling; to buzz about our heads; to torment us by their numbers, and by their very littleness; to be guarded against by every possible manœuvre, avoided as pests, slaughtered by hundreds, or poisoned by the thousand, such is their daily lot; yet, when an incautious straggler visits your breakfast-table, and, throwing itself on your chivalrous hospitality for a sip from the cream-bowl, falls a victim to its temerity, and you see it vainly struggling to escape from the white ocean of entanglement, how carefully, how tenderly do you lift it thence, lay it in the sunshine, and watch its heavy limbs regaining their vitality! how do you rejoice in its escape, and triumph in its flight, though so soon to return, ungrateful wretch! to torment by its vagaries the fingers that erst so daintily spared its life; to haunt your gilded cornices and picture-frames; and, finally, to fall unpitied, unlamented, a victim to the sugared snares which have proved fatal to the myriads of his brethren. Such is the episode in the life of many a fly, bringing it into contact with the tender humanities and inconsistent ruthlessness of mankind, and such the little wings that waken, or for the moment call into action, the dormant charities of your soul. May these hereafter find a larger scope, worthier objects, and less interrupted success!

Help ever the helpless, be it a drowning fly, or a brother floundering through the difficulties of life's first tasks; and down the long vista of life I see you, with little wings and slender strength, it may be, for it needs not vastness of resources or extent of power to minister such heart-help as the true-hearted can render,—I see you the friend of the friendless, it may be of the ungrateful and ungracious; the raiser of the fallen, though, perchance, only perversely to fall again; the cheerer of the cheerless, though it may be they droop again when your bright presence has passed away: but evermore, a little thing, perhaps, in all that men deem great, on little wings of small assistance, bearing little souls—and ever on that track yourself—heavenward, upward, and may it be to heaven itself, where true greatness and true littleness are mirrored in their real proportions. Yes, draw them upward, win them from the slough of despond; bring them, as best you may, to the sunshine of better days, and see to it that heart and strength revive for efforts in the right direction, which to men, as to flies, are the only safeguards against the poisoned footfalls that await the weak and unwary.

And what the return? Perhaps nought. In helping your helpless fellow-man, woman, or child, you may meet with that rarest of graceful things, a grateful heart, while yet expecting not; you act out the energies of a loving nature, as in saving the fly you simply gratify a benevolent instinct; for, in truth, no one *loves* a fly, nor may we suppose that, speaking in the language of men, and especially of women, that the fly has a *heart* at all. In aërial nature they are the *million*, the great untaught, unwashed, democratic, self-willed; destitute of the constructive, imitative, or ornamental faculties for which other winged things are noted. They congregate in myriads, when they please, *where* they please; or, if they please, they walk alone, defying our centres of gravity by a promenade feet upwards on the ceiling, or up and down the glittering surface of the window-pane, pursuing objects invisible to us, following behests unknown to all besides, and apparently responsible to no authority for the fulfilment of any of their duties in life.

To torment in various modes and degrees, from the gentle titillation of the bluebottle, who presumes to take his exercise on your cheek, to the maddening irritation caused by the fatal bite of the tropical Tzetze, appears to be a law of their nature: they seem to be creatures eminently unwelcome, and of whom in general a good riddance is the *summum bonum* of our wishes.

They are, in fact, very *negative* characters, and, like many human beings of whom we are apt to think that their room is better to have than their company, we have little idea what they save us from,—these *city* Arabs of the air, street-sweepers of the invisible atmosphere, scavengers of the ethereal poisons.

*Cause* and *effect* are so blundered over in all our hasty notions about things in general, that we can be in no way surprised if a strange jumble should often exist also in our judgments as to *cause* and *cure*, and if we should sometimes blame for the very evils he removes the unfortunate insect whose advent portends corruption, and especially our old friend the bluebottle, if in his efforts to withdraw from use that food which is becoming unfit for man, he calls to his aid the fecundity of his nature, and takes possession of our viands after a fashion peculiar to himself, and which we decline to dispute with him. The evil so abhorrent to our feelings can only be laid at his door in a sense which involves its speedy removal, and the conferring on ourselves a very real though a negative blessing.

Treated much in the same manner as the pariah dogs of the East, these and other flies are yet in their myriad-winged activities no less the scavengers of the air than those despised quadrupeds are of the earth; and happy may it be for ourselves that we see not the reverse of the picture at which we rail, nor find ourselves in some sad hour, when the spider proves too many for the fly, surrounded by the odours, the miasma, and the pestilences from which the presence of the fly relieves us.

We have heard of vast clouds of flies attending on the course taken by certain epidemics; and hence, say some, the flies have brought the fever, the cholera, &c., &c. Nay, truly, camp-followers they are on these fierce invaders, not predatory hordes themselves, gathering out from the air all that could offend, relieving the earth from many of the death-inviting miasmas which haunt its surface and prowl around the dwellings of careless men, inviting evermore the exterminating breath of the invader, and hatching a camp for the dwelling of disease.

Though to make friends with a fly were a vain attempt, we may yet bring it to terms of very close acquaintance, and, under the friendly light of the microscope, discover that it is, in its anatomy, one of the most interesting objects of scientific research, and in such close analogy to the larger animals, that a similarity, amounting almost to identity, has been discovered between the foot of the fly and the hoof of the rhinoceros. Adapted, doubtless, both these members are to the behests of their owners, though so different: the one being suited for the slow terrestrial locomotion of a heavy beast, the other to rambles in an inverted position on the ceiling above our heads, where, under the the aspect of antics and evolutions of the oddest description, it is engaged in the consumption and carrying away of many a fly-load of accretions on those surfaces which have met the ascending vapours of many noxious things.

*Musca vomitoria* (Blow Fly).—Natural size.

The scavenger life of a fly, in that phase of its existence, is yet a very short, though by us the most observed, period of its existence. We call them torments, crush them, poison them, set traps for them, and exult in their destruction, and then save the stragglers one by one, just when, in old age, they come to claim our hospitality, and a shelter from the chill winds of autumn, having deposited their eggs in the old crannies whence erst they crawled themselves, and spent their early days among the fruit and flowers, or (if they were bluebottles) among the animal decays which nourished their youth. The changes they have undergone are not less curious than those which attend the more interesting creatures whom we watch, as caterpillars, cocoons, or moths,—yet unobserved, save by the most ardent students of nature, let them, say we, remain; over the early life of a fly let us draw a veil. A busy, active, useful life it is, like that of all scavengers, from the pariah dog to the crossing-sweeper, and with much more to laugh and grow fat upon than the latter, poor fellow, ever enjoys; yet its diet is uninviting, its form repulsive; we shudder to think of it, we name it not. Like some of the more despised, yet useful, members of the human family, they far outnumber the more attractive individuals, and that they are designed to fulfil no insignificant part in nature may be gathered from the fact that from one mother not less than twenty thousand living sons and daughters are known to have descended.

Between the early life of the fly in this caterpillar form and its subsequent aërial existence there seems to be less connection than between the earlier and later life of any other insect. There is no point of resemblance either in outward form or in inward structure between the white, soft, pulpy, eyeless and legless body which has twisted and wriggled through its blind career, and the restless, buzzing tormentor which dances merrily in the sunshine. The one does not seem to grow out of the other, but to be suddenly transformed, when the little creature leaps forth to its new life. Its white skin becomes dark-coloured and hard as soon as it has

Larva of *Musca domestica*.
Magnified one-half.

*Musca domestica* (House Fly).
Magnified one-half.

reached its full size, and forms a case, within which the unseen change goes rapidly on. If we counted the rings of which the wriggling larva is composed, we should find that they are always seventeen in number. Now these rings or segments each go to make up some part of the perfect insect, but not always in the same order in which they are now arranged. In fact, the fly, in becoming one, turns itself partly inside out, after the fashion of the clever snake which the showman said could swallow himself, beginning by putting his tail into his mouth. The first three rings form the mouth, this important member taking up its fair share of the whole; the next two form the antennæ or feelers by which the fly exercises the sense of smell, and the eyes; the sixth, seventh, and eighth form the thorax or throat, by which it breathes, and to which are attached the wings and the three pairs of legs; while the nine last form that most important part of the body, the stomach.

But in the fly, while the mouth is formed from the first segments, the head is formed behind from the fourth to the eighth. The brain (for the fly really has a brain) is the only part that remains the same in substance though quite changed in form, while all the other parts of the new animal grow round it.

As soon as the perfect insect is ready to come forth, it splits the old dried shell in front, doubles itself up, and so cracks it along its whole length below, when it sallies forth with very stunted wings, which in a few minutes, while the newly-exposed skin is becoming dry, quickly grow, until after a little flapping and fanning the fly finds it can mount into air, and starts on its new life.

And now, if we proceed to examine the transformed insect, we shall

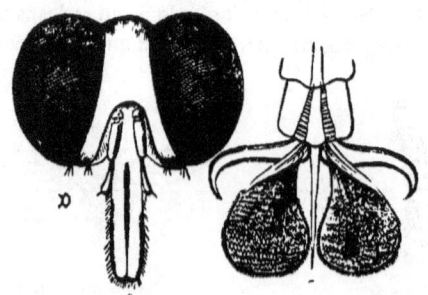

Head of *Musca domestica* (House Fly).

Foot (under-side) of *Musca vomitoria* (Blow Fly).

find wonders in its construction equal to any of those revealed to us in the anatomy of the largest and most complex quadrupeds. We have often seen a fly walking upon the ceiling, or upon any inverted surface, or running up a smooth pane of glass, and we may have wondered how it managed not only to hold on, but to run about so nimbly. An examination of that wonderful machine, a fly's foot, by a powerful microscope, will explain the whole of the very simple contrivance by which it seems to set the laws of gravitation at defiance. There have been several very clever guesses. Some have fancied that the hairs on its foot could take advantage of the slightest irregularity of surface; others that the foot was furnished with a natural air-pump, by which the air in its hollow was exhausted, and that it thus clung like a cupping-glass when applied to the flesh by the pressure of the atmosphere outside.

Now, if we examine the foot, we shall find it to be composed of a pair of pads with a pair of hooks above them, and the pads clothed with a number of very fine short hairs. Each pad is hollow, with a little nipple projecting into it. Behind the nipple is a bag connected with it, filled with a very clear, transparent gum. This gum, which is quite liquid, exudes from the nipple by the pressure of the insect in walking, and fills the hollow. The hairs are also hollow, with trumpet-shaped mouths; and these are also thus filled with the gum. This gum becomes hard the moment it is exposed to the air, and will not dissolve in water. Thus, at every step, the fly glues itself to the surface; and so tenacious is the gum that one foot is quite sufficient to bear the weight of the whole suspended body. If we examine the footprints of a fly on a window-pane by a powerful magnifier, we shall find that each foot-mark consists of rows of dots corresponding to the hairs on the foot-pads; in fact, the footprint is merely the traces of the gum that have been left behind. But how is it that the fly is not glued for life to the spot at the very first step it takes? It might be so, if it tried to lift its foot directly in a perpendicular direction; but it draws it up gently in a slanting direction, detaching the hairs in single rows, just as we might remove a moist postage-stamp, by beginning at one corner and gently drawing it back. When, however, the insect is diseased, the gum is very apt to harden, and at its death it at once becomes solid. Thus, we may often see a dead fly firmly attached to the wall, or to a window-pane, with a dull-coloured mark on the glass. This is caused by the fluid having glued the weak or sickly insect to its last resting-place, and having then hardened, the fly is cemented to the spot, till it decays away, leaving the legs behind. So very small are these trumpet-shaped hairs, that there are more than 1,000 on each foot-pad. We may add that moths, beetles, and all other insects have the same kind of gum secreted under their foot-pads.

Not less wonderful is the brain, or rather that which stands for it in the fly. We have spoken of the brain of a fly, but it must not be thought that insects have brains like the higher animals. In all these there is a large mass of brain protected by the skull, from which the spinal cord or marrow, which is a sort of continuation of the brain, extends to the extremity of the back-bone. Insects have nothing like this, though they have what answers the same purpose in their organization. They have what are called the *ganglions*, or large clusters of nerves, from which fine threads run in different directions. But instead of their being collected into one centre, there are different groups of them in different parts of the body; those of the head supplying the different organs of sense, the mouth the eyes, and the antennæ by which they smell; those of the thorax or middle section of the insect supplying the place of the heart, and being the nerve-centre of animal life, while another set supply the stomach or abdomen. From this separation it happens that the life of the insect chiefly depends on the thorax or middle part. If this is crushed the fly is instantly killed, and there is not the slightest motion afterwards; but if the head be cut off, while this ceases to move or to show any sensibility, the body will move for hours. If breathed upon or touched with a needle, there will be an attempt to run or fly; if dust or water be dropped either on the legs or abdomen, the feet will at once begin to rub it off. This seems to prove that these movements of the insect are at all times not the result of intelligence, but simply involuntary natural actions.

But yet there may be some intelligence even in a fly. Its substitutes for a brain, called the *cephalic ganglia*, are far larger than those of any other insect except the bees and ants, and are thirty times larger than the corresponding organs in a beetle of the same size. In bees, wasps, ants, and flies have been discovered, what have not yet been found in other insects, a pair of nerve-centres on the top of the cephalic (or head) ganglia, which anatomists suppose to answer to the brain lobes in higher animals. Perhaps this is the reason why the fly seems to show some intelligence, at least memory, in avoiding any one who has been chasing it. At any rate, from these centres proceed the nerves which run to its lips and enable it to taste, and to its eyes and antennæ and enable it to see and smell. All the senses are very highly developed in insects, more so than in higher creatures; and this renders it the more probable that their acts are for the most part caused by impressions from without, just as we shut our eyes when anything approaches them, or suddenly withdraw our hands from a burning substance.

Of all the organs of sense in the fly, the eye is the one most like the corresponding organ in other animals, and it is far more powerful than in any of them. To see as well as the fly, which can observe everything in four-fifths of the circle round it, we should require two more pair of eyes, in the side and at the back of our head. In the fly, no two facets or eye-discs look towards the same spot; and we must remember that the insect's eye is not a simple eye, but a vast collection of eyes in one head looking in all directions at the same time. There are between 4,000 and 5,000 of these little lenses, each of them the thousandth part of an inch in diameter, and set in a six-sided flat frame; and behind every facet or lens is a transparent cone, with a nerve from its point to the ganglion or brain.

We cannot be quite sure that flies smell by means of their antennæ, but if they are cut off the insect seems quite helpless, and from its uses of these feelers, it seems likely they are the organs of this sense. That they can hear we may conclude from their power of emitting sounds, and from the way in which they will disappear if one of their companions is caught and makes the sharp piteous drumming cry they send forth when in pain or fright. But no one has yet discerned the outward ear or organ of hearing in this insect, though it has been believed that their ears are just behind their wings, in the thorax instead of the head. These are what are called the *halteres*, and are in the place where the second pair of wings are fixed in bees, ants, and other hymenopterous insects.

We might go on page after page to describe the wonderful anatomy of the fly, as the microscope has revealed it to us. In fact, there is as much to be said about it as about the body of man, and perhaps more, for its organs are more numerous and more complex. Not only has it 4,000 eyes, instead of two; three sets of brain or nerve-centres, instead of one; 1,000 hairs and two claws instead of toes on its foot; it has also wings, which we have not; three pairs of legs, instead of one; a mouth which would bewilder any dentist; and a proboscis as far beyond that of an elephant in complexity of structure as a railway engine is beyond a wheelbarrow. But we hope we have said enough to excite your curiosity and your admiration of the works of God, which are as perfect and sublime in the tiniest thing that creeps as in the great worlds that whirl through space.

Of one thing we may be quite sure, that even flies were not made for nothing, and that a great many much larger beings would do very badly without them. They are often very troublesome, they may be, and have been, literally plagues, and certainly in many hot countries they are true plagues at some times of the year. But it is just when they are the greatest plagues that they at the same time are most useful. Of course if every fly mother reared 20,000 offspring, and every daughter of hers reared as many, there would soon be no room for anything else in the world but flies, and then the flies must die out of starvation. But it is not so intended. Living and dying the fly has its uses. Dying, it supplies the food not only of the spiders, who must live like other things, and would fare badly if no incautious fly glued its feet in their web; but it is the direct support and dependence of thousands of our special favourites. How dull would be our lanes and fields without the swallow and the martin skimming over them, or our hedgerows without the little song-birds which come back with the spring to refresh themselves and rear their young in our northern home after a winter's sojourn in the deserts or forests or marshes of Africa! Yet these feed almost entirely on flies. The swallow, whose appetite is as large as its mouth, needs myriads for its sustenance, and darts through the air for hours without a pause, snapping up the flies in its course. The little willow-wren darts up into the air, and as it hovers for a few minutes, snap! snap! goes its beak, while one fly after another is secured, till the swarm that buzzed over the hedgerow is scattered, and the little sportsman alights again, till very soon the incautious cloud is dancing over the fatal spot again, and invites him from his retreat to collect another savoury mouthful. The pretty graceful flycatcher of our gardens and shrubberies adopts a more dignified mode, and sits motionless on its perch at the end of a branch, or on the top of a pea-stick, its keen eye keeping an eager watch, till it sees an unsuspicious fly approach, when with a dart it shoots forth and seizes him, but always returns to its perch to swallow each morsel, as though it disdained to gobble up its food without due consideration.

Chameleons and many other of the innumerable lizards that swarm in all hot countries secure an ample sustenance with even less exertion than this, for they sit contentedly on a bough, and only shoot out their gummy tongue at the fly that buzzes near them. The toad, too, is fain to content himself with what he can catch on the tip of his tongue. But as he is not fond of bright sunny situations, it is well for him that his appetite is not very voracious, for he has to practise as much patience as an angler, before he can induce a curious incautious fly to come and examine his nose. Wasps, too, feed largely on flies, as we may see by the heaps of flies' wings strewn round a wasp's nest. And if the trout had not discovered what a savoury morsel is the fly that dances on the stream, what a very dull stupid amusement would fishing be! There would never have been the ingenuity which humbugs the fish by covering the hook with the feathers so neatly fastened together to imitate the living fly, and the skill that makes the little cheat dance so lively on the water, that the trout must come to look at it, and is very soon safely landed on the grassy bank. How many a schoolboy would lose the greatest treat of a summer's holiday, if there were no flies, and no trout that appreciated them!

If it were not for the bluebottle and other flies whose larvæ feed on decaying flesh, the carcases of animals might often create pestilence and disease. There are other flies whose larvæ are equally partial to decaying vegetable refuse. In fact, every substance that ought to be out of the way, and is a nuisance, is the object of the fly's research; and soon they can clear them off. They will make skeletons of a mouse or a little bird, very cleanly picked out for the student of anatomy, in a very few days.

But this habit of searching out and feeding on putrid matter causes one of the most serious injuries of which the fly is guilty in hot climates. Its taste or smell leads it to settle on any sore, and to feed on it. This would be all very well; but, unhappily for many, the fly, as we have seen, has a very gummy foot. One of the most common diseases in Egypt and other Oriental countries is ophthalmia, or running sores of the eyes, which often produce blindness when neglected. This complaint is highly contagious, but only by inoculation. If it were not for the flies, there would not be much danger in this, as people are not in the habit of kissing with their eyelids. But the fly perseveringly attacks the sufferer, and perches on the moist eyelid. Soon chased away, off he goes, and if with his wet feet he makes his next settlement on the eyelid of a healthy person, which too often happens, the result is certain to be an attack of ophthalmia. This is a kind of infection which no care or cleanliness can obviate.

There are other flies, very nearly akin to the house fly and the bluebottle, of which it is very hard to discover the use, except it be to act as scourges to man and beast. Who can ever say a good word for that thirsty little bloodsucker, the mosquito, as the large gnat is called, which

*Culex* (Mosquito).

murders sleep in Lapland and Labrador as much as in India or South America? Happily in England we know very little about them, excepting from the painful recollections of travellers. There is only one safeguard against them, and that is to be miles away from any water, for water is indispensable to the early life of the mosquito. The female lays her eggs, from 250 to 350 in number, always close together in the shape of a boat, on the edge of some leaf or substance floating on the water. The little raft sails away, and in two or three days the eggs are hatched, and the larvæ swarm by millions in every pool and stagnant ditch, and even in water-jugs or basins that have been allowed to stand a day or two. Like the larvæ of flies, they have no legs, and when not disturbed float on the top of the water with their head downwards, for they breathe air through the ends of their tails, which are shaped like a funnel. If the water is in the least degree disturbed, they dive or swim most rapidly, but they feed only while motionless. They do this by means of a circular fringe of hair round the mouth: the little creature twirls these hairs about, so as to cause tiny currents, which bring microscopic substances within their reach, which are thus drawn in. In another week or two it moults, and in a few days a second, and then a third time splitting up its old skin, and coming out with a fresh dress, when it changes to its third state, like the chrysalis of a butterfly, only that it still moves about, sometimes at the top, sometimes at the bottom of the water, but never eats. In eight or ten days more it comes to the surface, lies on its back, and after a few struggles splits its hard skin into the shape of a boat, in which it sits. It raises its head and then its body, till it stands upright like a mast in the floating boat; then it gets its feet clear of their shell; and when it has freed its third and hindmost pair of feet, it leans to one side, rests its fore feet on the water, waits a few instants while its body, which was quite white at first, becomes first greenish and then black with white rings; it unfolds, fans, and dries its wings, which were snugly folded very close within its old skin; and then off it darts, to disport for a day or two in the air, and, unless picked off by some passing swallow, to torment any human or other beings within reach.

It would be some excuse for these bloodthirsty little creatures, if it were a necessity of their existence to draw the blood of the giants they torment; but there are millions of them who lead innocent and happy lives, without ever having used their proboscis. It is only a bad habit, a mischievous luxury, in which they indulge when they have a chance. When they do find a victim, nothing but leather will keep them out; and besides their sting, their sharp stridulous note is so tormenting, that many sufferers find it worse than their bite. Their proboscis is a sort of hollow pipe, with a very sharp point; this tube they thrust into the skin, and as soon as it has penetrated to the veins, they shoot down through it several lancets with barbed points, notched like a saw, and then suck up the blood from the wounds. Not content with this, they eject a powerful

## ABOUT A FLY.

acid at the same time, which causes swellings and intolerable irritation—sometimes for several days—and often serious sores.

Sometimes they have appeared in such numbers that their clouds, at a distance, looked like volumes of smoke rising from a fire. We are told that Sapor, the King of Persia, was once compelled by them to raise the siege of a city; and that not only his soldiers were attacked, but his elephants and beasts of burden, till they became perfectly maddened and unmanageable. In some parts of South America the inhabitants have been compelled to sleep on the ground, buried in the sand, with only their heads out, and these covered with handkerchiefs, to secure rest. This was both witnessed and experienced by Humboldt. The traveller, Dr. Clarke, tells us that once in the Crimea he was attacked by a swarm at night, when not a breath of air was stirring. Driven from his quarters, he vainly took to the carriage outside for refuge. Almost suffocated with heat, he dared not venture to open the windows: still the torturing little creatures contrived to find their way in. He wrapped his head in a handkerchief in vain; they filled his mouth, nostrils, and ears. At length he succeeded in lighting a lamp; but it was extinguished in a moment by such a prodigious number of mosquitoes, that their carcases choked up the glass chimney, and formed a heap over the burner.

*Tabanus bovinis* (Gadfly).—Natural size.

*Slossina morsitans* (the Tzetze).

There is another fly which is, if possible, a yet greater torment of cattle than the mosquito is of man—the gadfly. When its tormenting buzz is recognized, we may see a whole herd of cattle rushing wildly about, with tails outstretched, in abject terror. The gadfly is far worse than the gnat, for it actually buries its eggs in the skin of the animal, and leaves them there to hatch, when they live upon the flesh for days, and often destroy the poor animal by the sores they create. The most terrible of the gadflies is the Tzetze of Abyssinia and other parts of Eastern Africa. It sometimes renders whole districts desolate for miles, by destroying all the animals, and thus reducing the poor inhabitants to a state of starvation; and Dr. Livingstone tells us of regions where for a part of the year no one is able either to live or travel, until the wet season destroys these terrific scourges.

This is indeed the dark side of fly life, so far as we are concerned; but we have said enough, we hope, to let our readers see that even flies have their uses, and that, whether we examine their history or their structure, they are not the least wonderful of the many wonderful things with which God has stored this prolific earth of ours.

---

## EARTHQUAKES.

MANY of you boys and girls have doubtless seen pictures representing the ruins of South American towns destroyed by earthquakes, and it has puzzled you and made you sad. You want to know why God killed all those people—mothers among them, too, and little children?

Alas! my dear children! who am I that I should answer you that?

Have you done wrong in asking me? No, my dear children; no. You have asked me because you are human beings and children of God, and not merely a cleverer sort of animal,—an ape who can read and write and cast accounts. Therefore it is that you cannot be content, and ought not to be content, with asking how things happen, but must go on to ask why. You cannot be content with knowing the causes of things; and if you knew all the natural science that ever was or ever will be known to men, that would not satisfy you; for it would only tell you the *causes* of things, while your souls want to know the *reason* of things besides; and though I cannot tell you the reasons of things, yet I believe that somehow, somewhen, somewhere, you will learn something of the *reason* of things. For that thirst to know *why* was put into the hearts of little children by God Himself; and I believe that God would never have given them that thirst, if He had not meant to satisfy it.

There—you do not understand me. I trust that you will understand me some day. Meanwhile, I think—I only say I *think*—that we may guess at something like a good reason for the terrible earthquakes in

South America. I do not wish to be hard upon poor people in great affliction; but I cannot help thinking that they have been doing for hundreds of years past something very like what the Bible calls "tempting God"—staking their property and their lives upon the chances of no earthquakes coming, while they ought to have known that an earthquake might come any day. They have fulfilled the parable of the nation of the Do-as-you-likes, who lived careless and happy at the foot of the burning mountain, and would not be warned by the smoke that came out of the top, or by the slag and the cinders which lay all about them; till the mountain blew up, and destroyed them miserably.

Then I think that they ought to have expected an earthquake?

Well, it is not for us to judge any one, especially if they live in a part of the world in which we have not been ourselves. But I think that we know, and that they ought to have known, enough about earthquakes to have been more prudent than they have been for many a year. At least we will hope that, though they would not learn their lesson before, they will learn it now, and will listen to the message, spoken in a voice of thunder, and written in letters of flame.

And what is that?

My dear children, if the landlord of our house was in the habit of pulling the roof down upon our heads, and putting gunpowder under the foundations to blow us up, do you not think we should know what he meant, even though he never spoke a word? He would be very wrong in behaving so, of course; but one thing would be certain,—that he did not intend us to live in his house any longer if he could help it, and was giving us, in a very rough fashion, notice to quit. And so it seems to me that these poor Spanish Americans have received from the Landlord of all landlords, who can do no wrong, such a notice to quit as perhaps no people ever had before; which says to them in unmistakable words, "You must leave this country, or perish." And I believe that that message is at heart a merciful and loving one; that if these Spaniards would leave the western coast of Peru, and cross the Andes into the green forests of the eastern side of their own land, they might not only live free from earthquakes, but (if they would only be good and industrious) become a great, rich, and happy nation, instead of the idle, and useless, and I am afraid not over-good, people which they have been. For in that eastern part of their own land God's gifts are waiting for them, in a Paradise such as I can neither describe nor you conceive; precious woods, fruits, drugs, and what not—boundless wealth, in one word,—waiting for

The Ruins of Arica, South America.

them to send it all down the waters of the mighty River Amazon, enriching us here in the Old World, and enriching themselves there in the New. If they would only go and use these gifts of God, instead of neglecting them, as they have been doing for now three hundred years, they would be a blessing to the earth, instead of being—that which they have been.

God grant, my dear children, that these poor people may take the warning that has been sent to them—" The voice of God revealed in facts," as the great Lord Bacon would have called it, and see not only that God has bidden them leave the place where they are now, but has prepared for them, ı their own land, a home a thousand times better than that in which they now live.

But you ask, How ought they to have known that an earthquake would come?

Well, to make you understand that, we must talk a little about earthquakes, and what makes them; and, in order to find out that, let us try the very simplest cause of which we can think. That is the wise and scientific plan.

Now, whatever makes these earthquakes must be enormously strong; that is certain. And what is the strongest thing you know of in the world? Think. . . . . .

Gunpowder?

Well, gunpowder is strong sometimes, but not always. You may carry it in a flask, or in your hand, and then it is weak enough. It only becomes strong by being turned into gas and steam. But steam is always strong. And if you look at a railway engine, still more if you had ever seen—which God forbid you should—a boiler explosion, you would agree with me that the strongest thing we know of in the world is steam.

Now I think that we can explain almost, if not quite, all that we know about earthquakes, if we believe that on the whole they are caused by steam and other gases expanding, that is, spreading out, with wonderful quickness and strength. Of course, there must be something to make them expand, and that is *heat*. But we will not talk of that yet.

Now, you may have heard this riddle?—"What had the rattling of the lid of the kettle to do with Hartford Bridge Flat being lifted out of the ancient sea?"

The answer to the riddle, I believe, is—Steam has done both. The lid of the kettle rattles, because the expanding steam escapes in little jets, and so causes a *lid-quake*. Now suppose that there was steam under the earth trying to escape, and the earth in one place was loose and yet hard, as the lid of the kettle is loose and yet hard, with cracks in it, it may be, like the crack between the edge of the lid and the edge of the kettle itself; might not the steam try to escape through the cracks, and rattle the surface of the earth, and so cause an *earth-quake?*

So the steam would escape generally easily, and would only make a passing rattle, like the earthquake of which the famous jester Charles Selwyn said, that it was quite a young one, so tame that you might have stroked it; like that which I myself once felt in the Pyrenees, which gave me very solemn thoughts after a while, though at first I did nothing but laugh at it; and I will tell you why.

I was travelling in the Pyrenees, and I came one evening to the loveliest spot; a glen, or rather a vast crack, in the mountains, so narrow that there was no room for anything at the bottom of it, save a torrent roaring between walls of polished rock. High above the torrent the road was cut out among the cliffs, and above the road rose more cliffs, with great black cavern-mouths, hundreds of feet above our heads, out of each of which poured in foaming waterfalls streams large enough to turn a mill, and above them mountains piled on mountains, all covered with woods of box, which smelt rich and hot and musky in the warm summer air. Among the box-trees and fallen boulders grew hepaticas, blue and white and red, such as you see in the garden; and little stars of gentian, more azure than the azure sky. But out of the box woods above rose giant silver firs, clothing the cliffs and glens with tall black spires, till they stood out at last in a jagged saw-edge against the purple evening sky, along the mountain ranges, thousands of feet aloft; and beyond them again, at the head of the valley, rose vast cones of virgin snow, miles away in reality, but looking so brilliant and so near that one fancied at the first moment that one could have touched them with one's hand. Snow-white they stood, the glorious things, 7,000 feet into the air; and I watched their beautiful white sides turn rose-colour in the evening sun, and when he set fade into dull cold grey, till the bright moon came out to light them up once more. When I was tired of wondering and admiring, I went into bed; and there I had a dream—such a dream as Alice had when she went into Wonderland—such a dream as I daresay you may have had ere now. Some noise or stir puts into your fancy as you sleep a whole long dream to account for it; and yet that dream, which seems to you to be hours long, has not taken up a second of time; for the very same noise which begins the dream wakes you at the end of it: and so it was with me. I dreamed that some English people had come into the hotel where I was, and were sleeping in the room underneath me; and that they had quarrelled and fought, and broke their bed down with a tremendous crash, and that I must get up, and stop the fight; and at that moment I woke, and heard coming up the valley from the north such a roar as I never heard before or since; as if a hundred railway trains were rolling underground; and just as it passed under my bed there was a tremendous thump, and I jumped out of bed quicker than I ever did in my life, and heard the roaring sound die away as it rolled up the valley towards the peaks of snow. Still I had in my head this notion of the Englishmen fighting in the room below. But then I recollected that no Englishmen had come in the night before, and that I had been in the room below, and that there was no bed in it. Then I opened my window—a woman screamed, a dog barked, some cocks and hens cackled in a very disturbed humour, and then I could hear nothing but the roaring of the torrent a hundred feet below. And then it flashed across me what all the noise was about; and I burst out laughing, and said, "It is only an earthquake;" and went to bed again.

Next morning I inquired whether any one had heard a noise. No, nobody had heard anything. And the driver who had brought me up the valley only winked, but did not choose to speak. At last at breakfast I asked the pretty little maid who waited what was the meaning of the noise I heard in the night, and she answered, to my intense amusement, "Ah! bah! *ce n'était qu'un tremblement de terre; il y en a ici toutes les six semaines.*" And now the secret was out. The little maid, I found, came from the lowland far away, and did not mind telling the truth; but the good people of the place were afraid to let out that they had earthquakes every six weeks, for fear of frightening visitors away; and because they were really very good people, and very kind to me, I shall not tell you what the name of the place is.

Of course after that I could do no less than ask Nature, very civilly, how she made earthquakes in that particular place, hundreds of miles away from any burning mountain? And this was the answer I *thought* she gave, though I am not so conceited as to say I am sure.

As I had come up the valley I had seen that the cliffs were all beautiful grey limestone marble; but just at this place they were replaced by granite, such as you may see in London Bridge or at Aberdeen. I do not mean that the limestone changed to granite, but that the granite had risen up out of the bottom of the valley, and had carried the limestone (I suppose) up on its back hundreds of feet into the air. Those caves with the waterfalls pouring from their mouths were all on one level, at the top of the granite and the bottom of the limestone. That was to be expected; for, as I will explain to you some day, water can make caves easily in limestone, but never, I think, in granite. But I knew that beside these cold springs which came out of the caves, there were hot springs also, full of curious chemical salts, just below the very house where I was in. And when I went to look at them, I found that they came out of the rock just where the limestone and the granite joined. "Ah, ah!" I said, "now I think I have Nature's answer. The lid of one of her great steam boilers is rather shaky and cracked just here, because the granite has broken and torn the limestone as it lifted it up; and here is the hot water out of the boiler actually oozing out of the crack; and the earthquake I heard last night was simply the steam rumbling and thumping inside, and trying to get out."

And then, my dear children, I fell into a more serious mood. I said to myself, "If that steam had been a little—only a little—stronger; or the rock above it only a little lighter or weaker, it would have been no laughing matter then; the village might have been shaken to the ground; the rocks hurled into the torrent; jets of steam and of hot water, mixed, it may be, with deadly gases, have roared out of the riven ground: that might have happened here, in short, which has happened and happens still in a hundred places in the world, whenever the rocks are too weak to stand the pressure of the steam below, and the solid earth bursts, as an engine boiler bursts when the steam within is too strong." And when those thoughts came into my mind, I was in no humour to jest any more about "young earthquakes," but rather to say with the wise men of old, "It is of the Lord's mercies that we are not consumed."

Most strange, but most terrible also, are the tricks which this underground steam plays. It will make the ground, which seems to us so hard and firm, roll and rock in waves, till people are sea-sick, as on board a ship; and that rocking motion (which is the most common) will often, when it is but slight, set the bells ringing in the steeples, or make the furniture and things on shelves jump about quaintly enough. It will make trees bend to and fro, as if a wind was blowing through them; open doors suddenly, and shut them again with a slam; make the timbers of the floors and roofs creak, as they do in a ship at sea; or give men such frights as one of the dock-keepers at Liverpool got, in the earthquake of 1863, when his watch-box rocked so that he thought some one was going to pitch him over into the dock. But these are only little hints and warnings of what it can do. When it is strong enough, it will rock down houses and churches into heaps of ruins, or, if it leaves them standing, crack them from top to bottom, so that they must be pulled down and rebuilt.

You see that picture of the ruins of Arica, and from it you can guess well enough for yourselves what a town looks like which has been ruined by an earthquake. Of the misery and the horror which follow such a ruin I will not talk to you, nor darken your young spirits with sad thoughts which grown people must face, and ought to face. But the strangeness of some of the tricks which the earthquake-shocks play is hardly to be explained, even by scientific men. Sometimes, it would seem, the force runs round, making the solid ground eddy, as water eddies in a brook. For it will make straight rows of trees crooked; it will twist whole walls round—or rather the ground on which the walls stand—without throwing them down; it will shift the stones of a pillar one on the other sideways, as if a giant had been trying to spin it like a teetotum, and so screwed it half in pieces. There is a story told by a wise man, who saw the place himself, of the whole furniture of one house being hurled away by the earthquake, and buried under the ruins of another house; and of things carried hundreds of yards off, so that the neighbours went to law to settle who was the true owner of them! Sometimes, again, the shock seems to come neither horizontally in waves, nor circularly in eddies, but vertically, that is, straight up from below; and then things—and people, alas! sometimes—are thrown up off the earth high into the air, just as things spring up off the table, if you strike it smartly enough underneath. By that same law (for there is a law for every sort of motion) it is that the earthquake-shock sometimes hurls great rocks off a cliff into the valley below. The shock runs through the mountain till it comes to the cliff at the end of it; and then the face of the cliff, if it be at all loose, flies off into the air. You may see the very same thing happen, if you will put marbles or billiard-balls in a row touching each other, and strike the one nearest you smartly in the line of the row. All the balls stand still, except the last one; and that flies off. The shock, like the earthquake-shock, has run through them all; but only the end one, which had nothing beyond it but soft air, has been moved; and when you grow older, and learn mathematics, you will know the law of motion according to which that happens, and learn to apply what the billiard-balls have taught you, to explain the wonders of an earthquake. For in this case, as in so many more, you must watch

Nature at work on little and common things, to find out how she works in great and rare ones. That is why Solomon says that "a fool's eyes are in the ends of the earth," because he is always looking out for strange things which he has not seen, and which he could not understand if he saw, instead of looking at the petty commonplace matters which are about his feet all day long, and getting from them sound knowledge, and the art of getting more sound knowledge still.

Another terrible destruction which the earthquake brings, when it is close to the seaside, is the wash of a great sea-wave, such as swept in some years ago upon the island of St. Thomas, in the West Indies; and also upon the coast of Peru. The sea moans and sinks back, leaving the dry shore; and then comes in from the offing a mighty wall of water, as high as, or higher than, many a tall house; sweeps far inland, washing away quays and houses, and carrying great ships in with it; and then sweeps back again, leaving the ships high and dry.

Now, how is that wave made? Let us think. Perhaps in many ways. But two of them I will tell you as simply as I can, because they seem the most likely, and probably the most common.

Suppose, as the earthquake-shock ran on, making the earth under the sea heave and fall in long earth-waves, the sea-bottom sank down. Then the water on it would sink down too, and leave the shore dry; till the sea-bottom rose again, and hurled the water up again against the land. This is one way of explaining it, and it may be true. For certain it is that earthquakes do move the bottom of the sea; and certain too that they move the water of the sea also, and with tremendous force. For ships at sea during an earthquake feel such a blow from it (though it does them no harm) that the sailors often rush upon deck, fancying that they have struck upon a rock; and the force which could give a ship floating in water such a blow as that, would be strong enough to hurl thousands of tons of water up the beach and on to the land.

But there is another way of accounting for this great sea-wave, which I fancy comes true sometimes. Suppose you put an empty india-rubber ball into water, and then blew into it through a pipe. Of course, you know, as the ball filled, the upper side of it would rise out of the water. Now, suppose there were a party of little ants moving about upon that ball, and fancying it a great island, or perhaps the whole world—what would they think of the ball's filling and growing bigger?

If they could see the sides of the basin or tub in which the ball was, and were sure that they did not move, then they would soon judge by them that they themselves were moving, and that the ball was rising out of the water. But if the ants were so short-sighted that they could not see the sides of the basin, they would be apt to make a mistake, because they would then be like men on an island out of sight of any other land. Then it would be impossible further to tell whether they were moving up, or whether the water was moving down; whether their ball was rising out of the water, or the water was sinking away from the ball. They would probably say, "The water is sinking, and leaving the ball dry."

Do you understand that? Then think what would happen if you pricked a hole in the ball. The air inside would come hissing out, and the ball would sink again into the water. But the ants would probably fancy the very opposite. Their little heads would be full of the notion that the ball was solid, and could not move, just as our heads are full of the notion that the earth is solid, and cannot move; and they would say, "Ah! here is the water rising again." Just so, I believe, when the sea seems to ebb away during the earthquake, the land is really being raised out of the sea, hundreds of miles of coast, perhaps, or a whole island, at once, by the force of the steam and gas imprisoned under the ground. That steam stretches and strains the solid rocks below, till they can bear no more, and snap, and crack with frightful roar and clang; then out of holes and chasms in the ground rush steam, gases—often foul and poisonous ones—hot water, mud, flame, strange stones—all signs that the great boiler down below has burst at last.

Then the strain is eased. The earth sinks together again, as the ball did when it was pricked; and sinks lower, perhaps, than it was before; and back rushes the sea, which the earth had thrust away while it rose, and sweeps in, destroying all before it.

Of course, there is a great deal more to be said about all this; but you you are too young to understand it. You will read it, I hope, for yourselves when you grow up, in the writings of far wiser men than I. Or perhaps you may feel for yourselves in foreign lands the actual shock of a great earthquake, or see its work fresh done around you. And if ever that happens, and you be preserved during the danger, you will learn for yourself, I trust, more about earthquakes than I can teach you, if you will only bear in mind the simple general rules for understanding the "how" of them which I have given you here.

But you do not seem satisfied yet? What is it that you want to know?

Oh! There was an earthquake here in England one night, while you were asleep; and that seems to you too near to be pleasant. Will there ever be earthquakes in England which will throw houses down, and bury people in the ruins?

My dear children, I think you may set your heart at rest upon that point. As far as the history of England goes back, which is more than a thousand years, there is no account of any earthquake which has done any serious damage, or killed, I believe, a single human being. The little earthquakes which are sometimes felt in England run generally up one line of country, from Devonshire through Wales, and up the Severn Valley into Cheshire and Lancashire, and the south-west of Scotland; and they are felt more smartly there, I believe, because the rocks are harder there than here, and more tossed about by earthquakes which happened ages and ages ago, long before man lived on the earth. I will show you the work of these earthquakes some day, in the tilting and twisting of the layers of rock, and in the cracks (*faults*, as they are called) which run through them in different directions. There are some of these cracks to be seen in the chalk cliff at Ramsgate—two sets of cracks, sloping opposite ways, which were made by two separate sets of earthquakes, long, long ago, perhaps while the chalk was still at the bottom of a deep sea. But even in the rocky parts of England the earthquake-force seems to have all but died out. Perhaps the crust of the earth has become too thick and solid there to be much shaken by the gases and steam below. In the eastern part of England, meanwhile, there is but little chance that an earthquake will ever do much harm, because the ground there, for thousands of feet down, is not hard and rocky, but soft —sands, clays, chalk, and sands again; clays, soft limestones, and clays again—which all act as buffers to deaden the earthquake-shocks, and deaden, too, the earthquake noise.

And how?

Put your ear to one end of a soft bolster, and let some one hit the other end. You will hear hardly any noise, and will not feel the blow at all. Put your ear to one end of a hard piece of wood, and let some one hit the other. You will hear a smart tap, and perhaps feel a smart tap, too. When you are older, and learn the laws of sound and of motion among the particles of bodies, you will know why. Meanwhile you may comfort yourself with the thought that there is prepared for this good people of Britain a safe soft bed—not that they may lie and sleep on it, but work and till, plant and build and manufacture, and thrive in peace and comfort, we will trust and pray, for many a hundred years to come. All that the steam inside the earth is likely to do to us, is to raise parts of this island so slowly, probably, that no man can tell whether they are rising or not.  Or again, the steam power may be even now dying out under our island, and letting parts of it sink slowly into the sea, as some wise friends of mine think that the fens in Norfolk and Cambridgeshire are sinking now.  You can see where that kind of work has gone on in Norfolk; how the brow of Sandringham Hill was once a sea-cliff, and Dersingham Bog at its foot a shallow sea, and therefore that the land has risen there.  How, again, at Hunstanton Station there is a beach of sea-shells twenty feet above high-water mark, showing that the land has risen there likewise.  And how, farther north again, at Brancaster, there are forests of oak, and fir, and alder, with their roots still in the soil, far below high-water mark, and only uncovered at low tide; which is a plain sign that there the land had sunk.  In the sunken forest at Brancaster beautiful shells may be picked up in the gullies, and there are millions of live Pholases boring into the clay and peat which once was firm dry land, fed over by the giant oxen, and giant stags likewise, and perhaps by the mammoth himself, the great woolly elephant whose teeth the fishermen dredge up in the sea outside?  Then remember that as that Norfolk shore has changed, so slowly but surely is the whole world changing around us.  Hartford Bridge Flats, in Hampshire, for instance, how has it changed!  Ages ago it was the gravelly bottom of a sea.  Then the steam power underground raised it up slowly, through long ages, till it became dry land.  And ages hence, perhaps, it will have become a sea-bottom once more.  Washed slowly away by the rain, or sunk by the dying out of the steam power underground, it will go down again to the place from whence it came.  Seas will roll where we stand now, and new lands will rise where seas now roll.  For all things on this earth, from the tiniest flower to the tallest mountain, change, and change all day long. Every atom of matter moves perpetually, and nothing "continues in one stay.  The solid-seeming earth on which you stand is but a heaving bubble, bursting ever and anon in this place and in that.  Only above all, and through all, and with all, is One who does not move nor change, but is the same yesterday, to-day, and for ever.  And on Him, my child, and not on this bubble of an earth, do you and I, and all mankind, depend

# FRUIT.

TO be suddenly in a strange country, surrounded by trees and fruits unlike any I had ever seen, would puzzle me considerably. After a time I should of course feel hungry or thirsty, and the simple thing to do would be to gather the ripe fruit. This would hang temptingly from the branches, or peep from under the creeping stems and glossy leaves, in such a provokingly delicious fashion, that I am nearly certain that I should at last eat some, quite forgetting that I might poison myself if by any chance I plucked a dangerous kind. You see, this is the difficulty; for being placed, without any previous knowledge of Botany, in the position of Robinson Crusoe, my adventures might come to a sudden conclusion, and would not make a story at all. For it is by no means a rule that fruit is always good to eat; indeed, there are many things in every garden that are of no use for their fruit. Potatoes and asparagus are illustrations of this, besides many wild berries that are extremely hurtful to the little folks who gather them. Ah, you are thinking, But people here always know which of the wild fruits are safe to eat, and there is no fear that you will be killed or even made ill by taking the bad ones. This is true, certainly, and is just the advantage of living in a country that has been hundreds of years growing up and educating itself into a nice convenient place for your benefit. But once upon a time things were very different, and some awkward misthkes were made before the various parts of either home or foreign plants found their present uses. For instance, in these enlightened days no one could possibly make you a tart of rhubarb-leaves instead of stalks, or send you the shells instead of the peas for dinner. But somebody must have made the first tart and the first tea. Puzzled enough they were with this last article, if all, or even half, the stories are true. I heard once of an old lady whose sailor son sent her a present of tea from China: she did not know what to do with the dark dry-looking stuff, but at last decided that leaves were only fit to eat when they were boiled. Boil them accordingly she did, and patiently waited for them to come to what she considered would be an eatable state. In spite of all her efforts, however, they remained tough, and, quite tired out, the poor old lady threw them away. I imagine her son must have been very angry when he found the precious leaves destroyed, for money could not buy more, and even the great East India Company considered they made the Queen of England a brilliant present when they sent her two pounds of tea.

If you had lived three hundred years ago, no one could have given you warm coffee for breakfast, or have called you from play to the pleasant tea-table, and for the best of reasons, neither of these nice things were then known to the people of England. It is almost impossible to realize such an uncosy existence; but it is even more difficult to believe that after coffee was known to the Arabs, they should have had conscientious scruples about using it. The Mohammedans, however, decided that the drinking of coffee was a wicked practice, and must not be encouraged, because it enabled those who used it to keep awake during their hours of prayer with less effort. That tea and coffee produce wakefulness, seems to have been discovered almost as soon as the use of the plants, for the Chinese have made the following curious story about the origin of the former. A pious hermit who was constantly overpowered by sleep during his watchings and prayers, became so exasperated with his eyelids that, in a fit of rage, he cut them off and threw them on the ground. But a god caused a tea-shrub to grow out of them, and its leaves resemble an eyelid bordered with lashes, and possess the gift of hindering sleep.

But stories will not help us to find out how the fruit grows on the trees, and it is quite time to think of this.

In the early spring, when the trees were full of blossom, you often heard it said that there would be plenty of fruit in the autumn, and you knew without telling that the people who prophesied about the fruit only did so because they saw the flowers. How the marvellous change from a pink blossom to a rosy apple was to come about no one explained; and even if you were curious enough to ask questions, I doubt if the answers satisfied you.

But let us look again at the lilies. The flower-leaves (petals) and stamens have fallen away, and only the green pistil remains. This is already passing into fruit, and the ovules at its base are growing up into true seed. Surrounding these are the walls of the ovary, whose business it is to grow into a safe home for them. But the changes that occur in the development of the seed-coverings cause the varieties of fruit that are known to us, and it is curious to examine the different methods taken by the plants for the same purpose. For of course, to protect the seed is the real reason fruit grows; it is not very likely the trees take any pains to make nice juicy currants and gooseberries only for our pleasure.

The pulpy mass that really represents our idea of fruit is generally like the cells of the boiled rhubarb; through this pass vascular bundles like fine thread, and the outer covering, at present known to you as the epiderm, encloses and protects the whole. Thus the little tender grain has, after all, only a delicate resting-place, and does not approve of the intrusive habits of the wasps and birds. In fact, it soon perishes if these impertinents appear before it has had time to ripen. Soft seeds, such as apple-pips, require more covering than those of peaches and apricots, and the thickened ovary is enclosed by the fleshy substance that develops from another part of the blossom.

To understand this thoroughly you must look carefully at the drawing,

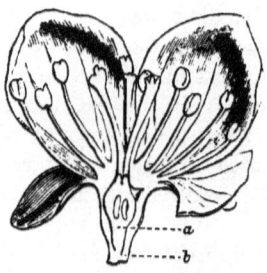

which shows you a vertical section of an apple-blossom. *a* is the true ovary, and *b* the cup or calyx that supported the leaves and stamens; after these have fallen off, the calyx, which is parted at its upper edge into little green points, fills up its cells and enlarges much quicker than the ovary. This, when the apple ripens, is called the core, and is the original birthplace of the pips or seeds, while the outer succulent layer is only the thickened calyx-tube. If you look at the top of the fruit, you will see the sepals and sometimes a few of the stamens remaining.

Gooseberries and currants also retain these leaves when they are fully grown, but this does not prove that they are very closely related to their companions the apples. We will cut one of these from the rose to the stalk, and another transversely, and we shall see how prettily the core divides to make separate cells for the seeds. Now cut a gooseberry, and notice how differently the seeds are placed; there are many more of them, but no separate partitions to keep them apart; there is no confusion, however, in this pretty transparent nursery, for every seed is attached to the side of the ovary, and the rest of the cavity is filled up with the juicy pulp. Fruits of this kind have single cells, and are the only true berries; raspberries, strawberries, and mulberries go about under false colours, and have no claim, as I will presently show you, to their surname. But the stone fruits, such as apricots and cherries, deserve to be looked at before these small people. Now it is clear enough there are no stones in the blossom, so this must also be a substance that develops after the fertilization of the ovule, when the flower-leaves have died away.

We have seen how carefully the apparatus for the production of seed is protected by the petals of the corolla and the sepals of the calyx; dur-

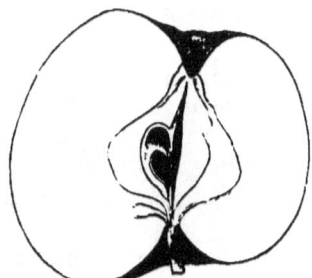
Vertical section.

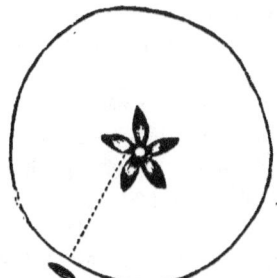
Transverse section of Apple.

ing the process of ripening it is equally necessary that the baby plant should be clothed and nourished. If you have comprehended how the apple-trees meet this difficulty, you know that the skin of this fruit corresponds to the epiderm of the calyx. But the outer skin of the stone fruits is only the covering of the ovary, and is now called the epicarp; the inner layer of cells next the ovule thickens and hardens into the shell. You are wondering how the little soft kernel ever gets free from its wooden case, and finds power to absorb moisture and get warmth to grow into a fresh plant. This, which is a sufficiently simple process in dehiscent fruits seems a mystery when we remember that apricots and peaches are planted enclosed in a case that appears hard enough to remain a stone for ever. In the damp earth this stubborn covering soon decays, and the liberated seed at once begins the work of germination.

This change of ovule and pollen into seed, and its combination with the ovary to form fruit, is a comparatively natural development when compared with the complex operations of the strawberries. This white flower, that looks too wild and uneducated to know any but the most ordinary way of bringing up seeds, chooses to find out a method of its own, and very much puzzled me to discover where the red fruit was coming from.

But let us study carefully the pistil of the lily, and learn how this is made before we go on to the strawberries.

By pulling or cutting this to pieces we shall find that ovary, style, and stigma are formed by the union of several leaves called carpels. These curl inwards, and at their base the ovules lie until fertilized by the pollen grain. But the carpels of strawberries and raspberries have no idea of binding themselves together, and forming one general home for the children; on the contrary, each of these leaves starts off in a different direction, carrying with it an already growing seed; the receptacle from which they spring also begins to develop rapidly, and it is this that grows into the juicy part of the fruit. The little yellow specks that lie in the dimples of the strawberries are the true seeds, and it is these that are gathered and planted when the gardeners want entirely fresh plants; they are enclosed in a pericarp, never become soft or juicy, and are called *achenes*.

Now the deceivers who are falsely called rasp*berries* and black*berries* have no greater claim to the name than the one we have just examined, but differ from strawberries by surrounding their seeds with a soft pulp this comes from the swelling of the carpellary leaf, and the white mass from which it springs supports the little juicy heads called drupels.

Mulberries show us another and entirely new way of growth, and may be distinguished from the others by being called a collective fruit. It has an undoubted right to this distinction, because it is not made up through the development of any one part of the blossom, but is composed of entire flowers growing from a single stem. The leaves of these become succulent and juicy as the pistil ripens; a marked contrast to some of those now known to us, whose petals take the earliest opportunity to retire from business.

I shall not attempt to describe any more varieties of fruit, for I have said enough to convince you that, useful as it is to us, it is of far greater importance to the plants, and like the root, stem, and foliage leaves, develops to perform distinct work for their benefit.

## WASPS AND PAPER-MAKING.

"YOU or I, fair damsel? you or I? For which of us is that sweet fruit ripening in the summer sun? Day after day you come before any one else is awake, and gaze upon it, and as you watch its delicate hue mellowing into the rich colouring of its perfection, you say to yourself, 'To-morrow! to-morrow I will pluck it.' But you shall find to your cost that one has been there before you. Look to it, for I am armed, and, once in possession, I know how to hold my own. Let every one, man or insect, look out for himself."

Thus hummed and buzzed to himself, in selfish soliloquy, a brilliant black and golden insect, hovering near the garden wall; now wheeling in airy circles, greatly to the terror of the gentle child who attempted to approach the tree; and now settling on the luscious fruit, to thrust in his proboscis, for the purpose of making his temporary abode inside the mellow peach, so soon as it appeared to his fastidious senses to be sufficiently ripe for the attack to be made. Sorely terrified was the little girl, who, standing on tiptoe, just ventured hastily to touch the fruit, and thought, "To-morrow! to-morrow it will be quite ripe, and then——Oh, dear poor Janie! and then——Oh, how she will like it!"

To-morrow! To-morrow! Alas! to-morrow brings with it many unlooked-for disappointments; and yet how strangely mixed sometimes with the signs of a worse evil averted! The wasp had his work set him by his own greedy nature, and, ensconced at length in a cavity scooped out by his own exertions in the side of the peach, settled himself in conscious security, armed, as he thought, against all intruders, and feasting on the luscious dainty so long coveted by him. Did any relentings beset him, as to the sting he was preparing for those fairy fingers, when they should be extended to pluck his dainty morsel? Could he not have wished it were some selfish greedy creature like himself?—perhaps some tyrannical schoolboy, at home for the holidays, full of the conviction so common among boys, that all good things ought to go but one road, or at most two, into their own pocket, or down their throat; and that but one crime exists in the world—that of depriving them thereof.

Whether or not the actual turn in the tide of affairs was any real satisfaction to the moral sense of the insect, it will doubtless be so to us, as we turn at the sound of rapid footsteps, not lightly tripping as we expected, but heavily pounding the gravel-walk, and, rushing up to the tree in hot haste, we recognize the identical being of our imagination. No, we will not wrong the British schoolboy: it is the spoilt child of a foolish home, who, having heard of his sister's secret and loving longings to carry the peach, as soon as it should be ripe, to the bedside of her suffering friend, resolved to be beforehand with her, and secure the fruit for his private eating.

Who can pity him? Who does not rejoice, and cry out, "Served him right!" "Glad it was he who got stung." Yes; listen to him, how he yells! See how he dances about, flings his fingers from him as it were! and then seizes his cap and makes a dart at the wasp, determined to kill the creature whose only crime was being there before *him*. And who will help him? Who is there to pity the pitiless boy? Who comes up at the moment, her face beaming with hope? She sees the fruit gone—dashed to the ground—her cherished pleasure snatched from her, and yet has hardly time for the tear to gather in her eye, or the grieved tenderness of her heart to find expression in her face, when the tone of anger and the flush of selfish disappointment on her brother's countenance distract at once her attention from her own trouble, and she hastens to soothe the rage, and to use such remedies as she can. Ah, little girl, we are so glad the wasp stung him instead of you! and, though you have lost the peach you were waiting for, see, there is another hanging close by among the leaves, which you may safely gather now, for I think *that* wasp has left its sting in the proper place, and is not likely to trouble any one again.

But why were wasps made with stings? and of what use are they in the world, even if they have no stings to make them a terror to the selfish and the cruel among human beings? They make no honey for the delectation of the lovers of sweetness—no mead for the libations of our ancestors was ever manufactured from the secretions of their industry. For what benefits to society from their labours are we indebted to these irascible creatures, whose very appearance at the breakfast-table has the effect of a bombshell in dispersing the company; but who seem to have hitherto failed to establish a claim to respect or affection, or even to have exhibited in their instincts any model on which more rational creatures may improve in their elaborate efforts after comfortable homes or luxurious living?

No honey, no wax, for they must die with the season which gave them birth, and the cells they inhabit for the brief space of their existence are of far less durable materials than those which we see in the structures of the honey-bee. Yet for delicacy of structure, for minute elaboration of its material, for exactness of adaptation to all their wants, the nest of a wasp will never suffer in comparison with that of any other living creature. Fragile in the extreme, it is protected from attack by its situation, or concealed underground at a distance of more than a foot from the surface; and, like the nest of the bee, is furnished with rows of cells for the habitation of the workers. A question seems to arise as the process is examined whether the wasp exists to build his nest or the nest is built for the existence of the wasp, since his life is extended so short a period beyond the completion of his work. But nests have been seen above-ground which served for several successive generations of wasps. One especially was long preserved under a glass case in the drawing-room of a house which had for years been infested by these irritating insects, whose resort no one could discover until the repair of a disused chimney led to the discovery of one of the largest known nests, which the wasps had inhabited, repaired, and enlarged for years.

The character of our wasps has greatly suffered at the hands of prejudice. They are doubtless armed against attack, and are justly feared when the ripened fruit hangs daintily in presence of the creatures, both of human and of insect life, for whom, doubtless in common, such feast is bountifully provided by the great Lord of Nature. But *who ever* saw a wasp attack a wasp? They never indulge their irascibility against one another; they all work in harmony; nay, more, they are very good neighbours to many other creatures. For instance, one of our ground-wasps always makes its nest in the close neighbourhood of one of the humble-bees, and no one ever knew the wasp, sweet though his tooth may be, help himself from his neighbour's store-room, but he honestly goes out and caters daily for himself. Indeed, the wasp has been hardly dealt with in being held up to hatred as a *waspish* creature. He is not so "waspish" in his temper as his cousin the bee; and though he has not the character of being a hard-working labourer, yet he is ordinarily far too busy in getting his daily bread to turn aside and waste his time by picking a quarrel with the passing stranger. When he comes into the breakfast-room, drawn by the savoury odours of jam and preserve, he only asks for a share; but when every handkerchief is flashed over his wings, and when ferocious attemps are made to crush him when, in his hurry to beat a retreat he has struck against the window-pane, his temper would be more angelic than even a wasp's can be expected to be if he did not retaliate by attempting to use the weapon with which Nature has endowed him. But we can examine a wasp's nest with far less risk than we can investigate a beehive, if only we do not irritate the inhabitants by too officious a curiosity. When a scorching summer's sun has quickened the energies and somewhat tried the temper of the busy colony as they pass to and fro, we must not stand in their way, and we must beware of treading on some wearied insect crawling home on the ground and too weak to fly, though quite strong enough to sting with effect.

There is, however, a great difference in the temper of the different species of wasps, as we are told by those who have studied them. The large British wasp, which builds in the ground, is said to have the sharpest sting; the wood-wasp (*Vespa sylvestris*), which hangs its home in the bushes and is very common in the north of England, is the most ready to use its sting, probably because, having so conspicuous a nest, it must be most alert in its defence; while the red wasp (*Vespa rufa*) is said to have the most amiable and inoffensive nature of all, and to hold the place among wasps that the humble-bee does among bees. Of course, when we attempt to take their nests, any wasps will sting if they have a chance. Their devotion to the home they have built by their own mandibles is very strong, and even the loss of their queen will not drive them from it. No wonder, then, that like other creatures they have great repugnance to being disestablished and disendowed. They will bear many things for peace sake, but they would not be wasps to stand this. When, however, the wasp-hunter, safely protected by his veil and strong leggings, has succeeded in digging up the nest of the ground-wasps or cutting down the branch with the nest of the tree-wasps, he may carry it where he will, and the little republic will cling to it still. He has only to place it where they can have easy access and work at leisure, and they will soon begin to repair damages and to feed with devoted attention the young larvæ in the combs.

But after all, if we do leave them alone, what is the use of these dangerous little creatures, who will *not* leave us alone if we disturb them in gathering the plum or the gooseberry we have been cultivating for ourselves?

We shall soon find, if we will only study the life of a wasp, that it deserves to be reckoned among our real benefactors. If they take toll of our gardens in autumn, they have been really working for us in spring.

First of all, they are active scavengers. No vegetable matter, no decaying garbage in which the vinous fermentation has commenced, comes amiss to them, and they clear off much that would otherwise taint the atmosphere. So also they gather a great deal of rotten wood. But besides this they have a very carnivorous appetite, and devour spiders, flies, and especially caterpillars, those enemies of the farmer and the gardener. Dr. Ormerod, the charming historian and champion of the wasp, brings forward instances in which the careful destruction of wasps has in a year or two resulted in infesting the place with Egyptian swarms of flies. About a wasp's nest the wings of flies and other insects may be gathered in handfuls; in fact, they form little insect kitchen-middens. Every one who has studied the subject is aware of the mischief which is done by killing off the larger animals, as for instance the kestrel and the owl, who destroy millions of rats and mice, lest they should occasionally fall in with a wounded partridge. The sea-birds, who used to be butchered by every stupid fellow that could borrow a gun when out for a holiday, were found to be so useful to the sailor and even the fisherman, though they eat fish, that at length they are protected by law. But in the case of these larger creatures man can by mouse-traps and other contrivances do something to supply their office. In the case of insects he can do nothing, though, happily for him, they are too small and too rapidly propagated to render his ignorant efforts for their destruction successful, and so the wasp still lives in spite of gardeners and their boys and helpers, whose whims and prejudices would long since have doomed him to the same fate which has befallen the noble peregrine falcon at the hand of the gamekeeper, if the power of gardeners had been equal to their will, and their decision final.

Now, if we can only persuade the gardener to observe the wasp in spring, at the very time he is most energetic in its destruction, we shall soon convince him that he has in it a true friend. With the bright sunny mornings in April, the old queen wasps that in some sheltered cranny have survived the frosts of winter, come forth, not like lone widows, but the royal foundresses of new kingdoms. We may see them then busily occupied on the slender twigs of the gooseberry-bush, or in the young wood of the apricot. There is no fruit for them there to pierce; they are feasting on the aphides and the mildew insects, which the gardener with all his syringes and decoctions of tobacco-water cannot subdue as the wasp can. Wherever these minute little pests most abound, there you may see the wasps' mandibles hard at work, carefully clearing off all the gummy exudation till the top of the young bough is reached. If the red spider or its eggs come in her way, the wasp uses them with much relish as her *sauce piquante*. One wasp unmolested will thus, in a day or two, clear the insect pests from a whole tree, and will secure the owner that crop which he never could have had without her aid, and on which surely she may put in a claim for her future family to take tithe. It is scarcely possible to calculate the number of aphides which a hungry queen wasp will thus devour in the spring and early summer months.

But it is rather as the original paper-maker that we would discourse of wasps, for here it is that their most marvellous instinct is fully displayed. Excepting in the shape of its cells, there is nothing in common between the nest of the wasp and the hive of the bee. The architecture, the material, the position, the arrangement, the uses all vary in the two families. All bees make their combs of wax collected by them from plants, and kneaded. All wasps are paper-makers, not wax-collectors. The comb of the bee is destined for various uses,—to be the home of eggs and larvæ, or young bees, to hold either honey or pollen, and is also intended to last for many seasons. The comb of the wasp is built but for one year and for a single purpose—to contain the young from the egg till it comes forth a perfect insect. Then, while the cells of the bees' comb are arranged back to back in the same comb, which hangs vertically from the roof of the hive, the wasps' comb is suspended horizontally by a pillar in its centre, and is composed of a single layer of cells, all opening downwards. There is one point in which the wasp shows greater architectural power than the bee. The latter trusts to nature or to man for its hive. The wild bee finds a hollow tree or a crevice in the rocks, in which the combs may hang protected from the weather. The wasp, not content with manufacturing its own household furniture, builds the house also for itself, and that of the same material, and relies upon her own exertions to defend herself from the effects of wind and rain.

The wasps, like the bees, comprise prolific females or queens, barren females or workers, and males or drones, which, among the former as among the latter, are stingless. But the lives of the queen wasp and the queen bee are very different. The queen bee, from that sunny morning on which, like some Viking of the North in olden time, she set forth to found a new empire on new soil, with the swarms of her attendant and devoted subjects, never again leaves her palace, far more closely immured within it than any Eastern Sultan or Japanese Tycoon, until she has done her life's work, having known no labour save that of depositing myriads of eggs. But then she is founding a dynasty, and her little kingdom may be handed down in the female line for many generations, unless prematurely extinguished by the hand of the spoiler—man. How different the life of the queen wasp! Like some hardy colonist, she goes forth in early spring into the wilderness, the lonely and solitary survivor of her family, with no obsequious damsels crowding round her such as those that throng the court of her more dusky cousin. She has no parental rooftree which is hers by succession. No inheritance has come down to her, but, like the human pioneer in a new land, she must cater for herself. She must be her own architect, her own mason, her own gardener and purveyor, and this too with the cares of a family coming on, and all her youthful progeny, swathed and helpless, dependent on her sole exertions for everything.

During the winter she has lain torpid behind some shutter or cornice, in the crevice of an old wall, or under the shelter of a roof, in the cranny of a chimney-stalk. With the first warm mornings of April she comes forth, very often to perish prematurely by the cutting spring frosts, and keenly chased by the hungry starlings and the gardener's boy. Few of her race have survived the blasts of winter; and of those that have, fewer still run unscathed the gauntlet of all their enemies. And still for weeks she must remain alone and unaided, with food to seek, a home to find, a nest to build, and then all the hungry grubs that soon follow to feed. She does not hurry about beginning her nest, but takes a long time in selecting her house.

There are six kinds of wasps in England besides the hornet, which is in reality only a large species of wasp; and of these some build in the ground, in holes, or in fissures of rocks; others hang their nests in trees or among bushes. Whatever be the situation, the nests of all our species have much in common. They are all built of paper, made by the insect itself, and whether hanging from a bough with the paper dome that shelters them from wind and rain, or snugly suspended under a roof-beam, hidden in a hollow tree, or excavated in a bank-side, the nests all begin and go on in the same way, adapted, of course, to circumstances. The tree-wasps take care to have their dome smooth and rounded to carry off the rain at once. The hornets make a stout case when their nest is exposed, a very slight one when they choose a hollow tree or similar shelter. The ground-wasps make a strong, rough, coarse, brown paper shield underground, but a much firmer and lighter shell when they build, as they sometimes do, from a rafter.

But how do the wasps get their paper? They manufactured it long before the Egyptians had discovered how to pare papyrus-stems into shavings to make their books. Before the Chinese had learned to squeeze and spread out the thin cotton pulp into sheets of paper, the wasps knew how to apply almost every substance which has been employed by our paper-makers to the fabrication of their dwellings. Grass-fibres, withered leaves, rotten wood, paper cuttings, bark scrapings, the thin coating of buds, vegetable down—all these and many other substances are worked up by the wasps, and laid on precisely in the same manner by all. If we watch a wasp on an old rail or gate-post, we shall see it diligently peeling off little strips of woody fibre, which it rolls into pellets and carries home in its mandibles. There are also many kinds of rushes and water-plants, the stems of which are covered with tough filaments, which the wasps peel off, and which make *papier-maché* of a stronger and superior quality, much more proof against the rain than the wood scrapings. As soon as the busy insect has rolled up a good-sized pellet of wood or grass parings, she tucks the burden in and under her mandible or large pincer jaws, and with outstretched neck flies home. Then on arrival she retires within the nest to rest for a minute, and coming out again, promptly sets to work. If what is required be the strengthening or enlarging of the outer walls, she gets astride the edge of the shell of the nest, takes hold of the pellet with her fore legs, presses it down firmly, and kneads the end of it, fastening it with her gummy saliva; and then slowly she walks backwards, unrolling the pellet as she goes, pounding and working it firmly down, while keeping it moist, and when she has come to the end she runs forward again, and commences to retrace her steps, drawing the edge through her mandibles, flattening and kneading it as she goes, and repeating the process several times till the little addition she has made is evenly and neatly welded on to the structure, and as soon as it is dried cannot be distinguished from the former work. The nests often have a striped appearance, caused by different wasps bringing materials of different colours, and working in their own quota as they find a vacant place on the edge of the nest.

But the first commencement was very humble. The queen began in spring by attaching a little cap of grey paper, of the shape of a tiny parasol, to a stalk of paper gummed securely to the under-side of a branch or stone. Below the cap this footstalk is extended and spread out to form the beginnings of four little octagon cells, hanging downwards, in each of which she drops an egg, and glues it into its place.

The lonely lady then begins to enlarge the cap, and adds other cells on each side of the first, strengthening the foundation pillar as she proceeds. Her labour grows upon her, for the first eggs hatch, and now she must feed her young and go on with her house-building at the same time. She busily flies backwards and forwards to the nearest bushes, and hurriedly gathers a supply of juicy aphides or well-fatted spiders to support her larvæ.

At length the first brood is hatched (though by far the greater number of nests begun never reached this stage, owing to the precarious fortunes of the mother), and then the queen begins to assert her dignity, and to rely upon the labours of her offspring. The nest is soon brought into shape, and the covering drawn down and completed underneath, so as to form a perfect sphere, with a small port-hole near the bottom for an entrance. But the original work must be rapidly undone. There is no room for enlargement within, and therefore one outside cover after another is added over the former, each quite independent of the preceding layer, which is removed from the inside, as the outer cover is completed. Meantime the comb inside grows apace, as fast as the walls expand. The four original cells grow into a comb, with six or seven combs hanging layer beneath layer, each perhaps six inches or more in breadth; and the pillar, which is the centre and key of the work, is proportionally strengthened, and the strips of paper which attach it above are doubled and trebled to bear the additional weight. As the comb grows, every day the inside of the case is cut away to make room for it. Thus the quantity of paper used is very great, for the cuttings of the old are not used again, or, if they are, it is only after they have been nibbled and reduced to pulp again by the jaws of the workers, and then mixed with new material. The floor of the nest is thus always strewn with scraps of used paper, as that of a beehive with waste wax-plates. Sometimes, too, the wasps scamp their work, and if they find leaves at hand that suit their purpose, they will work them into the nest without any previous manufacture.

Many foreign wasps—and the species of wasps are countless—differ much in their architecture from ours. Some, in countries exposed to much rain and wind, make their paper stout, thick, clean, and white as cardboard—so strong, that it may be knocked about and washed with impunity. The cardboard is made of the finest cotton down felted together: as many as sixteen or more layers may be counted forming the wall of the nest. This wasp lives in Demerara, where the sudden and violent rains would soon wash away the whitey-brown fabric of the British paper-maker. Then again in the East Indies, where rain hardly ever falls during the lifetime of the wasp republic, another large species is content with mud with a little straw mingled, like the bricks at which Israel had to toil in Egypt, and makes a huge clumsy structure, which one heavy thunderstorm would reduce to a hopeless wreck of mud. Others again use only leaves, and are tailors rather than paper-makers. Others make no roof at all, and some hang their combs with a paper umbrella over them, but no flooring or other protection. One species contents itself with using a great leaf for its cover, while it makes its cells of paper. In the Holy Land there is one species which hangs its combs in cavities in the sandy banks of rivers, and which, suspending its great comb from the roof, economizes labour by omitting all covering, while it prevents any injury to the comb by running a thin irregular sheet of very fine whitey brown paper along the under-side of the roof of the cave, so that no sand or pebbles can fall on to its nest.

Dr. Ormerod tells us also that wasps can foretell the weather with a preciseness far superior to that of the most skilled of almanac prophets. A gamekeeper who had spent his life in a land of brooks informed him that the height at which wasps make their nests above the water is a rough index of the amount of rain that is to be expected during the summer. In a wet season they choose the top of the bank, in a dry year they excavate nearer the water-level. Again, it is found that when a hedge-bank is selected, instead of the more ordinary situation, the edifice is much slighter than when wind and rain have to be provided against. But still, under whatever conditions it is built, we can always recognize the difference between the architecture of the different kinds of wasps. Besides the hornet, which is only a very large species of wasp, and must always be counted with them, there are three kinds of ground-wasps and three of tree-wasps in Britain. Each species makes a distinct sort of paper, and we have only to hold it up to the light, to read the water-mark of nature's impressing, and we can recognize the builder.

The hornet (*Vespa crabro*), for instance, who does everything on a large and coarse scale, makes its paper very thick and brittle, of a yellow colour, composed of fragments of decayed wood, bits of straw, and other rubbish glued up with sand into a coarse pulp. There is a good clear space, inside the hornet's nest, between the combs and the wall, like the open space that used to be kept in Vienna and other fortified towns between the houses and the walls.

The common ground-wasp (*Vespa vulgaris*) builds on the same plan, but its paper is a very superior sample. It is much finer, the fragments of wood are much more carefully beaten into pulp, and instead of being yellow, the colour is much more varied, generally with stripes of whitish-brown. This wasp will build anywhere where it can find shelter. Though generally underground, yet a good cottage roof, especially if it be thatched, never comes amiss. It has even been known to build attached to a sugar-loaf. This last was rather an extravagant use of the loaf for a wasp, since it usually prefers to take the sugar inside. At least we have read of a Government sugar store at Shahjehanpoor, in India, which was taken possession of by a swarm of hornets, and held by them in defiance of the order of the East India Company, till the end of the season, when on the commissariat officer claiming his charge at last, he found they had got through two thousand pounds of sugar! We must confess that the paper of the ground-wasp is not of the strongest quality, and would not be at all appreciated in the grocer's shop.

Our other common ground-wasp (*Vespa Germanica*) makes a similar nest, but it can always be distinguished by having no mottling or stripes of colour in its construction, but is of a uniform dull grey colour, and in texture is not more stout or durable than its cousins. It easily comes to pieces, though there is no stint of material in its construction; but the layers are heaped on overlapping each other, and without the neatly trimmed edges which mark many of the others. There is generally a neat little mouth, with a landing-place and porch at the entrance of this nest, which is always near the bottom of the building. Much prettier is the ground nest of the red wasp, a rather rare species, which lays on the plates of paper very neatly, and with the edges smoothly tacked down.

The tree-wasp makes much larger sheets of paper. *Vespa Britannica*, the commonest tree-wasp in the south of England, makes its nest of much better paper, for it uses stout vegetable fibre instead of rotten wood in its construction; and, indeed, it requires a much stronger material, for the nest hangs exposed in a bush or hedgerow, open to all the changes of weather. The paper is prettily mottled with white, brown, and yellow streaks. One other tree-wasp (*Vespa sylvestris*), so common in the north of England, makes its nest generally of paper of one colour, but very tough, and hanging loosely, like petticoat flounces, one over another, in a great many layers. The hole of this nest is always exactly at the bottom. This bell-shaped nest is, I think, the prettiest of all.

The internal domestic arrangements of all these species of paper-makers are the same. The eggs we have said are glued to the bottom of the cell by the mother wasp, or else, of course, they would drop at once out of the inverted cup. But when they are hatched, and are still helpless, the tail of the little infant remains glued to the top of the cell in its old egg-shell; and though it moults several times, still its tail remains glued until it has nearly reached its full size. But it often becomes detached, and then the workers, who have no toleration for untidiness, and treat everything that is out of its place as dirt, are sure to carry the little larva away without pity, and eject it with other scraps and rubbish —a fearful warning to other baby wasps to keep in their cradles. As soon as, after various moultings, the little larvæ have attained their full size, nearly large enough to fill the cell, but still able to turn round in it, they begin to weave a silk case, which is to protect them while they change to the pupa or chrysalis state. This done, they weave a white silk cap on the bottom of the cell, and then cast their skin a second time. It is a curious fact that the wasp larvæ have sharp mandibles, with which they mince for themselves the food brought them by their nurses. They get a new pair with their new moult, which are used at the end of their chrysalis existence to cut their way into the outer world. As soon as the newly-awakened insect has cut its way through this nightcap, it begins to feed itself, and actually eats its slight dress piecemeal directly it emerges from the cell a full-grown, pale-looking wasp. Soon its wings expand and dry, and it sets to work at once at paper-making, as if it had long since served its apprenticeship. Meantime all the old silk casing and other loose fragments are cut off the empty cell, and it is considered ready for a new-laid egg, though much dirt may be left at the top, which is never thoroughly cleaned out.

I must give the history of the growth and end of the wasp's nest in Dr. Ormerod's words:—"By the conjoint labours of all the busy workers, here a little and there a little, the nest grows. The work of one week may have to be renewed the next week, to make way for modern improvements and for the requirements of the growing city; and, as we have seen, it has nearly all to be done twice over. But wasps work very hard, and the nest grows visibly day by day. The little egg-shell in which it began is lost in the changes which the top of the nest undergoes. The slight strap from which it hung is now quite inadequate to sustain the daily increasing weight, and new points of attachment are sought to projecting roots, or stones, or branches. Sometimes a branch runs all through a nest. Or, failing these, the original point of support is strengthened by layer upon layer of paper rubbed smooth, and thickly coated with wasp-gum, to preserve so vital a point from all accidents of wind and weather. . . . .

"One thing more British wasps' nests have in common, viz., the end of all their labour, the wreck and ruin of their wonderful fabric. The history of the most long-lived swarm of wasps extends only over a few weeks. The end comes very speedily, as well as surely, whatever the cause, and the story of the decay of the nest, whose growth we have traced, may be traced in a few lines. Thus: no additions are made to the structure, the repairs are neglected, the loose ends are not neatly cut off and fastened down. A few idle wasps hang about, but the nest seems almost deserted. Perhaps a shake of the hedge will bring out a few fussy wasps for a minute, or a sunny afternoon will develop signs of life in the remains of the swarm, yet their strength is gone. A cold night or two, a few damp cold days, and all is over. Now the collector may take his prize safely; and he must be quick about it, for if he delays, the rain and wind will soon destroy whatever of this curious structure the moths and wood-lice and earwigs have spared. These are now its occupants. The little creatures who made it, and held it against all comers, have succumbed to cold, and disease, and old age, like other brave soldiers. They have skulked off to die, like old cats, away from home, and the most unlikely place to find a live wasp in is an old wasps' nest."*

So much for the story of paper-making and wasps. Much more remains to tell, for wasps yield not to bees in interest and in display of forethought and instinct, which can never be explained but by reference to Divine prevision. But I hope I have told enough to lead my young readers to look on a wasp as something better than a "horrid nasty thing," to be crushed on the window-pane or trodden underfoot whenever there is a chance.

* Ormerod, "Natural History of Wasps," p. 209.

## BESSIE'S CALENDAR.

To live in the midst of fields, and meadows, and pastures, wide-spreading woods, hop gardens, apple and cherry orchards, and nut groves—doesn't it sound delightful?

But little Bessie did not always find it so.

First let me tell you what kind of house she lived in.

Strangers from the county town, who drove past it in flys, used to exclaim, "How picturesque!" and sometimes made their drivers pull up in order that they might take a longer look at the cottage; and wandering artists from beneath the shade of their white umbrellas have sketched it in pencil, in chalk, and in water-colours. And it was a very pretty place to look at from the outside.

It had a thatched roof spotted with patches of moss-like bits of green velvet carpet, and walls of ripe and dim red and purple brick, parqueted with grey beams,—both variegated with many-coloured lichens and fungi.

But the leaky roof let in the rain, and the cracked chinky walls, the warped doors, and the often paneless windows, the draughts. Owing to the pig-packet-like crowding of the inmates, the air was never pure, and yet in winter the lower rooms, unevenly paved with worn, pitted, chipped bricks, were miserably cold. There was not a mite of comfort, and scarcely an approach to decency, about the "pretty" place.

Bessie, a dark-haired, dark-eyed little puss, whom exposure to all weathers had made almost as dark-skinned as a little gipsy, was an orphan. When her father and mother died of fever, her mother's married sister had adopted her. She did not receive a very warm welcome from her uncle by marriage, or her cousins—a swarming hungry brood, all older than herself. Then the aunt died, the uncle married again, and the new wife looked upon this little "anteloper," as she called Bessie, even more coldly than she had been looked on before. The child was not turned out of her uncle's house, but as soon as she could do anything she was set to work; all the little she earned she had to give to her stern step-aunt, and very scanty was the "keep" she received in return; whilst as for raiment, Bessie's clothes had been worn by all her girl cousins before they came down to her.

Early in the morning, sometimes almost as soon as it was light, Bessie was roused from her sleep and her dreams of "mother" and rest by day

LUSSIE.

as well as night, and sent forth to her field labours with the crust or two of stale dry bread, which, with water got anyhow, and now and then a cup of wishy-washy tea when she came home, or a scrap of hard cheese about as toothsome as horn, formed her provender for the day. On Sundays, in prosperous times, she just tasted "pig-meat" or "cag-mag," picked up cheap overnight in the nearest town. On other Sundays she and her cousins had to content themselves with bread, or potatoes, or "tea-kettle broth" for their dinner; and Bessie was grudged her scanty share of even that Lenten fare.

Bird-keeping, *i.e.*, scaring, was the work to which the little girl was first put,—from November to April in the fields, and again later on in the orchards. For hours the little girl was all alone, save for the big rooks and smaller black and other birds, whom it went to her heart to have to frighten off from the corn and fruit. Her lack of human company she did not mind. What she got of it was not so pleasant that she should regret its loss; but she wished that she might make friends with the birds. Sometimes, moreover, when she sat all Sunday long beneath a hedge or on the sun-and-shadow-chequered orchard grass, and heard the bells ringing out from the grey and red towers, half hidden in trees, of the village churches round, she could not help wishing that she too had nice Sunday clothes like the farmers' daughters, and could go to church hand-in-hand with mother in a silk dress.

But having no human being to love her, Bessie was obliged to make her friends of the so-called "inferior" creatures, animate and inanimate, in the midst of which she lived.

Her life was a calendar of nature. She did not always know the names of the things she loved to watch, she could not have given any but the baldest description of them, and yet she knew *them* well, and had their images photographed on her little brain. Her dark eyes looked as if they had been meant for laughing ones. As things were, a wistful yearning for smiles to awaken them to answer, was their chief expression when she was in company with the people of the village; but when she thought there was no one to see her watching or listening to her country favourites, light danced in those dark eyes. She welcomed, after the dead time of the year, the reappearance of the wagtails and the chaffinches, the new year's song or chirping of the redbreast, the thrush, the blackbird, the sparrows, the wren, the skylark, the woodlark, and the titmouse; the rooks returning to their nests, the jackdaws resuming their church-going, the nuthatch running up the trees; the first bat, the first butterfly; the white lambs on the brightening grass; the worms and snails crawling out of their winter quarters, bees booming from their long-hushed hives, gnats and flies again buzzing about; catkins hung upon the hazels, the red gold of the fresh crocuses, the pale gold of the rathe primrose, the snowdrop trembling upon its slender stalk, the nettles, honeysuckles, speedwell, dandelions in flower.

And as the year warmed, lovingly she noted the rooks and ravens building, and listened to the blackcaps, yellow-hammers, green and goldfinches singing, the drowsy coo of the ringdove, the startled cry of the curlew, even to the croak of the awakened frogs. The hoot of the wood-owl, the scream of the woodpecker, the pheasant's crow were more of her music. "Glad to see you again," her heart said to the wheatears, swallows, and martins. The only drawback was that the snakes had come back too. She was afraid of snakes, and had got a notion that they were made by the Devil. She had picked up a good deal more about the Devil than about God; but God Himself taught her that the flowering coltsfoot, the floating gossamer, the peacock butterflies, the daffodils, the violets, the crowfoot, the fruit-blossom, the catkins on the aspen and the filbert, the young leaves on the gooseberry-bush, the flowers upon the elm, the periwinkle, the wild hyacinths, and wood-anemones, were His handiwork.

Next she rejoiced in the season of the nightingale, the cuckoo, and the cowslip, the milk-white blossoms of the blackthorn, leaping trout, lady's-smock and ladybirds, harebells and hawthorn, dragon-flies and cabbage butterflies, glow-worms and guelder roses, lilac, and laburnum, and ragged Robin.

June Bessie loved because of its dog-roses starring the hedges and littering the lanes. Its hay-making was a very different thing to her from what it is to you. She was kept hard at work instead of being allowed to roll, and romp, and bury other children in the fragrant drying grass, as you do. Sheep-shearing she did not like, because the sheep struggled so when they were thrown down to be shorn, and bleated so piteously when they trotted off released,—scarred sometimes with nasty snips from clumsily-wielded shears.

In July the calling quail, the plump young brown partridges rising out of and dropping into the green wheat,—here and there beginning to turn yellow,—the scarlet pheasant's-eye, the bluebottle in the corn-fields, the straggling traveller's-joy in the hedges, the hemlock in the ditches, the white lilies and the hollyhocks and sunflowers in her uncle's cottage garden, were more of Bessie's friends; and in later summer and autumn the spotted foxgloves and nodding Canterbury bells, the floating thistledown and gossamer, and the second blossoms of the honeysuckle. She welcomed back the fieldfares, but felt sad when the swallows took their flight southwards, and the starlings mustered to follow them. But Bessie's calendar included labour from year's end to year's end, as well as pretty things in their season. Bird-keeping and hay-making were not the only work she did. She had to help in rag-cutting, hop dressing and tying, couching, thistling, weeding, fruit-picking, nut-gathering, hop-picking, acorn-picking. Except when she was birding, she got no rest on a week-day so long as she was awake, for in the cottage, too, she was made a little drudge. At first she thought it cruel that, slaving as she did, she could not get a word or look of love. No one praised her, however hard she worked. The kindest thing ever said of her was by her uncle one Sunday afternoon, as he came away from the little grave in which she had just been laid,—" Poor little gal! she didn't cost nothin', and we s hall feel the miss on 'er."

---

## SILK AND SILKWORMS.

" A LONG pull, and a strong pull, and a pull all together," was the practical language of the thousand and one slender fibres of the silken cord, strained and tightened from the bell-wire, which the master of the house pulled with a vigour and determination that bespoke attention, and with a result which might have awakened the Seven Sleepers.

"You'll break the rope!" was *our* exclamation, instinctively closing our ears with our hands: "you'll certainly break the rope, it's only *silk*."

"*Only* silk! and what more, and what better would you have for strength, for elasticity, for carrying weight, or bearing a pull? Look at little Madge at the table, winding away off her cocoon, and tell me what other material in the world for its size is half as strong? Talk of a hempen rope, an iron cable! you might as soon make use of a rope of sand, if either of them were taken in the slender form of that little thread; but *twisted, combined*, made into a cord, I think we have something far more telling than even the old fable of the bundle of sticks, as to the strength of *united* action."

"Oh for united action now!" Was it not the voice as of a plaintive and much-injured being, issuing from the glistening thread, that many times already had snapped under the impatient hands that attempted to wind it? "Had I but my ten sisters here to twist and twine their threads with mine as erst so lovingly we fed together on one mulberry-bough, I had not now to endure alone the impatient shocks that shiver my whole being and send a tremor from end to end of my thousand yards of length."

"Tiresome thing!" exclaimed little Madge, whose single thread of silk from the one cocoon which was her own especial property, had, notwithstanding all the elasticity so truly attributed to it, snapped now once too often for her impetuous spirit, and on whom the above valuable suggestion had been entirely thrown away, probably because entirely inaudible. "Tiresome thing! It *won't* go on! *I'm sure it breaks on purpose!*"

"And a good purpose too, if it were to teach you to twine a little patience and perseverance with the slender thread of a little objectless amusement; the threefold cord would stand a stronger pull than that which broke your silk, and that, slender as it was, would have given you no disappointment, if you had consented to the wish of the others, to wind the three cocoons together.

"But while your patience has time to recruit itself, come here and listen to a little silkworm memoir which I happen just now to have met with; just such a history as the little being inside that cocoon might have uttered, could it make itself understood by us:—

"Like all beings, clad not in the rough and borrowed garments of the flax or of cotton material, but in the luxurious folds of their own ancestral silk, *I* boast a *very* long line of ancestry. *My* native land possesses a history older than that of any other nation on the face of the globe— a land where also the theory of the transmigration of souls lent encouragement to the hopes even of a silkworm, that the soul which had so rapidly been transmitted through its various and exciting transformations, might one day fill the body of a mandarin, an empress, or an emperor, clad once more in its primitive raiment, and walking erect on two instead of crawling upon many legs.

"Comparatively recent, that is, not quite four thousand years ago, were the days when the great Emperor Hoang Ti cast the eye of appreciation upon the labours and the lives of my progenitors, and his Empress the inestimable Si Ling Chi, with the lilliputian feet and the fairy fingers, first caused to be assembled within the precincts of her Celestial garden, multitudes of the many-legged race, and gathered with her own hands the dainty leaves of viscid mulberry wherewith the voracious appetites of the mothers of millions might be appeased. Strange were the transformations of their bodies, strange also doubtless the transmigrations of their souls, prefiguring the heights to which patient industry, even embodied in a grub, may yet attain. Yet it is recorded concerning their work and the productions of their lives, that evermore that which was most hidden and nearest the centre of their body was the richest and most highly prized, while the showy exterior and lighter surroundings were thrown away as comparatively worthless.

"Nor was the Celestial lady content with the task of benevolence which consisted only in ministering to the hunger and pampering the appetites of the army of insatiables; but as in China and among our own race all things human are reversed, so was her chief work of benevolence, not that of clothing the naked, but of relieving the overcharged and sleeping bodies of my ancestors of their superabundant clothing.

"Up to that time the skins of slaughtered sheep sufficed to cover the human frame, and protect from the inclemency of wintry seasons a race who possessed neither the art of producing from their interior substance their external covering, nor yet the energy to condense within a single summer season the duties of a lifetime. But men were many and sheep were few, and the cradles of our race were used by the Empress and her attendant ladies to enclose their own august persons.

"Willingly we afforded to them the shelter no longer of use to ourselves, feeling abundantly requited by the provision so liberally made for successive generations of our family in the planting of extensive groves of that paternal tree whence we derive, not only the strength of our constitution, but the texture of our raiment. Well it was that such care was bestowed on their nourishment, as otherwise it might have become necessary to resort to emigration,—a step that would not only have seriously lowered the self-respect of a race whose welcome has in every land, and in all times, anticipated their arrival, and who have never had to wander in *search* of a settlement, but have nevertheless entirely baffled the boast of our great patroness Si Ling Chi, who reserved the best and richest of her silken fabrics for the great sacrifice of Chang Si, and suffered not the outside barbarians so much as to see our grandmothers, or to handle the delicate threads they spun.

"Then were we had in great esteem, then was silk worth its weight in gold, and then did the merchants trade with other lands for these precious things, making payment for the same in fabrics cunningly woven by secret arts from the many-threaded cocoons of the mulberry groves.

"Rough and hard were the men of old, and wool was for them the fittest covering—best suited both to their unclothed bodies and their sordid souls. And of them the roughest and the hardest were the Romans; and of the Romans, one stronger and braver than many yet saw and coveted the strange softness and dazzling brightness wherewith shine the garments of those who are clad in the cast-off raiments of our grandmothers. How must the great-grandsons of those noble old mulberry-eaters have shuddered could they have witnessed the scenes enacted before the first silken curtain which Julius Cæsar spread over his tent in the Colosseum, where human gladiators and savage animals fought together as fight the tigers in the jungle, and the outside barbarians shouted at the spectacle! The show was brave, the silk was a gorgeous prodigality, and—Cæsar was a great man.

"Yet did the Empress of the Celestials long outshine those of the West, for not even to the Empress Severina was the luxury accorded, so universally indulged in by the ladies of the little feet, of wearing a dress of a material so costly.

"Still the natural desire of that half of the human race for costly, soft, and splendid attire was destined to be gratified by means apparently the most unlikely.

"Clad in costume of dingy brown, the produce of the sheep or goat, and with no weapon but a staff, two men on foot invaded and succeeded in robbing of its precious monopoly the land that had hitherto cherished and protected us; and by ingenious concealment within those very staves, they imported in small numbers, and carried across the mountains and rivers of India, and the plains of Persia and Syria, the precious eggs whence should be hatched the successive generations of our now widely extended family. Since that epoch, our pride of family, our exclusiveness in social position, have rapidly given way before the revolutionary tide which has swept over Europe, and has even procured for the barbarians of the West a settlement in the early home of our race.

"No longer do emperors and empresses enjoy alone the privilege of wearing the produce of our labours. Not only does the blue ribbon sustain the star of honour that distinguishes the breast of the British mandarin, not only does the gorgeous train sweep gracefully round the person of the royal dame, the village maiden weaves in her golden hair a

4—2

tress of brilliant silk, and even the schoolboy, when he stoops to tie his shoe, fingers the ribbon that was once the work of a silkworm like myself.

"My personal history is but brief, for not to silkworms is it given to con again in recurring seasons the experience of former years. With the early sun of advancing spring I, who till then had been but an egg and had lain tranquilly on a shelf through the bleak storms of winter, crawled into being, a slender black thread of life. Larger I grew, for future greatness dawned on my distant horizon; and perceiving that the first, last, and only duties incumbent on a being like myself were to eat, to grow, and to cast my skin when too tight for my expanding body, I diligently pursued these avocations, and with a success that rivalled the largest, the most voracious, and most sluggish of my companions. Soon a longing for change seemed to oppress me, and as I raised my head to seek for new spheres of action, a torpor crept over my frame. I quitted the leaves on which hitherto I had feasted, and cast my lot as a dependent being on an isolated spot selected at random. But I soon felt the hour was come no longer to receive but to impart, and that in the process of giving forth of my substance I was myself enriched. I have now for some time dwelt in the midst of a golden abundance, never hungering for food, and possessed of that wherewith I may clothe the needy. I shall now soon end my career, a creature different far from my small beginning, feeble indeed in flight but prolific in eggs, and ready, after fluttering a few brief days a fair white moth and leaving innumerable hostages to posterity, to enter upon whatsoever stage of transmigration the theories of Confucius may point out as the future of a perfected *Bombyx mori*."

So much for the autobiography of the little silkworm; but by what little things may the history of nations be affected! When those two wandering Persian monks of whom we spoke contrived from the Indian Missions to penetrate the hitherto sealed empire of China, they discovered that the priceless tissues on which at that time the dainty dames of Byzantium expended princely fortunes were not combed from plants or distilled from Oriental dews, but were the produce of an unsightly caterpillar reared from a tiny egg. So important did they deem the discovery that, big with the secret, they traversed the breadth of Asia to lay it at the feet of the Emperor Justinian. Recognizing its importance, he persuaded them, by right imperial promises, to retrace their two years' journey and bring back the precious eggs. Rivalling in cunning the crafty Chinaman, they succeeded at length in filling their hollow canes; and those pilgrim staves, charged with a freight which has proved the seed of untold millions of wealth and has changed the fate and industries of nations, in A.D. 552 were safely landed on the Golden Horn. Long as they had remained concealed, the eggs were hatched at length, and fed and tended by the monks, who had carefully studied their culture.

From the little family which was landed at the Golden Horn have sprung, for thirteen hundred years, all the silkworms of Europe and of Western Asia. Long, however, did the Greeks retain the secret of their culture with a jealousy as vigilant as that of the Chinese; and it was not until eight hundred years had elapsed, when the Turk was thundering at the gates of Byzantium, and the fleets of Genoa and Venice were harrying the fairest provinces of the Greek empire, that Roger of Sicily carried off from the cities of Greece not only the silkworms, but the weavers, and compelled them to impart their mysteries to his subjects.

England had but small share in the silk of the East, for we read that the first time it was seen in this country was when the Emperor Charlemagne presented Offa, King of Mercia, with a royal gift of two silken vests. But from Sicily the culture soon spread over all the countries bordering on the Mediterranean where the mulberry-tree would flourish; and though Queen Mary forbade by law any person under the rank of an alderman's wife to indulge in a silken garment, and Queen Elizabeth was especially vain of the silken hose she received from Spain, the envy of her maids of honour; yet only a century later, in the reign of Charles II., it was the complaint of patriots, that every servant-maid in London spent half her wages in silk, to swell the revenue of the King of France.

It is strange, and almost unaccountable, how, for so many centuries, the origin and culture of silk remained so profound a mystery. For though it is not mentioned by Solomon, and we do not read that his ships of Tarshish brought bales of silk along with the ivory and peacocks, yet the very earliest writer on natural history whose works have come down to us, Aristotle, the tutor and friend of Alexander the Great, has given us (B.C. 325) a very accurate account of the origin of the precious tissue. He tells us it is spun by a horned worm, which passes through many transformations, and finally becomes a winged moth. But truth is often stranger than fiction. The story that so beautiful a texture could be produced from a creeping worm was too absurd to be believed; and until the cocoons were actually spun in the West, the tales of gossamers floating in the air, or combed down from silk-trees, were thought far more reasonable.

There is no country which those little eggs from within the pilgrims' staff have so wonderfully transformed as the old mountains of Lebanon. When Solomon was filling Jerusalem with all the strange curiosities of India, and the Holy Land was one vast garden studded with towns and villages, the long range of the Lebanon was one mighty cedar forest; very valuable indeed for building temples and palaces, but inhabited by bears and wild goats instead of by men. Now all has been changed. The cedar-trees have been cut down, and it is only here and there, in some wild corner, that the traveller can find them; the wild beasts have all been hunted away; and while the rich towns of Solomon have for the most part become desolate heaps, and the inhabitants of the villages have ceased in Israel, there are actually more people crowded amongst the valleys and rocks of Lebanon than are now to be found through the length and breadth of the Holy Land.

The silkworm has done it all. It was soon discovered by the industrious Syrians, that the Lebanon was exactly the country which suited the mulberry-tree, on the leaves of which alone the silkworms could be fed. In so hilly a country, garden ground is very precious; but the mulberry-tree strikes its roots so deep, that they do not interfere at all with the crops of carrots, cucumbers, and onions, which grow under the shade. Then, again, if the worms are to be healthy, they must be fed on leaves grown in dry places; for though they will eat very greedily of the large succulent leaves of trees grown in valleys and wet places, yet they often suffer from them, and the silk is not so good. The noble cedars have all been cleared away, and the homely mulberry has taken their place. Up and down the valleys the traveller passes for several days' journey along rocky mule-paths, that are more like staircases than roads, with villages curiously hidden in clefts of the rocks; and churches (for the people here are Christians) stuck on to the sides of the cliffs, with their flat roofs covered with turf, and grazed by kids, while rows of mulberry-trees swathe the mountains from top to bottom with closely-set waving strips of green. I call the mulberry a homely tree, for it is never allowed there to indulge its own taste, but is pollarded to the height of from six to eight feet, whence springs a dense thicket of small shoots, very useful and convenient, though not very ornamental.

It is a bright and cheerful scene to visit the Lebanon in the height of the silk season. There is no school then for either boys or girls; all are too busy in attending to the hungry little worms. As we ride along we are startled by the cuckoo cry of a little urchin, ensconced in the centre of the dumpy pollard. There he sits, busily engaged in shedding the leaves within his reach, and throwing them to the ground, and his merry face, with his red cap and black eyes, peers from out of the foliage, enjoying a saucy joke at the "howadji" as they pass. Beneath, the little sisters of the family are gathering up the leaves and heaping them into sheets, while even the little toddler of three years old looks proudly conscious of the dignity of labour and employment, as she stumbles along with her little contribution to the common stock. The elder girls are staggering home under their bulky but not oppressive loads, and one taller than the rest stands on a pair of rustic steps, and strips the twigs that are beyond the reach of the merry boy in his nest.

But it is when the leaves have been brought home that the most constant care is required. In the garden behind each cottage stands a large wooden erection, a sort of stage of laths, thatched to the height of about six feet with the green boughs of the oleander. The stage is full of trays from the top to the bottom, which slide in and out about six inches apart. On these trays the little worms are placed as soon as they are hatched. Here is the station of the housewife from morning till night. She draws out the trays one by one, carefully clears away the refuse, and picks out any diseased or dead insects, strewing the whole with the fresh-picked leaves which the children supply, and carefully screening the caterpillars from the sun, as they always feed on the under-side of the leaves. The fresh green roof and the open-latticed sides secure abundant ventilation and coolness, even under a Syrian summer sun. There, unlike the colder region of France, no artificial heat is required for the development of the eggs; and, from the first age to the fifth, the caterpillars continue to grow and thrive without any further care than air, food, and cleanliness, provided for them in this simple way. This feeding lasts for little more than a month, when the worms, as tired of eating as a schoolboy towards the end of his holidays, begin to leave the trays and creep up the sides of the lattices. They are then left alone, and allowed to spin in peace and quiet on tufts of grass which are placed at the corners of the trays. And now the silkworm's life is ended, for scarce one in a hundred is allowed to leave its little case alive. A few of the cocoons are laid aside to be pierced in due time by the chrysalis, to supply the eggs for next year. The others are gathered and baked, lest the insect should eat its way out before the family have had time to unwind the silk. This unwinding forms the employment of the Lebanon household during the early winter.

The cocoons are laid aside until the grapes and the olives have been gathered, and then the process begins. It is performed in a very simple fashion, just as children do it at home; about half a dozen cocoons being unwound as the children stand in a circle round the basin of hot water which contains a handfull of the golden balls. The little bobbins of silk

*a.* Oleanthus Moth. *b.* Tusseh Moth. *c.* Bombyx Moth (Common Silkworm). *d.* Cocoon of ditto. *e.* Chrysalis of ditto. *f.* Common Silkworm. Two-thirds of natural size.

are then duly weighed, and the village muleteer, when he hears that the roads are safe, and that there are no robbers in the neighbourhood, takes his precious freight to Damascus, and returns, if he is not plundered by the way, laden with the profits which are to provide all the simple luxuries of the village for the coming year. For strength, for toughness, for solidity, the ladies tell us there is no silk like that of Damascus, and it is all grown by the mountaineers of the Lebanon. The silk bazaars of Damascus are among the wonders of the Eastern world, and many little arched streets run out of them, covered over from the light of day, where hundreds of hand-looms are busily employed in weaving the beautiful shawls and girdles which every Turkish gentleman wears round his head and waist, in which every English traveller who visits that eldest of the cities of the world is sure to invest all the cash he has in his pocket,—if he have any sisters or daughters at home.

Of course, in more highly civilized countries, like France and Italy, the rearing of silkworms is carried on after a very much more scientific and artistic fashion; but I do not think it is nearly so interesting to watch as the happy industry of Hazrun, or any other village of the Lebanon. In England, though we rear no silkworms, yet the silk manufactory is a very important branch of industry. Two important towns, Coventry and Macclesfield, almost depend upon it; and many thousands of industrious artisans are employed in Spitalfields, in the east of London, in the same manufacture. Our silk is chiefly imported from Italy, for the French weave nearly the whole of their own produce; but our silk-weavers are, for the most part, the descendants of French Protestants, who were driven from their own country by Louis XIV. at the cruel revocation of the Edict of Nantes, which had promised toleration to the Protestants. That wicked revocation, whilst it deprived France of many of her best artisans, who fled in terror to other and freer lands, was the means of spreading a valuable industry, of which those who decreed that measure had little foresight. The exiles brought with them improvements in the manufacture into England, and have enabled our silks to hold their place in the markets of the world.

Though the little *Bombyx mori* is the silk moth with which we are best acquainted, yet there are many other silk-producing caterpillars. Some of these, which have long been cultivated in Japan, and are much larger than the mulberry moth, feed on the oak, and other trees more hardy than the mulberry. One of them, in the Himalayas, produces a silk which, under the name of Tusseh, is used largely for clothing in India. Another Japanese silkworm, which feeds on the oleanthus, has lately been introduced into Europe, and produces a coarse silk scarcely inferior to that of the mulberry moth.

But there are other moths which spin silk in great abundance, though, unfortunately, without any consideration for the needs or tastes of mankind. Thus, in our fields, we may often see, in early autumn, a whole network of glossy silk, stretching like a canopy from the heads of the taller stems of grass, and covering a space of two or three square feet. This is the umbrella of what are called the umbrella-spinning caterpillars. They particularly dislike to expose their bodies to a shower of rain, and so, when they have found an agreeable feeding-ground, they combine in spinning a screen which shall protect them from the sun and rain, while they devour at leisure the herbage beneath.

Akin to these, so far as they work in common, are what are called the sociable procession moths; the caterpillars of which may be often seen in oak coppices, marching by night, in regular order like files of soldiers, all moving in exact order, one after another, till they have found a feeding branch, and then returning before daybreak with the same regularity. These creatures hang to the stem or amongst the branches of the oak, large silken bags in which they remain secure during the day, lying heaped upon one another, till sunset calls them again to the march. But the silk of all these sociable moths is too short and scanty to be of use in commerce.

Silkworms, like larger beings, have many diseases, and many learned doctors have prescribed for their treatment. Some of these diseases appear to be very infectious, and three years ago an epidemic in Lombardy destroyed nearly the whole crop of the year.

They suffer most at the time when they change into their last moult, and it is then that the nurses are obliged to watch them most closely. Sometimes they writhe about, as if in acute agony, and at others they seem struck with paralysis. But I am afraid many of these diseases are the consequence of their own greediness, for it is the worms which have eaten most and become most fat, that fall victims to it. It appears that there is a curious microscopic fungus which takes root on their soft bodies, is nourished by their fat, and soon turns the living animal into a miniature mushroom-bed. The animal soon turns red and dies.

The silkworm is not the only caterpillar on which vegetables seem to grow. We have had sent home from New Zealand numbers of extraordinary specimens, each consisting of a large caterpillar, hard as wood, out of which rises a stem six inches in length, at the top of which is the fructification and seed of a sort of moss. This sphinx of nature is, after all, only like a gigantic *muscadine*, the name given to the silkworm fungus. Its little seed-spores, floating in the air, attach themselves to the back of the New Zealand caterpillar, which is in the habit of burying itself before it enters the chrysalis state for the winter. The unconscious insect, little knowing that he bears upon him the seeds of death, descends in due time to his living tomb. As soon as he is under the earth, the spore begins to germinate, and, drawing all its nourishment from its victim, sends its shoot to the surface, and fills the whole of his skin with a hard woody substance, which is its root, fed, not by the moisture of the earth, but by the flesh of the caterpillar. Its root never breaks the skin, and, as soon as the whole body is exhausted and transformed into hard fibre, the plant itself dies, and its seeds float in the air, till, perchance, one of them alights on the fostering back of another victim.

## WHAT HAPPENS IN GARDENS.

"ONE house is as good as another, and better too," said a domestic bee to a companion who was twice her size and more than three times as handsome.

"Buzz, buzz!" said the larger insect, "that is not true, for you have to take one that is given you; *I* build my own, and make sure to have it convenient, soft, comfortable, and *not over-crowded.*" And then the mis-called humble-bee went down into the bank by the lawn, and the poor snubbed little worker tried to hide herself in the sweet violets, feeling sorely the need of a big campanula, or blue gentian, to shield herself from a person of such independent tendencies.

It was just the grain of truth in the remark which made the poor bee feel hurt, for she and her relations were the victims of a new-fashioned owner, and they could not bring themselves to feel grateful for glass houses, patent slides, and other inventions intended for the development of honey rather than the comfort of bees.

"Indeed, it is very unpleasant," said the poor little thing when she went home at night, "to have these underground workers quizzing not only our houses but our motives, and the humble-bees quite despise us."

At this such an angry buzz went through the hive that the queen inquired what the disturbance was about. As soon as it was explained to her, she said that the statement was quite true, and that it had been on her mind for days how insulting and degrading it was to herself and her subjects to be put into houses where their domestic arrangements could be examined by every one, and she had been told that there was an enormous glass hive in the world, which contained many smaller ones of the kind they were then living in, and "Would you believe?" said the queen, making her great eyes flash red lights as she spoke,—"people actually pay to see us make cells and store honey!" Again the buzz of anger went through the hive, and a drone who was generally too much concerned with his own affairs to care for those of the community, said, "I cannot see why you should not live in the roof of the gardener's cottage." This magnificent idea was received with a general fluttering of wings, and was so earnestly desired that the queen decided to move before the sun disappeared over the hills. Such a fuss and commotion the moving caused! but no sooner did the queen pop into the hole of the roof—which she found with so little trouble that the common bees felt convinced she and the drone had arranged it beforehand—than the whole of her subjects followed, leaving the hive and its grand furniture just as clear as when they were put into it a week before. Happiness and hard work soon reigned in the home the bees had found for themselves, and they felt a comfortable conviction that they were established where they would not be ejected by burning. This has proved a very true conjecture, for they left the nice hives seven years ago, and have been undisturbed possessors of the space under the roof ever since, and it is likely they may remain as many more, for the patches on the thatch are renewed in the coldest of weather, no one caring for the privilege of mending it when the bees are buzzing round by thousands. It is impossible to get rid of the bees by burning, as the roof is so large, one could never make sure that they were all suffocated, besides the pleasant prospect of destroying the house if the wind carried a spark to the thatch; and therefore this lawless establishment is kept up in our own grounds, and fed with our own flowers, but we never partake of the honey. There are many straw hives and many bees living close by, but I suppose there has been no other insect so sensitive as the little worker who stirred up such a violent rebellion, or perhaps humble-bees are less arrogant, and have learned by experience not to throw stones at those who live in glass houses, knowing, as some of them must, the flaws in their own economy. Had the domestic bee who caused all the mischief known the whole story of her companion's life, she would have been better pleased with her own, for she had friends at home, and a monarchical government, whereas the humble-bee had only just found her way out of the ground after her winter sleep, and had at that moment neither home nor friends.

But she felt the power within her to make good her assertion, and her plunge into the lawn bank was her first effort towards it.

"I must have a home for myself," she thought, "and that hole where

Leaf, with Bee and Nest.

the croquet-stick went last night will save me a great deal of work." And she was just going to begin preparations when it occurred to her that a shower of rain would ruin her house, and that she could find no insurance office to remunerate her for damage by fire or water. Half ashamed of her idle thought, she had flown to the bank, and in the pride and pleasure of her first spring work had snubbed all other bees who liked gathering pollen and nectar better than building. How busy she was! she dug into the side of the lawn with her poor little front legs with as much energy as a dog who is scratching for rats. First with her legs, then with her head, and apparently with her whole body, she worked with an energy that would have made a steam-engine ashamed of the fuss it makes about working its way along an even road. As soon as she had made a hole large enough for two of her own size to enter, she made a transverse passage across the end of it, and then she gave her attention to the building of the nest. How many journeys she made carrying moss by instalments of one or two leaflets, and taking sand by the grain when such coarse material was needed, I could not tell, never seeming weary or tired of her work, and apparently never waiting to rest. Yet I think fatigue sometimes overtakes these bees, for I have picked up in the cold spring one or two fainting bumbledors, as they are sometimes called, who have soon revived by the warmth of my hand and flown away. After watching the little creatures as they crawl over my hand and arm, I get more puzzled than before to realize how such curious legs and jaws can build so cunningly such pretty habitations. The moss, the sand, scraps of straw, and tiny sticks were cemented together by a gummy secretion of the little lady's own manufacture, and the nest was finished in a very few days after her boast that it should be "warm, soft, and comfortable." But when all this was done a sense of loneliness came over the labourer; not a friend in the world had she, and to go out and seek companions did not accord with her notions of pride and independence. "I will have a colony, and companions and stores," said the energetic creature; and then in the transverse part of her home she constructed five or six cells with wax which she had already prepared. Then into each cell she put one small egg, which she thought a product of marvellous beauty; but unlike the hens who "cluck! cluck!" and invite every one to see their productions, she rushed to the other extreme, and putting into the cells a tiny portion of bee-bread, she fastened each of them down with a waxen lid. What a cruel mother she seems, to prevent the escape of her children by such a process! but with five or six eggs hatched perhaps all on the same day, she would have a worse time of it than the "old woman who lived in a shoe" if she had not taken the precaution to keep the wriggling young things in their proper places. But now is the happiest time of the bee's life, if she could only realize it, and she makes more cells and has more children, and gathers honey in a busy delightful way that is quite aggravating to the flowers who supply her with material. Each little caterpillar at last shows signs of having had food enough, and then the mother knows her work is nearly finished as a nurse. She again fastens down the cells and leaves the chrysalides, into which the children have developed, to work their own way out of them. This happens in a few days, and at last the lonely builder and mother has companions on her journeys and friends to share the labours of storing and gathering. It would be a happy state if this buzzing, humming life continued, and we can fancy how proud the mother must be of the new bright golden bands of coloured hair upon the bodies of her children. Her own have faded to pale dusky yellow since the work in the hot May suns; but she forgets this, and is proud of her children's gorgeous beauty. But presently the colony increases, each child builds cells and lays eggs, and these children as they grow older develop into bees and do the same, until the little family of five or six is sometimes increased to a colony of two or three hundred. There is no longer much peace, and the new and last comers, who are small male bees, try to rule in the nest; their mothers, who are also small, and the latest females of the broods, endeavour to support them in their claims; and then the older bees, who have worked for them and built for them, are obliged to fight to keep a place in their own homes.

As the season advances more quarrelling and fighting goes on between these rival mothers, and at last so many are killed, and so many driven out, that the nest with its stores and regular waxen cells is quite deserted. The foundress bee has been killed long since, destroyed by one of her own children perhaps, and her pretty home with all its wealth is left for a cunning field-mouse to break up, and carry off the effects to a family of sharp-nosed hungry babies, who sorely try her patience to find food for them. One would suppose that such rebellion and quarrelling would soon bring the reign of Bumbledom to an end; but it is not so: some of the larger early broods creep away and hide in the cracks of old willow-trees, in holes in the ground, or under the eaves of the thatched houses, and sleep away the winter. Then, when the warm spring days return, they creep from their curious crannies, and each one starts alone and unaided to do such work as I have described. A visit from a humble-bee is a sure sign of returning summer, and how many of us have heard the old nurse's tale that when one of these insects enters the house, strange visitors are sure to follow. Such stories were invented before railways and telegraphs brought people together so frequently as they meet now, and the story means nothing more than that the ground was dry and the bees at work, and so the roads were passable for foot or driving passengers. But these bees are not the only patient and solitary workers, although the others do not found such enormous colonies, and even those of humble-bees number generally only sixty or a hundred inhabitants. Perhaps the leaf-cutter bee, who is at this moment destroying the leaves of my rose-trees, is a more interesting designer than these I have described. This little artificer despises such material as the others delight in; not for her are the dead stalks of the curly moss, or the bruised petals of flowers, or scraps of hay which may have fallen from somewhere unknown and unclean. This little lady (and she is really such an atom!) prefers the crisp and just full-blown leaves of the choicest roses, and she will never touch those of the dog-rose, if certain cultivated kinds are to be had for the trouble of flying over the hedge where it is growing. She is not at all an easy bee to capture, and her mind is not so much occupied with her work that she is off her guard. The movement of a bird or the rustle of a dress disturbs her, and she will fly off, leaving her scrap of leaf hanging by a thread, which she comes back and cuts off as soon as she thinks the danger over. But it is a pity to catch her and stop the progress of her pretty mischief; it is far better to watch and to follow her when possible to the abode she is building with your property. There, under a fallen stick, or quite likely under your window-sill, or behind your shutters, you will see the cunning nest, which is built with such regularity as to the size of the pieces employed, that one might suppose the bees were taught mathematics and brick-making. There she puts in separate cells her tiny eggs, and she lives to see the family metamorphose result in producing less vicious insects than those of the bumbledors.

The other British rivals in insect architecture are the mason bees, who make their nests of sand, which they carry, a grain at a time, into some little nook which forms the outer wall of the structure. The sand is made into a kind of paste by the insect, and put in ridges to divide the separate cells, and it soon hardens by exposure to the air. Sometimes an empty snail-shell is converted into a house, whose chambers will contain an egg, a caterpillar, or a chrysalis, but never an idle bee. Such are the spring visitors to all our gardens, but they are only known to those who care for their acquaintance; and prying into their affairs, from any motive but that of an earnest love for them, is much disliked by these tiny aristocrats. If by an accident they intrude upon us, they do their best to escape, and one reason that they quarrel so much with their discarded relations, the wasps, is that these latter have not at present tact enough to know when they are not wanted. This vulgar habit, and the very dishonest tricks which the wasps have a way of playing upon the bees, render them very troublesome company, and now and then one is attacked by a united party of bees in a way which is quite certain to end in the death of the intruder. This result we certainly cannot regret, nor even blame the murderers, when we with our clumsy apparatus and enormous size are delighted if we succeed in crushing to death an insect whose only crime, so far as we are concerned, is that she has lost her way.

---

## A LUMP OF COAL.

EVERY one likes a good fire in bleak chilly weather,—not the huge block of black cold coal, with a backing of dross, with which the thrifty housewife fills the grate, and only a thin line of dull red glimmering between the lowest bars; but a little active volcano, whose cheery crackling does one's heart good. It is pleasant to sit beside such a fire when the short wintry day is fading into the gloomy night. The eye can no longer see the printed page, and the weary book is laid aside. It is the time for thinking. Gazing into the bright embers, the fancy is busy with all sorts of dreamy notions. Strange faces are seen in the centre of the pure white heat; and the shapes of the burning coals look like those castles and rocks, the abodes of giants and dragons, of which the young mind is so full. But the things that we see in the fire are not all fancies—airy nothings. From the ashes of the fire the man of science can raise before his mind's eye the shapes of the old trees whose remains formed the coal that is burning beside him. For, strange to say, the coal is not a mineral but a vegetable substance. It looks like a stone—a piece of black marble; but it is in reality made of the relics of plants, just as the limestone that is often found with it in the earth is made up of the remains of animals, shells, and corals, whose figures we see in the marble of almost every mantelpiece.

The page of the earth's story-book that tells us the history of coal is a very extraordinary one. It is to the familiar appearance of the world of he present day what the fairy story-books of childhood are to the sober duties and enjoyments of grown-up men. The earth has its ages just like a human being. It has its childhood and youth, and it is of the fresh green youth of the earth that the coal burning in the grate speaks. It tells us of forests growing where no trees are now found, and where everlasting winter reigns, covering the earth with a dark green mantle from pole to pole. The plants that composed these old forests are different from any that now exist. They passed away many ages before man came into the world—they were the first planting on the fresh soil raised above the ocean. But we see, at times, in sandstone quarries, pieces of their stems and branches; and on the roofs of coal-mines, and on the upper layers of coal-seams, we find impressions of leaves and other parts, and from these fragments we know what kind of appearance they had when green and growing.

Between three and four hundred different kinds of plants are found in the coal area of our own country. The great mass of them are ferns and pines, and belong to a lower order of vegetation than our own oak and beech forests. The most common coal plants are called *Sigillarias*, on account of the seal-like impressions which occur on their stems. Large fragments of them are frequently found in coal-pits, remarkably well preserved, and filled in the inside with sand. There are furrows or grooves running beside each other from one end of the trunk to the other, and along these the seal-like marks are arranged in rows. Each stem resembles a fluted Doric column, beautifully but variously chiselled —the pattern changing with the species. The seal-like impressions are evidently the scars left behind by old leaves, like the horse-shoe marks on the young twig of a horse-chestnut-tree when its leaves have fallen in autumn.

Another very common coal plant is called *Lepidodendron*, a long word which signifies a *scaly tree*. Its trunk was covered with scales, somewhat like a pine-cone. It grew in a fork-shaped form to a considerable height, and was covered with needle-like leaves like a larch or fir-tree. The wood of the stem was very soft, with a pith in the centre, and was therefore worthless for timber. Both the Sigillarias and the Lepidodendrons were intermediate between the pine-trees and the club-mosses now growing on our moors and hills, partaking of the characters of both. They supply the missing link connecting these two classes of plants; the Sigillarias most nearly resembling the pine-trees, and the Lepidodendrons most nearly the club-mosses. Every one knows what pine-trees are like;

and most persons have seen the club-mosses, often called fox-fetters which creep over the ground among the heather; so that by uniting the shapes of these two kinds of familiar plants in the mind's eye, any one may have a very good idea of the appearance of the coal plants. We are struck with their enormous size in comparison with the humble plants which represent them at the present day. They frequently attained the amazing height of seventy to one hundred feet, and a girth of five to six feet.

Another ancient plant still whose substance formed our coal is called *Calamites*, signifying a *reed*. It is found in jointed fragments which were originally round and hollow, but which by the pressure of the rocks above them have been crushed and flattened. The stem appears to have been branched; and both stem and branches are ribbed and furrowed, and present quite an elegant appearance. We know less about this plant than about the two others; but it is supposed to have been a kind of ancient horse-tail. Our modern horse-tails are among our humblest plants, a foot being about the usual height; but the ancient horse-tails or Calamites had stems fourteen or fifteen inches round, and grew to a height of thirty or forty feet.

Scattered among these huge club-mosses and horse-tails were many beautiful tree-ferns. Their wide umbrellas of richly-cut leaves waved high above the ground, upborne on tall dark stems. Beneath their shadow, forming the underwood of the forest, grew many thick clusters of graceful ferns, whose impressions are seen as delicately marked on the black slate of coal-pits as though they had been printed from copper plates.

There are many other plants that enter into the composition of coal, but those I have mentioned are the four principal kinds. And it is not a little strange that we owe our coal to plants so low in the scale of nature instead of to the lordly oaks and the princely palms. God shows us in this the value and importance of humble things, and proves to us that He is greatest even in His least productions.

The ancient coal forests composed of these trees must have looked very dark and sombre. The colour of the whole was one dull green, without the many delicate shades which we see in our woods. And all the year round there was no change of hue any more than in a pine wood. The stillness of these old forests was awful: only the grand sound of the wind among the tops of the trees seemed at intervals like the distant roar of the sea. Animal life was scarce. Only huge reptiles frogs, serpents, and crocodiles glided silently in and out among the fern clumps, devouring each other. Had man lived among these woods, he would have been sorely pinched for food and fuel and implements of work. He could not have cultivated the trees for their fruit or their blossoms or their shade; their wood was too soft and spongy to have formed his furniture or his instruments, and it was too wet and pithy to serve even for fuel. The

only form in which they could have been made useful was in making coal; and accordingly it is in this form alone that the great Creator has preserved them to us.

What the climate of the globe was at the time that these coal plants grew may be found out from the fact that ferns and club-mosses thrive best at the present day in moist sheltered islands. Some think that the presence of tree-ferns shows a tropical warmth; but in New Zealand, where they still occur, the climate is very much like our own, and tree-ferns are found even at a height of upwards of a thousand feet, beside the end of glaciers, waving their green feathery fronds over the never-melting ice. Ferns and club-mosses can therefore endure a wide range of temperature, so that their occurrence is no proof of a particular climate. In all probability, however, the climate of the coal period was warmer, moister, and more uniform than now obtains; and the huge size of the coal plants and the vast quantity of coal which they formed seem to prove that the atmosphere was highly charged with that carbonic acid gas which plants take in and work into all their structures. The sameness of the climate is shown by the general character of the coal plants being the same over every portion of the earth's surface, the same kinds being found in the most distant countries.

How coal was formed from these plants it is very difficult indeed to tell, because there is no process precisely of the same nature now going on anywhere. Most men of science believe that the plants have been swept down from the places where they grew by rivers or currents, and left in basins and firths of the sea or in fresh-water lakes. Sand and mud were heaped in alternate layers over them there, and thus formed the different strata of coal and sandstone which we now find in a coal basin. We can trace the gradual change between perfect wood and perfect coal in such situations, from the blackened tree-trunks of our peat-bogs, through the *lignites* or brown coal, up through bituminous coal to the true coals which we burn in our grates. After a long interment of the heaped-up plants beneath the water, gradually undergoing there the chemical changes necessary to convert them into the mineral condition, subterranean fires at last elevated the beds of coal above the waters nearer the surface of the earth. Were it not for this the coal would have been buried far beyond the reach of man.

Molten matter ran through the coal basin in different directions, like the lava that flows down the sides of Vesuvius, and this hardened into what is called trap-rocks. By these fiery eruptions of trap the seams of coal were broken up and divided into parts that are easily worked, and the coal itself was brought from the profound depths within reach of man. We find many wonderful proofs of God's wisdom and care for man in thus preparing and arranging the coal beds. Had they been formed on the surface, exposed to the air, they would have crumbled away into dross. But the precious treasure was safely hid deep down in the earth under beds of rocks, and yet not so deep as to be beyond the industry of man to get it by digging. It was not covered with hard rocks like granite and quartz, which it would be very difficult to blast and penetrate; but with limestone, sandstone, shale, and clay ironstone, which can be easily pierced, and yet afford a sufficiently safe roof for the mine, and which are very valuable in themselves for man's uses.

A lump of coal, it is often said, is made up of sunbeams. We could believe this more readily of the diamond, which is just a crystal of coal, for it is so bright and sparkling, and makes brilliant sunshine in a shady place. But even the dull black coal has been formed of the sunshine of long-forgotten summers. Every sunbeam that fell upon the club-mosses and ferns of the old coal forests, enabled them to withdraw the minute unseen carbon from the air, and form out of it their own solid tissue. They thus caged and imprisoned the floating light itself, and wrought its bright threads in their loom into the beautiful patterns of stem and leaf which they showed. To form one of the little rings of wood in the trunk of one of the old pines took the sunshine of a long summer falling upon all its thousand leaves; and who can tell how much sunshine has been worked up in all the stores of coal that lie concealed under our feet? This prisoned sunshine we set free whenever we kindle a fire of coals. When the sun ceases to shine upon us in cold misty wintry days, we draw upon the sunshine of a million years ago to drive away the frost and make us comfortable. The source of all labour is the sun; and we get the benefit of his labour when we burn the coal or the wood in which he has condensed and preserved it. No ray of sunlight has ever been wasted or thrown away. It is because Nature has been so thrifty in her household ways that we are enabled to be so prodigal of our resources to-day, spending upwards of one hundred millions of tons of coal every year, and with that vast consumption of sun-labour producing all the varied and extraordinary work that we do under the sun.

Why is a lump of coal black if it is composed of sunbeams, which every one knows contain all the colours of the rainbow? Why is it black if it is made up of the green stems and branches and leaves of plants? It is because its particles are so formed and arranged as to take in all the light that falls upon it without giving back any portion. A white object reflects all the light, and a black object absorbs all the light. What becomes, then, of the colours which the black coal has withdrawn from the sunshine? Are they lost? No! nothing in this world is lost. Everything is accounted for. When anything has served its purpose in one form, it seems to vanish altogether, but it reappears in another form, and in it works anew. There is everywhere change, but not loss. A growing plant absorbs some of the colours of the sunbeams that nourish it, and reflects others in its prevailing hue—yellow or blue or red or purple. But the colours that it absorbs are not lost; they generally reappear in some other or after part of the plant. So the colours of the sunshine that are absorbed in the black coal come out in the coloured flames of the blazing fire. The red and

yellow flame over which you warm your hands, is just the flower into which the sunshine, concealed and stored up in the coal for ages, has blossomed. But more than this: the lost colours of the rainbow in coal are brought out still more strikingly by our modern manufactures. Every one has heard of and most persons have seen what are called the coal-tar colours. Richest and brightest hues of blue and green, and mauve and magenta, and rose and yellow, are obtained from tar, and tar is obtained from coal. It would take too long to describe the process, but it is very curious, and is one of the many triumphs of this extraordinary age.

Were I disposed to draw a moral from this little essay, I could show that there are hidden beauties in everything and every one, however ugly and unpromising, which it would be well worth our while to find and bring out. But I shall merely throw out this as a hint, and simply ask those who have accompanied me thus far without sleepy eyes and yawns as round as O's, if it is not true what I have said, that there are wonders in the fire stranger than in any fairy story, stranger than any faces, and castles, and pictures that the young fancy sees in its glowing heat in the twilight hours, —wonders, the half of which has not been told them?

## THE SPIDER AND ITS WEBS.

ONCE upon a time there was a spider. Not one of those happy creatures who spin their gossamer webs in green lanes or shady forests, where they glance in the sunshine or glisten with dew to make captives of the bright unwary creatures passing by in the free summer air; nor yet one of those unfortunate Penelopes who day by day weave afresh their webs, to be as often torn away by the unwearied housemaid, who with her Turk's-head brush, poking into corners, peeping behind shutters and into the recesses of closets, would sweep to destruction the homes and hopes of the spider race. It was none of these, but a plain brown spider of homely birth and habits, suited to a cottage home, though possibly with hidden aspirations within him, such as may have swelled the breast of many a village Hampden who lived and died inglorious.

Well, now, to return to the spider. His home was a hovel, the rafters of which at wide intervals supported the thatch, through which the smoke found scanty exit. Without splendour, there was spider comfort, and no rude winds tore his web to atoms, no officious housemaid with her ruthless broom brushed away his larder well stocked with flies. But the spider was restless. He dreamed of a lot cast in a higher region, where his web might curtain the arches and festoon the pillars of a king's palace. Had he ever heard what Agur, the son of Jakeh, says of the spider, who "taketh hold with her hands and is in king's palaces"? (Prov. xxx. 28) (though philosophers *do* say that the wise man had a lizard of some kind in his mind when he made that allusion). "Alas!" said our cottage friend, "that my life should pass so uneventful and so unobserved on this retired rafter! that far from courting observation, or even attracting the notice of the sordid giants of the human race who inhabit this smoky den, I am never thought worth looking after, nor does my web, weighed down with dust and skeletons, demand of them the trouble of sweeping it away. A palace were a better sphere for me. My industry, my taste, and the fineness of my fabric might there win the admiration of all who had time to examine the ingenuity of its construction; while history might record my successful capture of a bluebottle about to settle on the nose of the monarch!"

Happily, a good night's rest, after a plentiful supper, restored the spider to a more complacent state of mind with respect to his existing circumstances and his native home; and, balancing himself on the delicate meshes of his dusty web, he took measure of his present position, and of his actual powers, with a view to their exercise and improvement, in case the palace for which he longed should ever open its portals for his reception. "Could I," he mused, "all untrained as I am in this lonely shed, weave like those lighter spiders whose slender bodies and lengthened limbs dart almost unseen through those lofty corridors, and weave in those elevated niches for which I so often sigh,—I, whose body has become heavy with the gross fare of this confined cottage,—my fate might be that foreshadowed by the poet, that 'vaulting ambition doth sometimes o'erleap itself, and fall on t' other side.' No. But I will fit myself. I will cultivate my powers, and extend my efforts, till my perseverance at least shall be worthy of imitation. Hitherto I have been content with spinning my web in this dark corner, where no effort is needed to stretch it from rafter to rafter. At least I will deserve, if I do not gain, the gratitude of the giants who sleep below me. How often have I watched them tormented in their sleep by some pertinacious fly, or some vicious little imp of a gnat, which settled on their eyelid, or buzzed in their ear, while they winked and winced, and turned from side to side to elude their plagues! But as they moved, their nimble tormentor darted out of reach, and came down again in an instant. Ah! if I had woven a nobler web, had I stretched it just over their heads, I might soon have trapped their plagues, and laid up a good breakfast for myself at the same time."

From this musing he proceeded to action, and from the inner recesses of his self-consciousness he evolved a long silken thread on which to hang not his argument, but himself. Dropping a little gum from his spinneret, he proceeded gently to draw out from his body a long viscous thread. The end of it he carefully glued to his rafter, and then gently letting himself down, he rapidly but very carefully combed out this gum into a long thread between the claws of his hind legs, taking care that it was strong enough to support his weight, when, meeting the air, it instantly dried into hard fine silk. Thus he transformed himself into a species of pendulum, the vibrations of which should affect, if not the history of the world, at least that of his own particular ambition.

Ignorant of the higher mathematics, or even of the simpler laws which govern gravitation, knowing nothing of the movements of the radii of circles, or of the arcs which they describe, he yet, in his vain attempts to reach the next rafter, and in the constant return of his little cargo of legs to the other side of his beat, described an orbit regulated by mathematical rules as exact as those which were at the same moment carrying him, his web, his hovel, and the earth on which it stood, with unerring force through the regions of space. The point at which he aimed was a distant one, but to his ardent imagination it seemed not unattainable. "What spider has done, spider may do," thought he, as he swung disconsolately back from his second unsuccessful attempt. Again he applied his hind legs to the end of his abdomen, and gently teazed out a little more of the glutinous fluid, and carefully combed it into long silken hairs. "Perhaps," he reflected within himself, "with a little longer web I may succeed." Pausing a moment, he looked down and contrasted himself favourably with a human being who had lately entered the hovel, and had flung himself down on the pallet bed with the hopeless and worn expression of one with whom effort and success had not gone hand in hand. The spider saw he was a stranger, and began to criticise somewhat contemptuously his soiled and tattered appearance. But vanity had a large share in our spider's composition, and when he noted how the wanderer's eyes, which at first roved vacantly from object to object with a weary stare, at length rested on himself, he felt spurred on to more vigorous efforts. Never before had one of the familiar faces of that cottage been fixed on him or his work.

"I will prove myself worthy," resolved he, "of human notice. That man shall learn now how even a cottage spider can float in air, where he cannot follow." Again, again he tried: with a sudden motion, stretching out his legs, he pushed himself back, and swung once more towards the rafter. It seemed to retreat still further from his unequal efforts. Again he tried. "England, Scotland, Spiderland expects every one to do his duty," was perhaps the thought that animated the renewed effort of the spirited creature; and it seemed in some mysterious manner to communicate its energy to the lustreless eyes of the wayworn traveller. Warrior he seemed as well as huntsman, for his bugle and his sword hung side by side from his loosened belt. His eyes became riveted on the little animal, so unremitting, so dauntless, yet so unsuccessful. A fellow-feeling touched, perhaps, a sympathetic chord, and kindled the almost extinguished embers of hope that had well-nigh given way to despair, alternating with sad and revengeful thoughts.

Beneath that spider lay Robert Bruce, the defeated and almost hopeless hero of Scotland. Ah! little spider, had you taken hold with your hands of a king's palace, instead of the rafters of a cotter's hut, your lot might have been a more brilliant, probably a shorter, most certainly a less useful one.

Four times, five times, six times, swings the living pendulum from side to side, and failing to gain his own object, rebounds again, baffled, but not disheartened. The seventh time it gathers up its energies and repeats the effort. It has won at last, and "Never say die" is the watchword that thrills the giant heart of the champion of Scottish liberties, as he recalls how six times he has been defeated, and beholds the little animal safely resting on the rafter it has scaled at last with so much effort. Little did it think how a mighty courage had been rekindled by its tiny struggles, and that a page in history would ennoble the memory of the *cottage spider*.

But it is not every spider that can expect a place in story. Yet there is many another spider, which, if we watched it, would teach us a lesson, if not as grand as that which Robert Bruce learned, yet one very useful or interesting.

There are nearly three hundred kinds of British spiders, living not only in cottages and halls, but in lanes and hedges, or trees, or in fields, but some burying themselves in the ground, and others, stranger still, living under water—not in it, like fishes or reptiles, but actually bottling the air, taking it down with them and keeping enough about them to breathe, and then, when that is exhausted, coming up again for a fresh supply. But all these spiders weave webs, and the webs are almost as various as the spiders. If there are near three hundred species of spiders in this country, there are as many different patterns of webs. Just as silk is woven into

sarsenet, or satin, or velvet. or net, so the fairy gossamer of the spider's web is spun sometimes to form the brown dust-catching silk which festoons the neglected corners of a room; sometimes those beautiful patterns of network we see jewelled with dewdrops on a summer's morning in the hedges, or the fine threads which stretch from tree to tree, or the light hairs we catch up with our feet as we walk across a field in early spring. But all spiders spin, though all do not spin nets. Some content themselves with spinning houses for their young ones, and very tight and tough houses those white and yellow silk bags are. Other subterranean spiders make silk hinges for the doors of their houses, of which we may have something more to say further on. And others make literal fishing-nets, for the water-spiders of which we spoke actually spin webs in the water and catch the water-insects.

There is one spider, the tarantula, not an English animal (insect we must not call it, for spiders, small as they are, are not insects, but far more like crabs or lobsters), about which strange stories are told not quite so pleasing as that of Robert Bruce's spider, for it is said to have a poisonous bite, which forces people not to try again like Bruce, but to dance like maniacs. The bite is not, however, very serious, and I have often caught the tarantula in warm countries without being hurt by it.

But there is another kind of spider, which by candle-light looks as large as a mouse running across a room, which is a very old friend of mine. I once had one of these spiders, a sort of *Mygale*, as it would be called in books of natural history, which I kept tame in my bed for a year and a half, and which I think was quite as noble a spider as Robert Bruce's friend. It was in the island of Bermuda, which swarms with every kind of disagreeable insects, and where the mosquitoes, gigantic bloodthirsty gnats, not only murder sleep by their sharp shrieking buzz in the ear all night long, but thrust their long lancets through the skin and suck out the blood, raising great sores which are often very troublesome. No one can sleep there in peace without a mosquito-net or large bag made of bobbin net, which is hung from a hook in the ceiling and covers the whole bed to the ground like a huge gauze nightcap. But the mosquitoes are very active, and when you lift up the net to get into bed, some of them are sure to be nimble enough to get in with you to keep you company. Now, my bedfellows were very troublesome, and would neither sleep themselves nor let me sleep. Sometimes they tasted the tip of my nose, then they bored my ears, then they ran their lancets into my eyelids, singing all the time most hideously. At last I determined to make friends with a large spider. I caught him one evening as he was jumping after the flies in the window curtains, and put him into a little bag which I fastened inside my net at the very top. Then I fed him with large flies for a few days, until he began to find himself in very comfortable quarters, and thought of spinning a nest and making his home. I then cut a hole in the bag, and my spider soon spun a beautiful nest as large as a wine-glass for himself, winding himself round and round as he combed out the silk from the end of his tail. In this nest he sat perfectly motionless, for these spiders do not weave nests, but only homes for themselves and their young ones, and catch their prey by leaping upon them with amazing speed. There at the top of the net sat my friend, and often have I watched him when a fly or mosquito got inside our gauzy tent. I could fancy I saw his eyes twinkle as his victims buzzed about, till when they were within a yard or so of the top, one spring, and the fly was in his forceps or nippers, and another leap took him back to his den, where he soon finished the savoury mouthful. Sometimes he would bound from side to side of the bed, and seize a mosquito at every spring, resting only a moment on the net to swallow it. In another corner of the room was the nest of a female *Mygale* of the same species. She was not content with so small a house as her husband, but added some beautiful little silk bags or cocoons larger than a thimble of very tough yellow silk made by herself, in each of which she laid more than a dozen spider's eggs, which used to sit on her back when hatched, but which all disappeared as soon as they were old enough to hunt and leap for themselves. I kept my useful friend in bed for nearly a year and a half, when unfortunately one day a new housemaid spied his pretty brown house, pulled it down, and crushed under her black feet my poor companion.

There was another kind of spider in Bermuda, much more handsome than my bedfellow, but not nearly so great a favourite of mine, about an inch long without measuring its long legs, and with a bright yellow and black body painted in beautiful patterns. This spider did not weave nets, but nooses of bright yellow silk. It spun them in the woods from tree to tree, sitting at the extremity of a branch, and then, taking advantage of a breath of wind, it would sail out into the air, carrying its thread behind it, till it reached the next tree, where it fastened it, and then started back again with another thread. These spiders generally choose the trees on each side of a pathway for their operations, and the silken threads hang across it in myriads. When the large beautiful butterflies come fluttering down the avenue in the sunlight, they often get their wings entangled in these cords. If the cord breaks at once the butterfly escapes, but if not, in its struggles it would soon touch two or three more lines, and as soon as it was completely entangled, the spider would come running along its thread from the tree, and rapidly moving round and round its lovely prey, would spin its gummy silk till the butterfly was completely fettered, when it devoured its captive on the spot. I once saw two of these spiders together capture a little bird, a greenlet, about the size of a wren, in this way. The threads had got so entangled round its wings that the spiders were able to seize it as it struggled in the snare, and had bitten its throat so severely that, though I freed it after watching the battle for a minute or two, the poor little bird died in my hand.

An ingenious American tried to make use of this silk, and once exhibited at a show in Bermuda a yellow silk handkerchief of spiders' webs. But though it was far finer than silkworms' silk, it was so troublesome to collect that no one attempted the manufacture afterwards.

There is another spider which I have often watched in Greece and the Holy Land, which is, I think, the most wonderful of all in its architecture. It is also a *Mygale*, but much smaller than those of which we have been speaking, and is commonly known as the Mason Spider. This spider is entirely nocturnal in its habits, and never either hunts or feeds in daylight, but makes itself a most comfortable house, where it is perfectly safe and locked up till sunset. It bores a circular hole in the side of a bank, or any sloping ground, about the size of a man's middle finger. The tunnel is most exactly rounded, and from two to four inches deep. To rake up the earth and shovel it away, it has a row of hard points on its head, like the teeth of a rake. As soon as it has scooped out the soil, it lines the tunnel with silk, through which no damp can penetrate; and no drawing-room was ever so beautifully plastered, and papered with damask, as the mason spider's sitting-room. But the door is the most wonderful part of this mansion. The spider does not like draughts, and cannot bear having the door left open, so it contrives that it shall shut itself. The door is perfectly round and flat, about the size of a sixpence, but very thick, made of thin layers of fine earth moistened and worked together with fine silk, so that it is very tough and elastic, and cannot crumble; with a wonderful silk hinge at the top. The hinge is elastic silk, very springy, and so tight that when the door is opened it closes immediately with a sharp snap. But the door does not fit on to the house, but into it. It has a beautifully hard socket, bound with silk, into which it fits very tightly, while the outside is covered with bits of moss or other things glued on, so that no one can possibly detect it. The only way of opening it from without is by a pin, and even then I have often seen the spider keeping tight hold of the bottom of the door with her claws, while holding on to the walls of her cell with her whole force. Here the little architect remains all day, and at night spins a few threads among the grass near her home, in which she catches her prey; but she also hunts for food by leaping upon beetles, and carrying them into her tunnel. So attached is she to her cellar, that I have often cut the nests out of the earth and brought them away in my pocket with the inhabitant within; and I have now before me a row of these nests, all with their doors fitting exactly alike. I once cut off the door of a nest near my tent, and next day found that a new one had already been hung on its hinges.

One more spider I should like to say a word upon, because it is one we may often see in this country, and is very little known. It is the water-spider. It has a very long Latin name, *Argyroneta aquatica*, *i.e.*, the water silver spider, and it is very interesting, because—as we said some time ago—it bottles up air and takes it under water to breathe with. In fact, had people only watched water-spiders as Robert Bruce watched the cottage spider, diving-bells would have been discovered hundreds of years ago, and people might have learnt how to go to the bottom of the sea and save the treasures of wrecks. We know there are two ways in which divers descend and work under water. One is by the diving-bell, which is like a great bell dropped into the water, so that the air cannot escape; the other is by a diving dress, in which there is a supply of air inside the clothes of the diver. The spider uses both these methods. It lives in ditches and stagnant pools, near the bottom, and weaves a strong silken cup of the shape of a bell, which it fastens by long cords stretched on all sides to the stems or water-weeds, and which is filled with air. As the bag is always kept mouth downwards by the cords, the air cannot escape; and here the spider lives and deposits its eggs in little capsules or bags, where its submarine cradle keeps them perfectly safe. Its body is covered with long hairs, and these hairs hold the atmosphere all around it, so that when it swims lying on its back—which is its regular method of moving about—it looks like a silvery bubble of air. It often comes to the surface to replenish its supply. The walls of its nest are very thin, composed of a tissue of fine white silk, to which is attached quite a fringe of threads to anchor it to the weeds. Here the spider lives, with his head downwards, ready to pounce upon any unwary insect. In winter, when it sleeps for many weeks together, it weaves a flooring to its nest to secure it from any accidental entrance of water.

I could tell of many other wondrous kinds of spiders' webs, but my readers will see from the few here mentioned how full of marvels is even the little spider's world, and how much there is to instruct any one who would rather go through life with eyes than with no eyes. The spider will teach us not only the lesson of perseverance which Robert Bruce learnt when he was nearly giving way to despair—it will teach us how to spin and how to weave, how to hunt and how to snare. It gives lessons in gymnastics, in swimming, and in leaping, and it has solved many a problem in mathematics before Euclid was born. Look at the spider's web, and see whether "any hand of man, with all the fine appliances of art, and twenty years' apprenticeship to boot, could weave us such another."

## VOLCANOES.

YOU want to know why the Spaniards in Peru and Ecuador should have expected an earthquake.

Because they had had so many already. The shaking of the ground in their country had gone on perpetually, till they had almost ceased to care about it, always hoping that no very heavy shock would come; and being, now and then, terribly mistaken.

For instance, in the province of Quito, in the year 1797, from thirty to forty thousand people were killed at once by an earthquake. One would have thought that warning enough; but the warning was not taken; and since then thousands more have been killed in the very same country, in the very same way.

They might have expected as much. For their towns are built, most of them, close to volcanoes—some of the highest and most terrible in the world. And wherever there are volcanoes there will be earthquakes. You may have earthquakes without volcanoes, now and then; but volcanoes without earthquakes, seldom or never.

How does that come to pass? Does a volcano make earthquakes? No; we may rather say, that earthquakes are trying to make volcanoes.

For volcanoes are the holes which the steam underground has burst open, that it may escape into the air above. They are the chimneys of the great blast-furnaces underground, in which Nature pounds and melts up the old rocks, to make them into new ones, and spread them out over the land above.

And are there many volcanoes in the world? You have heard of Vesuvius, of course, in Italy; and Etna, in Sicily; and Hecla, in Iceland. And you have heard, too, of Kilauea, in the Sandwich Islands, and of Pele's Hair—the yellow threads of lava, like fine spun glass, which are blown from off its pools of fire, and which the Sandwich Islanders believed to be the hair of a goddess who lived in the crater;—and you have read, too, perhaps, the noble story of the Christian chieftainess who, in order to persuade her subjects to become Christians also, went down into the crater and defied the goddess of the volcano, and came back unhurt and triumphant.

But if you look at the map, you will see that there are many, many more. Get Keith Johnston's Physical Atlas from the school-room—of course it is there (for a school-room without a physical atlas is like a needle without an eye)—and look at the map which is called "Phenomena of Volcanic Action."

You will see in it many red dots, which mark the volcanoes which are still burning; and black dots, which mark those which have been burning at some time or other, not very long ago, scattered about the world. Sometimes they are single, like the red dot at Otaheite, or at Easter Island in the Pacific. Sometimes they are in groups, or clusters, like the cluster at the Sandwich Islands, or in the Friendly Islands, or in New Zealand. And if we look in the Atlantic, we shall see four clusters: one in poor half-destroyed Iceland, in the far north, one in the Azores, one in the Canaries, and one in the Cape de Verdes. And there is one dot in those Canaries which we must not overlook, for it is no other than the famous Peak of Teneriffe, a volcano which is hardly burnt out yet, and may burn up again any day, standing up out of the sea more than 12,000 feet high still, and once it must have been double that height. Some think that it is perhaps the true Mount Atlas, which the old Greeks named when first they ventured out of the Straits of Gibraltar down the coast of Africa, and saw the great peak far to the westward, with the clouds cutting off its top; and said that it was a mighty giant, the brother of the Evening Star, who held up the sky upon his shoulders, in the midst of the Fortunate Islands, the gardens of the daughter of the Evening Star, full of strange golden fruits; and that Perseus had turned him into stone, when he passed him with the Gorgon's head.

But you will see, too, that most of these red and black dots run in crooked lines, and that many of the clusters run in lines likewise.

Look at one line: by far the largest on the earth. You will learn a good deal of geography from it.

The red dots begin at a place called the Terribles, on the east side of the Bay of Bengal. They run on, here and there, along the islands of Sumatra and Java, and through the Spice Islands; and at New Guinea the line of red dots forks. One branch runs south-east, through islands whose names you never heard, to the Friendly Islands and to New Zealand. The other runs north, through the Philippines, through Japan, through Kamschatka; and then there is a little break of sea, between Asia and America; but beyond it, the red dots begin again in the Aleutian Islands, and then turn down the whole west coast of America, down from Mount Elias towards British Columbia. Then, after a long gap, there are one or two in Lower California (and we must not forget the terrible earthquake which some years ago shook San Francisco, between those two last places); and when we come down to Mexico we find the red dots again plentiful; and only too plentiful, for they mark the great volcanic line of Mexico, of which you will read, I hope, some day, in Humboldt's works. But the line does not stop there. After the little gap of the Isthmus of Panama, it begins again in Quito, the country in which stand the huge volcanoes Chimborazo, Pasto, Antisanas, Cotopaxi, Pichincha, Tunguragua,—smooth cones from 15,000 to 20,000 feet high, shining white with snow, till the heat inside melts it off, and leaves the cinders of which the peaks are made all black and ugly among the clouds, ready to burst out in smoke and fire. South of them, again, there is a long gap, and then another line of red dots—Arequipa, Chipicani, Gualatieri, Atacama,—as high or higher than those of Quito. On the sea-shore below those volcanoes stood the hapless city of Arica, whose ruins we saw in the picture (p. 16). Then comes another gap; and then a line of more volcanoes in Chili, at the foot of which happened the fearful earthquake of 1835 (besides many more) of which you will read some day in that noble book "The Voyage of the Beagle;" and so the line of dots runs down to the southernmost point of America.

What a line we have traced! Long enough to go round the world if it were straight. A line of holes, out of which steam, and heat, and cinders, and melted stones are rushing up perpetually, in one place and another. Now the holes in this line which are near each other have certainly something to do with each other. For instance, when the earth shakes round the volcanoes of Quito, it shakes also round the volcanoes of Peru, though they are 600 miles away. And there are many stories of earthquakes being felt, or awful underground thunder heard, while volcanoes were breaking out hundreds of miles away. I will give you a very curious instance of that.

If you look at the West Indies on the map, you will see a line of red dots run through the Windward Islands. There are two volcanoes in them, one in Guadaloupe and one in St. Vincent (I will tell you a curious story presently about that last), and a little volcano, which now only sends out mud, in Trinidad. There the red dots stop; but then begins along the north coast of South America (which you must learn to call the Spanish Main) a line of mountain country called Cumana and Caraccas, which has often been horribly shaken by earthquakes. Now once, when the volcano in St. Vincent began to pour out a vast stream of melted lava, a noise like thunder was heard underground, over thousands of square miles beyond those mountains, in the plains of Calabozo, and on the banks of the Apure, more than 600 miles away from the volcano, a plain sign that there was something underground which joined them together, perhaps a long crack in the earth. Look for yourselves at the places, and you will see that (as Humboldt says) it is as strange as if an eruption of Mount Vesuvius was heard in the north of France.

So it seems as if these lines of volcanoes stood along cracks in the rind of the earth, through which the melted stuff inside was for ever trying to force its way, and that, as the crack got choked up in one place by the melted stuff cooling and hardening again into stone, it was burst in another place and a fresh volcano made, or an old one reopened.

Now we can understand why earthquakes should be most common round volcanoes; and we can understand, too, why they would be worst before a volcano breaks out, because then the steam is trying to escape; and we can understand, too, why people who live near volcanoes are glad to see them blazing and spouting, because then they have hope that the steam has found its way out, and will not make earthquakes any more for a while. But still that is merely foolish speculation on chance. Volcanoes can never be trusted. No one knows when one will break out, or what it will do; and those who live close to them—as the city of Naples is close to Mount Vesuvius—must not be astonished if they are blown up or swallowed up, as that great and beautiful city of Naples may be without a warning any day.

For what happened to that same Mount Vesuvius nearly 1800 years ago in the old Roman times? For ages and ages it had been lying quiet like any other hill. Beautiful cities were built at its foot, filled with people who were as handsome and as comfortable and (I am afraid) as wicked as people ever were on earth. Fair gardens, vineyards, oliveyards, covered the mountain slopes. It was held to be one of the Paradises of the world. As for the mountain's being a burning mountain, who ever thought of

that? To be sure, on the top of it was a great round crater or cup, a mile or more across and a few hundred yards deep. But that was all overgrown with bushes and wild vines, full of boars and deer. What sign of fire was there in that? To be sure, also, there was an ugly place below by the sea-shore called the Phlegræan Fields, where smoke and brimstone came out of the ground, and a lake called Avernus, over which poisonous gases hung, and which (old stories told) was one of the mouths of the Nether Pit. But what of that? It had never harmed any one, and how could it harm them?

So they all lived on, merrily and happily enough, till, in the year A.D. 79 (that was eight years, you know, after the Emperor Titus destroyed Jerusalem), there was stationed in the Bay of Naples a Roman admiral, called Pliny, who was also a very studious and learned man, and author of a famous old book on natural history. He was staying on shore with his sister; and as he sat in his study, she called him out to see a strange cloud which had been hanging for some time over the top of Mount Vesuvius. It was in shape just like a pine-tree (see p. 85); not, of course, like one of our branching Scotch firs here, but like an Italian stone-pine, with a long straight stem and a flat parasol-shaped top. Sometimes it was blackish, sometimes spotted; and the good Admiral Pliny, who was always curious about natural science, ordered his cutter and went away across the bay to see what it could be. Earthquake-shocks had been very common for the last few days; but I do not suppose that Pliny had any notion that the earthquakes and the cloud had aught to do with each other. However, he soon found out that they had; and to his cost. When he got near the opposite shore, some of the sailors met him and entreated him to turn back. Cinders and pumice-stones were falling down from the sky, and flames breaking out of the mountain above. But Pliny would go on: he said that if people were in danger it was his duty to help them; and that he must see this strange cloud, and note down the different shapes into which it changed. But the hot ashes fell faster and faster; the sea ebbed out suddenly, and left them nearly dry, and Pliny turned away to a place called Stabiæ, to the house of his friend Pomponianus, who was just going to escape in a boat. Brave Pliny told him not to be afraid; ordered his bath like a true Roman gentleman; and then went in to dinner with a cheerful face. Flames came down from the mountain, nearer and nearer as the night drew on. But Pliny persuaded his friend that they were only fires in some villages from which the peasants had fled; and then went to bed and slept soundly. However, in the middle of the night they found the courtyard being fast filled with cinders, and, if they had not woke up the Admiral in time, he would never have been able to get out of the house. The earthquake-shocks grew stronger and fiercer, till the house was ready to fall; and Pliny and his friend, and the sailors and the slaves, all fled into the open fields, amid a shower of stones and cinders, tying pillows over their heads to prevent their being beaten down. The day had come by this time, but not the dawn; for it was still pitch dark as night. They went down to their boats upon the shore; but the sea raged so horribly that there was no getting on board of them. Then Pliny grew tired, and made his men spread a sail for him, and lay down on it. But there came down upon them a rush of flames and a horrible smell of sulphur, and all ran for their lives. Some of the slaves tried to help the Admiral upon his legs; but he sank down again overpowered with the brimstone-fumes, and so was left behind. When they came back again, there he lay dead; but with his clothes in order, and his face as quiet as if he had been only sleeping. And that was the end of a brave and learned man—a martyr to duty and to the love of science.

But what was going on in the meantime? Under clouds of ashes, cinders, mud, lava, three of those happy cities were buried at once—Herculaneum, Pompeii, Stabiæ. They were buried just as the people had fled from them, leaving the furniture and the earthenware, even often jewels and gold, behind, and here and there among them a human being who had not had time to escape from the dreadful deluge of dust. The ruins of Herculaneum and Pompeii have been dug into since; and the paintings, especially in Pompeii, are found upon the walls still fresh, preserved from the air by the ashes which have covered them in. When you are older, you perhaps will go to Naples, and see in its famous museum the curiosities which have been dug out of the ruined cities; and you will walk, I suppose, along the streets of Pompeii, and see the wheel-tracks in the pavement, along which carts and chariots rumbled 2,000 years ago. Meanwhile, if you go nearer home, to the Crystal Palace, and to the Pompeian Court, as it is called, you will see an exact model of one of these old buried houses, copied even to the very paintings on the walls; and judge for yourself, as far as a little boy can judge, what sort of life these thoughtless, luckless people lived 2,000 years ago.

And what had become of Vesuvius, the treacherous mountain? Half or more than half of the side of the old crater had been blown away; and what was left, which is now called the Monte Somma, stands in a half-circle round the new cone and new crater which is burning at this very day. True, after that eruption which killed Pliny, Vesuvius fell asleep again, and did not wake for 134 years, and then again for 269 years; but it has been growing more and more restless as the ages have passed on, and now hardly a year passes without its sending out smoke and stones from its crater, and streams of lava from its sides.

And now, I suppose, you will want to know what a volcano is like, and what a cone, and a crater, and lava are?

What a volcano is like, it is easy enough to show you; for they are the most simply and beautifully shaped of all mountains, and they are alike all over the world, whether they be large or small. Almost every volcano in the world, I believe, is, or has been once, of the shape which you see in the drawing on next page; even those volcanoes in the Sandwich Islands, of which you have often heard, which are now great lakes of boiling fire upon flat downs, without any cone to them at all. They, I believe, are volcanoes which have fallen in ages ago : just as in Java a whole burning mountain fell in on the night of the 11th of August, in the year 1772. Then, after a short and terrible earthquake, a bright cloud suddenly covered the whole mountain. The people who dwelt around it tried to escape; but before the poor souls could get away the earth sank beneath their feet, and the whole mountain fell in, and was swallowed up, with a noise as if great cannon were being fired. Forty villages and nearly three thousand people were destroyed, and where the mountain had been was only a plain of red-hot stones. In the same way in the year 1698, the top of a mountain in Quito fell in a single night, leaving only two immense peaks of rock behind,

and pouring out great floods of mud mixed with dead fish; for there are underground lakes among those volcanoes, which swarm with little fish which never see the light.

But most volcanoes, as I say, are, or have been, the shape of the one which you see here. This is Cotopaxi, in Quito; more than 19,000 feet in height. All those sloping sides are made of cinders and ashes, braced together, I suppose, by bars of solid lava-stone inside, which prevent the whole from crumbling down. The upper part, you see, is white with snow, as far down as a line which is 15,000 feet above the sea. For the mountain is in the tropics, close to the equator, and the snow will not lie in that hot climate any lower down. But now and then the snow melts off, and rushes down the mountain-side in floods of water and of mud, and the cindery cone of Cotopaxi stands out black and dreadful against the clear blue sky, and then the people of that country know what is coming. The mountain is growing so hot inside that it melts off its snowy covering; and soon it will burst forth with smoke and steam, the red-hot stones, and earthquakes which will shake the ground, and roars that will be heard, it may be, hundreds of miles away.

And now for the words, cone, crater, lava. If I can make you understand those words, you will see why volcanoes must be in general of the shape of Cotopaxi.

Cone, crater, lava: those words make up the alphabet of volcano learning. The cone is the outside of a huge chimney. The crater is the mouth of it. The lava is the ore which is being melted in the furnace below, that it may flow out over the surface of the old land, and make new land instead.

And where is the furnace itself? Who can tell that? Under the roots of the mountains, under the depths of the sea; down "the path which no fowl knoweth, and which the vulture's eye hath not seen: the lion's whelp hath not trodden it, nor the fierce lion passed by it. There He putteth forth His hand upon the rock; He overturneth the mountains by the roots: He cutteth out rivers among the rocks; and His eye seeth every precious thing"—while we, like little ants, run up and down outside the earth, scratching, like ants, a few feet down, and calling that a deep ravine; or peeping a few feet down into the crater of a volcano, unable to guess what precious things may lie below—below even the fire which blazes and roars up through the thin crust of the earth. For of the inside of this earth we know nothing whatsoever. We only know that it is, on an average, several times as heavy as solid rock; but how that can be we know not.

So let us look at the chimney, and what comes out of it; for we can see very little more.

Why is a volcano like a cone?

For the same cause for which a molehill is like a cone, though a very rough one; and that the little heaps which the burrowing beetles make on the moor, or which the ant-lions in France make in the sand, are all something of the shape of a cone, with a hole like a crater in the middle. What the beetle and the ant-lion do on a very little scale, the steam inside the earth does on a great scale. When once it has forced a vent into the outside air, it tears out the rocks underground, grinds them small against each other, often into the finest dust, and blasts them out of the hole which it has made. Some of them fall back into the hole, and are shot out again; but most of them fall round the hole, most of them close to it, and fewer of them farther off, till they are piled up in a ring round it, just as the sand is piled up round a beetle's burrow. For days, and weeks, and months this goes on, even it may be for hundreds of years, till a great cone is formed round the steam vent, hundreds or thousands of feet in height, of dust and stones, and of cinders likewise. For recollect, that when the steam has blown away the cold earth and rock near the surface of the ground, it begins blowing out the hot rocks down below, red-hot, white-hot, and at last actually melted. But these, as they are hurled into the cool air above, become ashes, cinders, and blocks of stone again, making the hill on which they fall bigger and bigger continually. And thus does wise Nature stand in no need of bricklayers, but makes her chimneys build themselves.

And why is the mouth of the chimney called a crater?

Crater, as you know, is Greek for a cup. And the mouths of these chimneys, when they have become choked and stopped working, are often just the shape of a cup, or, as the Germans call them, *kessels*, which means kettles, or caldrons. I have seen some of them as beautifully and exactly rounded as if a cunning engineer had planned them, and had them dug out with the spade. At first, of course, their sides and bottom are nothing but loose stones, cinders, slag, ashes, such as would be thrown out of a furnace. But Nature, who, whenever she makes an ugly desolate place, always tries to cover over its ugliness, and set something green to grow over it, and make it pretty once more, does so often and often by her worn-out craters. I have seen them covered with short sweet turf, like so many chalk downs. I have seen them, too, filled with bushes, which held woodcocks and wild boars. Once I came on a beautiful round crater on the top of a mountain, which was filled at the bottom with a splendid crop of potatoes. Though Nature had not put them there herself, she had at least taught the honest Germans to put them there. And often she turns her worn-out craters into beautiful lakes. There are many such crater-lakes in Italy, as you will see if ever you go there; as you may see in English galleries painted by Wilson, a famous artist, who died before you were born. You recollect Lord Macaulay's ballad, "The Battle of the Lake Regillus"? Then that Lake Regillus (if I recollect right) is one of these round crater-lakes. Many such deep clear blue lakes have I seen in the Eifel, in Germany; and many a curious plant have I picked on their shores, where once the steam blasted, and the earthquake roared, and the ash-clouds rushed up high into the heaven, and buried all the land around in dust, which is now fertile soil. And long did I puzzle to find out why the water stood in some craters, while others, within a mile of them perhaps, were perfectly dry. That I never found out for myself. But learned men tell me that the ashes which fall back into the crater, if the bottom of it be wet from rain, will sometimes "set" (as it is called) into a hard cement, and so make the bottom of the great bowl waterproof as if it were made of earthenware.

But what gives the craters this cup-shape at first?

Think——While the steam and stones are being blown out, the crater is an open funnel, with more or less upright walls inside. As the steam grows weaker, fewer and fewer stones fall outside, and more and more fall back again inside. At last they quite choke up the bottom of the great round hole. Perhaps, too, the lava or melted rock underneath cools and grows hard, and that chokes up the hole lower down. Then, down from the round edge of the crater the stones and cinders roll inward more and more. The rains wash them down, the wind blows them down. They roll to the middle, and meet each other, and stop. And so gradually the steep funnel becomes a round cup. You may prove for yourself that it must be so, if you will try. Do you not know that if you dig a round hole in the ground, and leave it to crumble in, it is sure to become cup-shaped at last, though at first its sides may have been quite upright, like those of a bucket? If you do not know, get a trowel and make your little experiment.

And now you ought to understand what "cone" and "crater" mean. And more, if you will think for yourself, you may guess what would come out of a volcano when it broke out "in an eruption," as it is usually called. First, clouds of steam and dust (what you would call smoke); then volleys of stones, some cool, some burning hot; and at the last, because it lies lowest of all, the melted rock itself, which is called lava.

And where would that come out? At the top of the chimney? At the top of the cone?

No. Nature, as I told you, usually makes things make themselves. She has made the chimney of the furnace make itself; and next she will make the furnace-door make itself. The melted lava rises in the crater—the funnel inside the cone—but it never gets to the top. It is so enormously heavy that the sides of the cone cannot bear its weight, and give way low down. And then, through ashes and cinders, the melted lava burrows out, twisting and twirling like an enormous fiery earth-worm, till it gets to the air outside, and runs off down the mountain in a stream of fire. And so you may see (as are to be seen on Vesuvius now) two eruptions at once—one of burning stones above, and one of melted lava below.

And what is lava?

That, I think, I must tell you another time. But if you want to know (as I dare say you do) what the eruption of a volcano is like, you may read what follows. I did not see it happen; for I had never the good fortune of seeing a mountain burning, though I have seen many and many a one which has been burnt—extinct volcanoes, as they are called.

The man who saw it—a very good friend of mine, and a very good man of science also—went last year to see an eruption on Vesuvius, not from the main crater, but from a small one which had risen up suddenly on the outside of it; and he gave me leave (when I told him that I was writing for boys and girls) to tell them what he saw.

This new cone, he said, was about two hundred feet high, and perhaps eighty or one hundred feet across at the top. And as he stood below it (it was not safe to go up it), smoke rolled out from its top, "rosy pink below," from the glare of the caldron, and above "faint greenish or bluish silver of indescribable beauty, from the light of the moon."

But more——By good chance, the cone began to send out, not smoke only, but brilliant burning stones. "Each explosion," he says, "was like a vast girandole of rockets, with a noise (such as rockets would make) like the waves on a beach, or the wind blowing through shrouds. The mountain was trembling the whole time. So it went on for two hours and more; sometimes eight or ten explosions in a minute, and more than 1,000 stones in each, some as large as two bricks end to end. The largest ones mostly fell back into the crater; but the smaller ones, being thrown higher, and more acted on by the wind, fell in immense numbers on the leeward slope of the cone" (of course, making it bigger and bigger, as I have explained already to you), and of course, as they were intensely hot and bright, making the cone look as if it too were red hot. But it was not so, he says, really. The colour of the stones was rather "golden, and they spotted the black cone over with their golden showers, the smaller ones stopping still, the bigger ones rolling down, and jumping along just like hares." "A wonderful pedestal," he says, "for the explosion which surmounted it." How high the stones flew up he could not tell. "There was generally one which went much higher than the rest, and pierced upwards towards the moon, who looked calmly down, mocking such vain attempts to reach her." The large stones, of course, did not rise so high; and some, he says, "only just appeared over the rim of the cone, above which they came floating leisurely up, to show their brilliant forms and intense white light for an instant, and then subside again."

Try and picture that to yourselves, remembering that this was only a little side eruption, of no more importance to the whole mountain than the fall of a slate off the roof is of importance to the whole house. And then think how mean and weak man's fireworks, and even man's heaviest artillery, are, compared with the terrible beauty and terrible strength of Nature's artillery underneath our feet.

Now look at this figure. It represents a section of a volcano; that is,

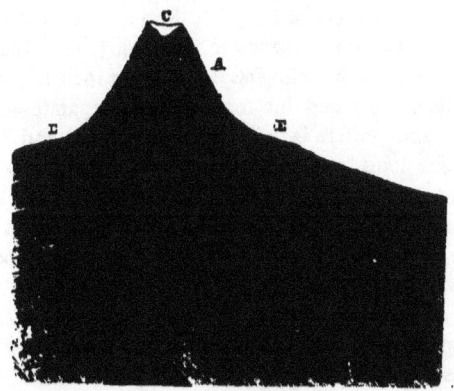

one cut in half to show you the inside. A is the cone of cinders; B B, the black line up through the middle, is the funnel, or crack, through which steam, ashes, lava, and everything else rises; C is the crater mouth; D D D, which looks broken, are the old rocks which the steam heaved up and burst before it could get out. And what are the black lines across, marked E E E? They are the streams of lava which have burrowed out, some covered up again in cinders, some lying bare in the open air, some still inside the cone, bracing it together, holding it up. Something like this is the inside of a volcano.

## BEAVERS AND BUILDING.

"ALAS for old home! Desolation, emptiness, the work of a lifetime swept away!" groaned a beaver to his wife, with a flap of his broad tail, expressive of mingled scorn and indignation. "Torn to pieces, to make room for the new mill-dam that these humans think they know how to put together. And the friends, 'our happy tribe,' together we gnawed the logs, together we fed on the bank, and dwelt side by side; and now, whither are they fled? Scattered, like the wreck of our happy dwelling, which the waters have carried beyond our depth. Times are changed for the worse since our young days; and the older the world grows, my dear, and the longer the rivers run, the fewer beavers and the more fools dwell upon their banks.

"Fools did I say?" cried he, rousing himself at the thought of the wrongs of his race, while another decided flap accompanied his words, and his partner listened with attentive ear and gleaming eye to his animated language. "Fools did I say? Idiots! Beings whose utter incapacity of mind and body glories in destroying what they can never imitate! And what are these inventions of theirs, their implements, their edifices? Why, a man and his tools are two separate affairs. If he leaves them behind—where is he? If he breaks one and has no smith at hand with *his* tools—where is he? If he blunts them and has no grindstone, and no stonecutter with *his* tools to make one—where is he? If he cannot find limestone for his mortar, and cow's hair and what not —where is he, and where is his work? And when the floods come— where is he, and where are all his fine constructions? Ay, my dear, let the west wind and the cataract answer me that. But a beaver *is* a beaver, worth a human and his tools put together ten times over. Whatever happens, there he is. Wherever he is driven by the draining of his pools, or the levelling of his dam, there he is. The beaver is a beaver still: give him his tail and his teeth, and back him against the finest and strongest tools that ever built a dam against wind and water, and plastered it over with compost of lime and hair! Yes! *wherever* we may go,— and go we must, for we beavers have had our day,—change and emigration are now the cry. Humans emigrate to beaver-land—beavers must emigrate too; but whither? Are there yet rivers to be found where no beaver ever yet laid his brain and his tail together, and made the water alive with his dwellings and his work? Thither let us go, and, gathering our scattered forces, prove ourselves superior to the adversities of civilization!"

"You do well, my love," replied a gentle voice beside him, while a feminine flap of another broad tail was heard in a subdued plash under the water;—"you do well to withdraw from a region where your superior example and endowments have so little benefited mankind. Vain, apparently, have been all the attempts of human beings to absorb into their own brain the wisdom that dwells in ours. Generations of beaver hats have been worn out—our parents, our kindred, our sons and our daughters sacrificed in the cause—and men are no wiser! The fur from our bodies, felted and fitted to the human hand, has been likewise adopted, and with a like result to their practical utility. Nor can I anticipate that the newest covering to the human head, so pretentiously styled a '*wide-awake*,' should indicate or stimulate an intelligence superior, or even equal to that of our little ones, *born* as they are with their eyes wide open. Might we hope for better days: that wisdom should not die with you; that men might live and learn; a glorious future might yet be before you—to educate the human head and hand——Ah! let us remain—my heart clings to this old reedy bank!"

"Never! my darling, never!" and he plashed a despondent plash as he spoke. "The days of 'go-ahead' are not ended yet. Men have yet to learn what strength, what vigour, what resources of every kind reside in the tail. Deprived by niggard nature of that most useful, that essential appendage, how can they understand its value, or appreciate the living virtue of that which when cooked for food they only esteem as the richest diet? No! let others go *a-head* through life; but while this old heart beats warm beneath the beaver fur, the motto of my existence shall be, 'Make sure you're right, then go *a-tail*.'"

Thus saying, he plunged into the stream, and for a few seconds disappeared, but was soon to be descried making his way down stream with little apparent movement of the feet, but with a conscious self-assertion of the member he so highly prized, which acted as a rudder, while he neared the rapids, and was soon concealed by the turnings of the stream from the view of his solitary mate, who wept in silence over the shattered home of her early loves.

Soon, however—for inactivity under misfortune forms no part of the beaver creed—soon she discerns, first by the distant sound of the ever-useful organ which so greatly assisted his movements, not least by the odour of *Castoreum*, by which beavers are so successfully traced by the hunters, and, finally, by the vision of the well-beloved nasal rotundity which appeared above the water, that her mate had neither spoken, resolved, nor departed in vain. Would that my tale had the breadth and energy which so greatly assisted him while he led the way to new waters, flowing through a yet untrodden forest, whose gigantic shade sheltered the growth of young and succulent boughs, and seemed to invite the disestablished and disheartened pair to begin life again! Happy were they that no prognostications of the revolutionary progress of a so-called civilization disturbed their minds or discouraged their efforts! Intent on the duties of the present, they left anxiety for the future, until that future should in its turn become the present.

Happy pair! whom destruction and disappointment only nerved to fresh effort; and who shall say that the sagacity expressed in action by those broad tails of yours, did not surpass the wisdom of many a human brain, toiling to create for himself appliances of existence alike foreign to his better nature and to his highest needs?

Happy pair! I think I see you on the deep shade of that virgin forest, where the weary sun looks in aslant each evening, and the cool shade protects your labours through the day. Deep among the ooze and the tangled reeds, by the near margin of the running water, you lay the foundation sound and strong, for you know by experience that tempests blow, and that floods can sweep away the best of surface work. Pursue your task as generations of your ancestors have done before you, so perfectly, so soundly, and without a flaw, that posterity shall find as little scope for improvement on your plans, and your work, as you have on those of your progenitors.

Not long did our friend Castor and his mate work alone on the new ground—I should rather say the new waters—they had selected for their home. Speedily joined by some of the scattered members of their former colony, they proceeded to organize their operations on that extended scale which had produced in the vast Canadian streams, and probably at remoter periods in the nearer waters of our Welsh and English rivers, those remarkable constructions, the beaver-dams and beaver-huts, whence doubtless were originated the first practical ideas in the mind of man, applicable to weirs and earthworks, for social or defensible purposes.

Shall we penetrate yet further into the past history of our race and theirs, and feel compelled to repudiate, even for those primitive inhabi-

A BEAVER-DAM.

tants of now civilized countries, whose only records are found in the remains of their dwellings, and their simple implements, the attribute of a pure originality of design? Side by side with the lake dwellings of a bygone age dwelt the beavers, whose bones and skulls are now disinterred in company with the flint implements and other relics of the untaught industry of that epoch. Shall we say that *they* learned no lesson in constructive ingenuity and engineering art from the creatures whose simple necessities differed chiefly from theirs in the fact that the object of the man was to raise his dwelling above the water, and that of the beaver was to raise the water to the level of his dwelling; and therefore, while the latter laid his logs in a horizontal position, the former drove in his piles vertically, and raised his platform above them? The appearance of the dwellings of the man and the animal—I had almost said of the animal and the man—must have been very similar, for the description of the two runs thus :—

"The dwellings of the Gauls are described as having been circular huts, built of wood and lined with mud. The huts of the pile works were probably of a similar nature, pieces of the clay used in the lining having been preserved. These fragments bear on the one side the marks of interlaced branches, while on the other, which apparently formed the wall of the cabin, they are quite smooth. Some of those found at Wangen are so large and regular, that Mr. Troyes feels justified in concluding that the cabins were circular, and about fifteen feet in diameter. Supposing on an average each cavern was inhabited by four persons . . ."*

"Every man had a hut on the planks in which he dwells, with a trap-door closely fitted on the planks, and leading down to the lake. Of fish there is such an abundance that when a man has opened his trap-door, he lets down an empty basket by a cord into the lake, and, after waiting a short time, draws it up full of fish."†

"The lake dwellers followed two different systems in the construction of their dwellings, in the second of which the support consisted not of piles only, but of a solid mass of mud and stones, with layers of horizontal and perpendicular stakes, the latter serving to bind the mass firmly together. They were from three to nine inches in diameter."‡

"The beaver lodges are composed chiefly of branches, moss, and mud, and will accommodate five or six beavers together. The form of an ordinarily sized beaver's hut is circular, and its cavity is about seven feet

* Lubbock, "Prehistoric Man." † Herodotus. ‡ Dr. Keller.

in diameter by about three feet in length. The walls of this structure are extremely thick, so that the external measurement of the same lodge will be fifteen or twenty feet in diameter, and seven or eight feet in height. The roofs are finished off with a thick layer of mud, laid on with marvellous smoothness, and carefully renewed every year."

"In order to secure a store of winter food, the beavers take a vast number of small logs, and carefully fasten them under water in the close vicinity of their lodges. When a beaver feels hungry, he dives to the store heap, drags out a suitable log, carries it to a sheltered and dry spot, and nibbles the bark away, and then either permits the stripped log to float down the stream, or applies it to the dam."

"In forming the dam, the beaver does not thrust the ends of the stakes into the bed of the river, but lays them down horizontally, and keeps them in their places by heaping stones and mud upon them. The logs of which the dam is composed are generally about seven inches in diameter."

Such a dwelling, with a barrier constructed on similar principles, was erected by our energetic pair of friends, assisted by the companions who shortly thronged around them, for the social instinct is strong in these wonderful animals, and the company of their fellows seems to call out into fuller exercise the instincts and powers which in solitude seem to lie comparatively useless and inactive, except so far as to supply their individual wants. Brought up together, these animals live in perfect harmony, and labour in concert; but, removed from such society, each can live no longer but for himself alone, and even the instinct of building suffers total extinction as long as the animal is kept in captivity; while the attempt to introduce two of them under these circumstances to one another is attended with the danger of their violent combats, and the severe wounds they inflict on each other, terminating fatally for one or both. The condition of isolation and of captivity, so utterly foreign to their nature, seems to paralyse every natural instinct, especially if adopted in early youth. The necessity for exertion in procuring food, and in protecting its dwelling from attack, and from the effects of climate and weather, seem to be absolute requisites, as they are found to be in the human race, for drawing out and educating to a full perfection not only the subtler sensibilities of their nature, but even their ordinary instincts.

The first and apparently the most arduous operation was to secure the most suitable material, not in the first place for their dwelling, but for the immense dam which was to be carried across the stream, and to secure for their base of operations a clear and undisturbed depth of water. No settler had yet made a clearing in that native forest; no axe had rung through its silent arcades, nor had a tree fallen, save on the height where the storms had from time to time uprooted the aged giants of the wood. But now the younger trees, the saplings, were to bow their waving crests, and succumb to the slow and sure inroads made by the teeth of the four-footed invader. These more recent growths were found in luxuriant abundance, rising chiefly from one of those strange clearings in the forest which testify to the force of the invisible agents, which from past ages to the present time have performed in a few hours the work which now costs the axe of the backwoodsman months to accomplish. And whether it be the whirlwind, gathering force as it sweeps over miles of forest and agitates the ocean of verdure, till it meets the obstruction on a rising ground of a few aged and gigantic trunks, or whether the lightning force of an electric current, devastating the line of its viewless march, the effect is one which enhances greatly the silent sublimity of those vast solitudes—for solitudes they now are, the trail of the Mohawk, the Oneida, or the Mohican having almost vanished from among them, and the human settler having not yet made a permanent home there. But on the spot where lay the vast trunks of uptorn monsters, heaped in wild confusion,—the noble elm, with its graceful and weeping top laid low; the maple in its rich variety; the sturdy oak, stretching its roots into the air, and mingling its giant boughs with those of the broad-leaved linden,—space was thereby afforded for the struggling growth towards sunshine and upper air of many an inferior tree, which essayed to lift its modest head to a level with the surrounding surface of undulating verdure. The silvery-stemmed birch, the quivering aspen, the useful and shrubby nut-bearing bushes, which, thrown in this manner into the company of the stately and the great, proved themselves, by their utility for the purposes which these skilful little architects had in hand, to be neither ignoble nor vulgar cumberers of the ground,—these in their turn were destined to a fate differing from that which their own ambition, could its utterances have been heard, would have asked, of supplying the vacuum created, not by the lightning and the hurricane, but by the encroaching hand of advancing civilization, which had commenced the work whereby some of the most interesting specimens of instinctive ingenuity have by this time been almost improved off the face of the earth.

Here the beavers fixed their abode, here in increasing numbers they began to nibble—sharp were their teeth and quick the fall, as tree after tree crashed down and interlaced their branches on their way. Cut afterwards into shorter lengths, they became more portable, and were conveyed, by the united strength and ingenuity of the builders, to the spot beside the waters where the wondrous edifice was soon to rise. Not ten, nor even a hundred, sufficed for the work—higher and higher up the course of the stream they plied their task, never following its downward flow, but in parts of the forest whence with vast labour they rolled the nibbled branches to the stream, which floated them to the destined spot, until the foundation was laid of a barrier three hundred yards in length. Slowly rose the superstructure, tapering from a thickness at the bottom of twelve feet, to a summit of from two to three feet. But why that strange variety in the direction of the work? Is it possible that the master-mind which seemed to guide the whole has hesitated, paused, changed its intention, and failed at length in producing that symmetry which so much united action seemed to promise? Here it runs boldly out into the river. There it takes a bend, and curves outwardly against the stream. Ah! at that very point they know that the current is the strongest—and though they have never visited an artificial harbour, or inspected a modern breakwater, they have adopted their principle with the most consummate engineering skill. Strengthened with stones and plastered over with mud, they have now completed the breastwork, behind which, in the artificial depths which they have created, they may calmly erect the dwellings which are to form for their growing colony, not only a cluster of compact and comfortable dwellings, but an invaluable defence against the attacks of their enemies.

Yes, experience is a stern teacher, and the new dwelling must be both commodious within and invincible without. The thickness of their conical mansions is such, that an external diameter of fifteen feet allows but seven feet within, and the walls are so curiously plastered within and without, that the outer surface, hardened by the action of frost, is a model for an iron-plated man-of-war. Danger there may be beyond the walls; for the savage foe to the beaver, as to the smaller quadruped, the wolverine, prowls through the woods in search of prey, with its jet-black fur, gigantic paws, and ivory claws; and, but for the thickness and hardness of the roofing of the beaver-huts, would even invade the privacy of their dwelling, where, in neatly arranged beds, separately placed against the wall, each family of beavers spends the winter months. A deep ditch affords additional facility of entrance and egress to the beavers, the doors of whose huts are under water, and to whom travelling by land is always less acceptable than aquatic exercise; and the store of bark stripped from the logs they have used in building, is kept in the lower stores of their dwelling, for the supply of their winter food. Notwithstanding these provident arrangements, the beavers generally emerge from their winter quarters greatly reduced by hunger and extremely thin, and find it prudent during the summer to separate from each other and to seek their food alone, until the necessity of shelter during the severe season re-unites them for another winter.

Not all of the beaver tribe are equally distinguished for their ingenuity and industry. Among them, as among the bees, are to be found a class, called by the Canadian trappers *Les Paresseux*, or the *Idlers*. These, the rejected and disappointed bachelors of the race, retire gloomily into burrows or tunnels on the river-banks, and, having no family affections to call out their natural powers in providing homes or provision for the young, they build no dams and construct no houses, and unsuspiciously fall an easy prey to the trappers, doubtless living unrespected, and dying unlamented.

Perhaps now, as men have failed in absorbing constructive instinct through their hats, and have descended for their fabrics to the insect world, the beaver in North America may linger at least as long as his compatriot, the Red Indian, whose wigwam is indeed but a miserable parody on the comfort of the beaver's hut. Time was when the beaver was the first architect in the British Isles. Long before round towers had arisen to mystify future antiquarians, the beaver was modelling his dome in the fens of Yorkshire and the mountains of Wales. But skill is not always victorious against brute force. The beaver's tail struggled in vain against the flint hatchet, and when the Roman came, the beaver and the Briton alike withdrew to Scotland and to Wales. They had had their day—the day when the great Irish deer, the wild boar, the roebuck, the stag, the goat, and the wolf, disputed with the Briton the fens of Lincolnshire and the wolds of Berks. They were the successors of the elephant, the woolly rhinoceros, the hippopotamus, the hyæna, and the cave tiger, which had ceased to exist before the peat began to grow on the swampy plains where the beavers built their dams. It is in the peat mosses that we find the only story of the English beaver.

But in Scotland and in Wales the beaver finds a place in written history. An old monk who writes his travels in Wales tells of the beavers in Cardiganshire, and three hundred years later, about A.D. 1490, we are told they abounded about Loch Ness, whence their furs were exported. Still later lingered the tradition among the Highlanders about Lochaber, of the former abundance of the "broad-tailed otter" there, the very same name by which it was known in Wales. But the beaver's coat was too precious for him to be allowed to wear it in peace, for Howel the Good, when he fixed the price of furs by law in the ninth century for the Welshmen, while he rated an otter's skin at 12$d$., estimated the beaver's as worth 120$d$.

We know not whether the Crusaders wore them for their cloaks; but when the Archbishop of Canterbury, A.D. 1180, went to the Principality to beat for recruits, his secretary and biographer was so delighted with the beavers of Cardiganshire that, forgetting the Crusades, he can only tell us about their huts, their tails, and their teeth, and how their habitations, formed of willow-stumps, so soon as the boughs begin to shoot, look like groves of trees, rude and natural without, but artfully constructed within.

But this was long ago, and all the traces the beaver has left are his name, still attached to some waters in the Principality, telling us of the home of the old family; just as in Yorkshire, Beverley, "Beaver's Legh," by its name and its coat of arms, remind us of an inhabitant more ancient than the monks and the minster.

But as there were heroes before Ajax, beaverdom, like man, had its giants of a yet older time. All over the northern world, from Siberia to Britain, have been found the remains of a gigantic beaver, buried in the clay with the bones of the mammoth and the rhinoceros, with teeth in comparison with which the incisors of our beavers are puny indeed; teeth which, instead of confining themselves to willows and alders, may have felled the huge pine for a morning's work, to dam some ancient and long-forgotten river.

Let us hope that our children may sometimes have the opportunity of seeing a beaver-dam even nearer home than the rivers of Arctic America. One enterprising proprietor proposes to re-introduce the beaver in Staffordshire, where, let us hope, the intelligent builders will learn that plantations are not forests, and will exercise no right of free forestry beyond the domain of their kindly patron. A few beavers still linger on the Vistula and other rivers of Russian Poland. There are many living who remember several colonies in Norway; but it was supposed that the Norwegian Parliament, when thirty years ago it passed a law imposing a heavy fine on any one who should kill a beaver, was acting on the old maxim of locking the stable door when the steed was stolen, for years had passed since a beaver had been seen in the land. It seems, however, that one family, perhaps only a single pair, had contrived to escape observation in a sequestered forest of Southern Norway. Unknown and unobserved, the family increased; they judiciously selected almost the only vast estate left in that democratic land. Its proprietor kept the secret, and three years ago introduced a delighted and astonished English professor to a thriving colony in a sequestered *Dahl*. In spite of his enthusiasm, it is to be feared that the professor's zeal for his University Museum might have overcome his respect for Norwegian law—at least, the temptation would have been strong—had he not, by strange good fortune, lighted on the unburied remains of an aged beaver, who had died in peace by the side of his village, and whose skeleton is now honourably entombed, with other gaunt specimens, in a palace of science.

But the living beaver may be studied still nearer home. The interesing Canadian family in the Zoological Gardens, though not allowed to indulge their taste by building a dam across the Regent's Canal,—a feat on which they seemed to have set their hearts, having more than once surreptitiously made nocturnal surveys, rudely interrupted by London barges,— have had to content themselves with building a hut of most inferior materials in a mere paltry pond. Even here, however, they contrived to elude the pertinacious inquisitiveness of keepers and sight-seers into their domestic privacy. Safely housed under that cumbrous dome, which they had been obliged to construct,—not of the neat willows their native taste would have selected, but of such rubbish as was within a captive's reach, —they have reared a little beaver family, of the existence of which no one was aware, until the necessity for extra-mural sepulture compelled them to deposit the remains of a little one on the edge of their prison; and a few days afterwards the watchful mother introduced her surviving offspring to the glories of sunrise on a summer morning.

# LIGHT.

A SINGLE ray of light is a marvellous thing, whether it is regarded in itself or in its influence on other objects in nature. Even every child knows that the sun is the earth's grand source of light, as it also is to the other planets of our system. The rays of the sun, however, are not simply rays of light. Associated with the luminous rays there are others which give out heat and chemical power. These various descriptions of rays are always found in combination with each other.

They fly with almost inconceivable swiftness. A sunbeam darts in a single minute to a distance of more than a million of miles. In a second it passes through a hundred and ninety-two thousand miles. When these rays fall upon any surface, they either pass through it, or they pass into it and are lost, or they are reflected from it. Bodies which permit the rays of light to pass freely through them are said to be transparent. Glass, water, air, and various gases, are instances of transparent bodies; but some of them are more transparent and others less so. In passing through a transparent medium, the sun will go in a straight line; but if it passes from one transparent body into another, that other being of a different density, it no longer pursues a straight course, but becomes more or less bent. The familiar experiment of the basin and shilling very satisfactorily illustrates this law. An empty basin is taken, and at the bottom of it a shilling is placed: the observer then goes back a step or two from the basin, until the side next to him just hides the shilling from his eye. If another person now pours water gradually into the basin, it will be found that the shilling becomes visible to the first observer, although he has not in the least altered his position. The reason is simply this: the rays of light reflected from the shilling, in passing from the water into the air, which is a transparent body of a different density, are bent in such a manner as to make them meet the eye at a point at which they would not otherwise have reached it. This property of the sunbeam—the property of bending—is called, or rather the effect is called, the refraction of light.

It is upon this property of the sunbeam's bending aside out of a straight

ISAAC NEWTON EXPERIMENTING WITH LIGHT.

line on its entering into bodies of different density to that of the air, that the effect of a lens on light depends. When the rays fall on the surface of the lens otherwise than in straight lines, they pass through it in a direction still different from that in which they reached it, and emerge from it in a direction also different from that which they followed in passing through its substance. In the case of a convex lens, or bulged and magnifying-glass, the rays of light are so bent in passing through it, as to meet in a common point at a little distance from the lens, and this point is called the focus. Thus a lens may be employed to collect or gather together the rays of light proceeding from any luminous body, such as the sun or a lamp, and to bring them to a focus upon any surface towards which we may choose to direct them. With this process all boys are well acquainted. All the lenses which are employed in the artificial experiments on light and shadow are varieties of the convex lens, the object being to collect the rays proceeding from various external bodies, and to collect them in such a manner upon a smooth white surface, as shall render those bodies visible as distinct images. Lenses are sometimes convex or rounded only on one side, the other being flat. In that case, the lens is called a plano-convex lens. Sometimes the one side is convex, and the other concave, or hollowed—and this sort of lens is called a *meniscus*. The most frequently employed lens, however, is that which has two convex surfaces. This is called a double convex lens.

Rays of light have also the property of being capable of reflection, or of being thrown back from various surfaces. Indeed, were it not for this property, all the objects around us would be invisible, and light could be perceived only when we looked at the sun, or at some other luminous body. In fact, we do not *see* external objects at all, but the image of them which is reflected and pictured upon what is called the *retina* of our own eye. The reflecting power of different objects upon the rays of light is intimately connected with the state of their surface, and is closely associated with the production of many remarkable effects which are especially worthy of notice.

To light also belongs the property of being absorbed, or lost, when it falls upon various kinds of surfaces—as black velvet, for example. It seems as if it really entered into such substances, and there became annihilated. It is, at all events, no longer to be detected.

The decomposition of light, as it is called, is another of its remarkable properties or capabilities. The process by which this is accomplished is well known. When a glass prism is held in the sunbeams, immediately may be seen a beautiful streak of colours upon the wall or on the ceiling. This is in consequence of the breaking up of the rays into those colours, which were previously united in the form of pure white light. It is generally believed that any coloured body, such as a red one, for instance, appears so coloured to the eye because of its decomposing power on the light—in other words, it absorbs all the other rays, and reflects only the red ones, which become impressed on the organ of vision. The three primary rays of light, as they are called, which by their various combinations produce all other colours, and which, mingled in due proportion, constitute white light, are red, blue, and yellow.

These remarks on the properties of light may help to explain certain appearances which we now proceed to notice.

One evening in the summer of the year 1743 a gentleman and his servant, as they were sitting at the door of a house in Cumberland, saw the figures of a man and a dog pursuing some horses along the side of a mountain. They knew the place to be so precipitous that a horse could scarcely walk upon it at all; but as these figures appeared to them they seemed to run at an amazing pace, until they disappeared from view. The servant and his master resolved to go next morning in search of the person they had seen, and accordingly, after daybreak, ascended the steep side of the mountain, fully expecting to find the man dead, and quite satisfied that the horses also must have been dashed to pieces. But there was no trace of either, nor could they find in the turf the slightest mark to indicate that the place had been recently visited by any living creature. They returned in great amazement, and on telling what they had seen were laughed at for their credulity, and most probably began themselves to believe that their own senses had deceived them. In the following year one of these same men was walking in the evening about half a mile from the same mountain, when, to his astonishment, he saw a troop of horsemen riding on the steep mountain side, in close ranks and at a quick pace. Not having forgotten the ridicule with which his neighbours had received his former statement, he continued, on this occasion, to observe the figures for some time in silence; but becoming convinced that there could be no deception in the matter, he went to the house and informed another person. They accordingly went out together, and in a short time the whole family had assembled to witness the strange spectacle. The mounted men advanced in regular troops along the side of the mountain, and finally disappeared from view by crossing over it. They

THE SPECTRAL BALLOON.

continued to be seen for upwards of two hours. Many troops were seen in succession, and frequently the last but one in the troop quitted his position, and, galloping to the front, fell into the same pace with those who were there. All this was seen by every person at every cottage within the distance of a mile; but no explanation presented itself to the beholders.

The loftiest of the Hartz Mountains in Hanover is called the Brocken, and from the earliest periods of authentic history this mountain has always been the seat of the marvellous. The scenery is weird and wild, and heathen rites and sacrifices were long ago to be witnessed in the midst of it. But the most remarkable spectacle there to be seen in certain states of the atmosphere is a gigantic figure which is popularly known as the "Spectre of the Brocken." A gentleman having ascended the mountain on a fine morning in spring when the sun was just rising, was astonished to see at some distance before him a colossal representation of himself. His hat having been suddenly almost carried away by a gust of wind, he raised his hand to his head, and was amazed to find that the figure did so likewise. He then bent forwards, and the figure repeated his movements. It then disappeared. Soon, however, it appeared again; when it mimicked his gestures as before. Another person having joined him, there very soon appeared two colossal spectres in the clouds, which imitated the motions of the wondering observers in a remarkable degree. After this the figures disappeared.

Hogg, the Ettrick Shepherd, remarks a similar fact, and describes the astonishment which he felt when on one occasion he observed a spectre in the clouds which imitated all his movements, and which he even found to be attended by a spectral dog like his own, but of enormous magnitude.

Some remarkable appearances of a corresponding character were witnessed by Baron Humboldt in his travels in South America. On one occasion he observed small fishing-boats swimming in the air during more than three or four minutes above the well-defined horizon of the sea. When residing at Cumana, he frequently saw the islands of Picuita and Baracha suspended in the air, and sometimes with an inverted image. At another place he beheld the extraordinary spectacle of cows suspended in the air at a considerable height above the soil. The same voluminous author says, "The well-known phenomenon of the mirage is called in Sanscrit 'the thirst of the gazelle.' The objects appear to float in the air, while their forms are reflected in the lower strata of the atmosphere.

At such times the whole desert resembles a vast lake, whose surface undulates like waves. Palm-trees, cattle, and camels sometimes appear inverted in the horizon. In the French expedition to Egypt this optical illusion often nearly drove the faint and parched soldiers to distraction. This phenomenon has been observed in all parts of the world."

Phantom ships, as they have been called, have repeatedly been seen by various observers. Mr. Scoresby, in his voyage to Greenland, in 1822, saw an inverted image of a ship in the air, so well defined that he could distinguish by a telescope every sail, the peculiar rig of the ship, and its whole general character, insomuch that he confidently pronounced it to be his father's ship, the *Fame*, which it afterwards proved to be. On subsequently comparing notes with his father, he found that their relative position at the time gave a distance of the one from the other of nearly thirty miles. But the most magnificent of these spectacles was seen by Mr. Scoresby after he had further penetrated the Arctic regions. On this occasion, the sky being clear, there was a tremulous and perfectly transparent vapour, and the land on the coast was particularly distinct and bold. The general telescopic appearance of the coast was that of an extensive ancient city, abounding with the ruins of castles, obelisks, churches, and monuments, with other large and conspicuous buildings. Some of the hills seemed to be surmounted by turrets, battlements, spires, and pinnacles; while others exhibited large masses of rock, apparently suspended in the air, at a considerable elevation above the actual termination of the mountains to which they referred. The whole magnificent scene was continually changing. Scarcely was any particular portion sketched before it assumed another form, and entirely differed in its whole aspect. But notwithstanding these many and repeated changes, the various figures had all the distinctness of reality; and not only the different strata, but also the veins of the rocks, with the wreaths of snow occupying the ravines and fissures, formed sharp and distinct lines, and exhibited every appearance of the most perfect solidity. The Fata Morgana, seen in the Straits of Messina, between Sicily and the coast of Italy, presents an appearance somewhat resembling that observed by Scoresby.

All the appearances just described depend upon those properties of light of which we before gave an account, and many of a similar character can, without difficulty, be provided on a small scale by artificial means. The spectres seen in the clouds, as at the Brocken, are merely shadows of the observer, or of other persons, projected on dense vapour or thin fleecy clouds, which have the power of reflecting much light. Vapours or clouds are necessary for their production, and they can be seen only when the sun is throwing his rays in a horizontal direction. The other phenomena depend upon the unequal refraction of light. If, in calm weather, the surface of the sea is much colder than the air which overlies it, that air will gradually become colder by giving out its heat to the water, and the air immediately above will give out its heat to the cooler air immediately below it, so that the air from the surface of the sea to a considerable height upwards will gradually diminish in density, and, therefore, everything requisite to produce the phenomena in question is present. Just as, in the shilling and basin experiment, the shilling is seen in consequence of the alteration in the path of the light proceeding from it, or rather reflected by it, by its having to pass through a denser into a rarer medium, so the ship below the horizon is rendered visible to the eye by a similar disturbance of the direction of the light proceeding from it, in consequence of the unequal density of the overlying air. All the objects seen by Mr. Scoresby were produced in the same way—that is, by the unequal refraction of the rays of light proceeding from real objects on the coast.

The spectacle of the troopers seen in Cumberland may be explained in this way: a number of troopers were probably being drilled on the other side of the hill to that on which the spectral troopers were seen. At the time of their being so engaged, the air near the ground was more dense than at a short distance above it. Such being the case, the light, just as in the experiment which has been mentioned, on passing from a medium of greater into one of less density, underwent refraction, and its course was so far changed as to bend the rays over the hill, and produce the impression that the troopers were actually on the one side of the hill instead of being on the other. The Fata Morgana are to be explained upon the same principle.

Thus nature is full of wonders, and the simplest causes, such as the production of a shadow, or the bending of a ray of light, may sometimes combine so as to produce effects the most remarkable. The light of science dispels the mystery in which ignorance or superstition would envelop such occurrences, and teaches us that those very spectacles which terrify the unenlightened are mere illustrations on the great scale of the properties of bodies, the less imposing examples of which are, at every moment, presented to our notice. The same law which makes the rod of the angler appear bent when thrust into the clear stream in which he

THE SPECTRE OF THE HARTZ.

is fishing, is that which, in other circumstances and in a different medium, produces all the splendid and fantastic scenery of the Enchanted Coast, the Fata Morgana, and the Phantom Ship.

When light is reflected from a polished surface at a particular angle, it becomes possessed of certain properties which it did not before possess. In this state, light is said to be polarized. By means of an ingenious apparatus, and with the assistance of the oxy-hydrogen light, or with that of the sun, the most splendid phenomena can be produced. Figures of various kinds, illuminated with a depth and brilliancy of colour otherwise unattainable, can be displayed, and shine with all the lustre of precious gems. But the arrangements by which these phenomena are produced are too complicated for intelligible description in a work like the present.

Still, very singular effects can be realized in a simpler manner. Let us suppose a room to be darkened by the exclusion of all light from without, and that in this room a lamp is then lighted which emits rays of only one colour—say yellow or red. By dissolving a little salt, or some other compound containing soda, and applying it to a sponge, or by merely sprinkling the surface of a sponge dipped in spirits of wine with some powdered carbonate of soda, the flame, when this sponge is ignited, will be almost wholly of a yellow colour. The effect of this experiment is most extraordinary. The dresses of ladies, which before the introduction of the yellow light had displayed every variety of tint, now appear simply white or a ghastly yellow, and black or purple. Neither is the effect on the countenance less appalling—every face wears the aspect of disease, and looks pale and ghastly. So is it with the furniture—so with everything. If a shutter is opened at the other end of the room, and the pure daylight is allowed to stream in, the contrast between the one division of the apartment and the other is exceedingly remarkable. The same phenomena are sometimes produced at scientific lectures, by means of trays of ribbons, and one or two sponges prepared in the manner which has just been described. In such an experiment, a solar lamp, within a dark lantern, is employed. When the lamp is allowed to pour its rays on the ribbons, they appear in their proper colours; but when the slide of the lantern is closed, and the light of the sponges falls upon the trays, all is instantly changed. The same experiment has been effectively made with a landscape painting. With the yellow light the whole scene is dismal in the extreme, but when the white light is permitted to fall upon it, it is at once restored to its former beauty.

These effects are all due to the properties of light considered in itself; but there are many interesting experiments which are related to the manner in which light is conveyed to the eye. These belong to the class of what are called optical delusions. If a lighted stick be twisted rapidly round, the appearance of a single moving point of flame is lost to the eye, and a circle of light takes its place. Many readers must be familiar with the thaumatrope. This instrument consists of an upright frame of wood, resembling the frame of a looking-glass; but the place of the mirror is occupied by a flat board, covered on each of its sides with white paper. This board is conected with a band and a pulley, and can be made to revolve with great swiftness. Upon one side it is usual to paint a rat-trap without a rat in it, and upon the other a rat without a trap. By causing this board to revolve at a certain rate, the rat and the trap cease to be seen separately, and the rat is distinctly beheld within the trap. Of course, many varieties of objects may be used in connection with this instrument—all depending upon what is painted on the revolving board. Another is a figure of a body without a head on the one side, and a head without a body on the other. By the revolutions the head is seen as if restored to its proper place.

Among the various instruments which are associated with light and its uses, must first be named the *Camera Obscura*. This was discovered by a celebrated natural philosopher of the sixteenth century, John Baptista Porta, a Neapolitan gentleman, born about the year 1540. The name, camera obscura, means the dark chamber, and in giving an account of his instrument, and his invention of it, Porta tells us that all the windows of an apartment must be closed by shutters, and that it will be well if even the chimney also is stopped up, so as to exclude every possible ray of light. A single round aperture, of the size of the little finger, must then be made in a shutter, and upon the white wall opposite will be seen images of external objects, but all upside down, and their movements reversed. This must often have been seen before Porta's time, but none had so carefully observed the fact as he. He informs us that he placed a lens of glass in the aperture, and immediately every object appeared with a far superior brilliancy. The countenances of the passers-by, the colours of their dresses, their movements, all shone forth with great lustre, and formed a spectacle of such beauty that all who witnessed it were never weary of looking at it. This was the origin of the camera obscura. Porta continued to experiment upon his discovery, and that especially with the view of restoring the images, which were all reversed, to their proper position. In this he finally succeeded, by the assistance of two concave mirrors. He now took portraits of his friends, and also produced pictures of hunting scenes, and battles, and other spectacles, which he produced from paintings partly, and partly from living figures associated with paintings. The uses to which the camera has been devoted in connection with photography are well known. The instrument, as we now find it, is constructed on precisely the same principles as those of Porta, of a great variety of sizes, and of course is much improved in its mechanism. It consists essentially of a dark space, into which the light proceeding from external objects is permitted to enter through a lens, adapted to a tube in one of its sides. The rays thus entering are thrown upon a white surface of paper, or of any other material, and produce the images of external objects in all their natural colours.

The most general and ordinary use of the camera, in our time, is in connection with photography, and the form of the instrument is so well known that it need not here be described. Photographic cameras are generally required for one of three purposes—portraits, landscapes, or copying, and for each of them it used to be necessary to make suitable modifications in the construction of the instrument. But a camera has been recently contrived which contains within itself all the conditions required for these purposes. Literally, photography means "light-writing," or "writing by light," and it really is so. This art has combined the various discoveries in reference to the nature and properties of light made by investigators at different periods in such a manner as has secured for it a far more rapid progress than has been the experience of most of the sciences. Like other branches of chemistry, it owes its origin to the alchemists, who, in their fruitless researches after the Philosopher's Stone and *Elixir Vitæ*, produced a substance to which they gave the name of *Luna Cornea*, or horn silver, which was observed to blacken on exposure to light. This property of the substance constitutes the leading fact upon which the science of photography is based. There are great names connected with the progress of photography, such as Scheele, Wedgwood, Davy, Niepce, Daguerre, Talbot, and many others. Thomas Wedgwood was the first to produce pictures by the action of light on what is called a sensitive surface—this was in 1802—but immense progress has been made in the art since his time. The action of light has, by arrangements not widely different from those of photography, been successfully employed in engraving on stone, and on wood, and in various other ways; and altogether, this class of effects produced by light is among the most interesting and surprising in the circle of the sciences.

The *Diorama* and *Panorama* are pictorial representatives of the whole of a surrounding landscape as seen from one point. The invention is claimed by the Germans for Professor Breisig, of Danzig, but he never constructed one. The real inventor was Mr. Barker, an ingenious artist of Edinburgh, to whom the idea occurred when he was taking a sketch of the city from the top of Arthur's Seat. The paintings must be very carefully executed, and each sketch is the section of a circle. The position from which the picture is viewed is in the centre of a circular room, and the light is admitted from a concealed aperture in the roof. The effect of light on such pictures is such that the real scene appears to be before the spectator. So much is this the case, that several years ago, in London, when a panorama was being exhibited, which included a representation of the wreck of a ship's boat, with sailors struggling in the waves, a dog belonging to one of the spectators at once leaped over the handrail to rescue the men who were supposed to be drowning.

The *Magic Lantern* is an instrument which consists of a lens which orms on the wall an enlarged image of any object which is placed before it. The objects used in the magic lantern are pictures painted with transparent varnishes, and the magnifying power of the instrument may be diminished or increased by bringing the white screen which receives the image either nearer to the lantern, or removing it to a greater distance. The instrument, which was formerly employed very much for mere amusement, is now largely used in almost every branch of scientific instruction. In connection with the use of this lantern, spectres have been apparently raised by the management of the lights. These are called *phantasmagoria*, and this process very forcibly illustrates the power and quality of light. The same may be said of the *chromatrope* and of *dissolving views*, all of which class of experiments depend much upon the use of the magic lantern.

Ignorance and superstition were, ages ago, practised upon by designing men who knew the properties of light. Ghastly images, and other appalling sights, were called up to terrify the credulous. But the advance of science has happily brought such impostures to an end.

# AMONG THE BUTTERFLIES.

## I.

IT was just half-past three in the afternoon of Friday, July the 9th, that we reached the Loudwater Station on the Western Railway, after rather a hot but pleasant journey from London. We had been just enough to fill the carriage, eight of us, besides a little rough brown terrier called "Skye," several carpet-bags, two baskets full of prog, a bundle of fishing-rods, sticks, umbrellas, and butterfly-nets; and a very jolly time of it we had, I can tell you, all the way down. It was like a very long, cosy picnic, without any of the insects that *will* creep up one's back, or wasps that *will* taste the sandwiches and sip the beer. The two elder boys were just home from school, and had appetites that were not to be satisfied, and a thirst that nothing could quench, though they tried every possible variety of liquid they could lay hands on. We were all in the best of spirits, for, after six months' hard work, we were getting away from the smoke, dirt, and noise of the great city to a quiet country vicarage, in the very heart of the summer woods, with green fields, lanes, and heathy downs on all sides of it; and there we were going to rest, and drink the fresh pure air for seven weeks. But our great business was to hunt for and catch Butterflies; and I am now going to try and tell you how we spent our holiday, and found out how full of beauty and of good healthy pleasure God has made the green fields, lanes, and woods, and everywhere left traces of His goodness and wisdom among the least and lowest of His creatures.

Well, in spite of our long and jolly picnic, we were all glad enough to get out of the train, and take a breath of pure, bright, Hampshire air, after that last long tunnel, in which we seemed to taste nothing but brick-dust and smoky fog, though we shut up both windows as we rushed into it. Loudwater was a quiet little roadside station, with fields of yellow waving corn all round it, and white lilies growing up in the broad sunshine in little patches of earth from the platform, and already we seemed to have got a thousand miles away from smoky Babylon.

"Now, boys," said I, "come with me and count the packages and boxes as they come out of the van." There were just ten of them, besides all the baskets, carpet-bags, sticks, umbrellas, etc., etc.; but at last all were counted. Mamma and the two little ones were packed into the pony carriage; the two servants, Skye, and the legion of boxes into the carrier's cart, and away they went to Loudwater Vicarage, to get all unpacked and ready for us who were going to make the journey on foot, a matter of four miles, through fields, over the down, and along the edge of some woods. It was rather late in the day to think of doing much in the way of butterflies; but, not to miss a chance, we resolved to take our nets and see what could be done.

Away therefore, we went, Henry, Mary, Cecil, and I, each having a light stiff bamboo cane, between five and six feet long, at the end of which was fastened a bag of soft green * gauze, called a bag-net.

The sun was shining brightly as we set out, and the wind was from the west, so that by the time we got to the top of our first hill we found out that, if not quite warm enough for butterflies to be out, it certainly was quite warm enough for us who had to chase them. Our way at first lay through some bright sunny corn-fields, but the flowers were few, and we saw nothing worthy of our notice;—"Nothing," as the boys said, " but Meadow Browns, that we could catch at any time." Then we came to a piece of hot dusty road between a double row of firs, where there was plenty of blackberry blossom, wild scabious, and a few early sprays of yellow bed-straw by the side of the ragged hedges; but still not a butterfly to be seen. All this time we were gradually mounting up to higher and higher ground, and finding out that a west wind could blow very fiercely when it pleased; and then we all at once came out on the open down, covered with patches of furze, wild thyme, and purple heath, and stopped for a moment to turn our faces to the breeze, and drink in the glorious fresh air, that seemed to penetrate to every dusty corner of our lungs and fill us with new life.

Of butterflies, however, there was scarcely one, though we knew the ground well, and only two years before had spent many a long hour on this very down, away under those two tall Barrows, chasing Chalkhill Blues, Graylings, Clouded Sulphurs, Skippers, and small Heaths, till we were hardly able to stir hand or foot, and our boxes were nearly full. There was one strip of chalky ground, especially, along by the edge of a cornfield, where we had seen and caught clouds of the smaller butterflies, when busily feeding, a score of times; but even this was now all but deserted, and the freshening blast of wind had clearly swept them all away

---

* Our nets were all of *green* gauze, but I strongly advise my young readers to try *white;* through which they will far more easily examine a butterfly, see what he is, and whether he is worth keeping, before killing him.

among the stalks of barley, the long grass, and into the quietest corners in the hedges, till to-morrow morning.

Still, on we went, across the smooth short grass of the chalky down, watching narrowly every cluster of flowers and bunch of thistles, every patch of thyme, and bush of hazel and hawthorn, that fringed the edge of the common, and cut us off from the gritty road. But not a butterfly could be seen.

"Well," said I at last, "it is not much matter, after all, for in such a wind as this, on the open down, there would be little chance of catching it, if we saw anything worth getting."

Just at this moment, up from a clump of long grass, near a furze-bush, sprang a greyish-white-looking butterfly, and flew away as hard as he could go before the wind. In a moment Cecil was after him, as hard as *he* could go. They started fairly enough, and Cecil was a good runner; but in less than two minutes he was fairly distanced, and came back panting and out of breath.

"I don't mind the run a bit, but what was that butterfly?"

"It wasn't a Common White," said I; "there was too much black about the wings, and there was not enough yellow for a Clouded Sulphur. The only butterfly I can fancy at all like him is the Marbled White, whose wings are marbled with black and greenish-white, and when he flies fast, are apt to look of a dingy colour. But I never found the Marbled White on a chalky down, his favourite haunts being along the edge of woods, in grassy brakes, and meadows. However, he is now quite beyond our reach, and no other butterflies appear to be out; so, as we have still a couple of miles to walk, let me tell you something about the ten different families of butterflies, all having Latin as well as English names, which it is well that you should know.

"Family I. is that of the PAPILIONIDÆ, or *Swallow-tails*, of which only One species is found in England, chiefly in the fenny districts of Norfolk and Cambridgeshire; colour chiefly yellow, lower wings sharply tailed, hence the name *swallow*-tails; having six legs, and straight antennæ knobbed at the point.

"Family II. PIERIDÆ, or *Whites*, of which we have Seven species in England, all having six legs, straight knobbed antennæ, and rounded lower wings; colour chiefly white.

"Oddly enough, the Marbled White has only four walking legs, and does not belong to this class, but to the Satyridæ. (All butterflies, in fact, have six legs, but a few have only four which they can use for walking, the two front ones being short and imperfect, used, perhaps, as the common fly uses his fore legs—for cleaning his proboscis.)

"Family III. RHODOCERIDÆ, or *Red-horns*, a very lovely class of butterflies, chiefly of a yellow colour (Sulphurs and Clouded Sulphurs), having six true legs, short clubbed antennæ of a reddish colour, and a very downy fringe at the edge of the wing.

"Family IV. ARGYNNIDÆ, or *Fritillaries*, a beautiful class of spotted butterflies, containing Seven species, all but one (the Queen of Spain Fritillary) common in England, all having four legs only, rounded wings of a rich bright brown spotted with black, and on the under-side washed or dotted with silver. The antennæ are knobbed and of a fine dark brown colour.

"Family V. VANESSIDÆ, or *Angle-wings;* that is, having wings which run to sharp points at the upper and lower extremity, and containing some of the rarest and most beautiful of the larger English butterflies, such as the Camberwell Beauty, the White Admiral, the Comma, the Red Admiral. All this family (Eight in number) are decked with brilliant colours, and have fine knobbed antennæ, but only four true legs.

"Family VI. NYMPHALIDÆ, or *Nymphs*, the smallest of all the English families, and containing only One species found in England. But then that one is the Purple Emperor, and he, by virtue of his size, strength, and splendour, is entitled to a royal name and place. The Emperor has but four true legs; his wings are angled at the lower extremity, and his long antennæ taper to a thick point. His throne, as we shall see by-and-bye, is generally fixed at the top of some lofty oak, from which he takes broad sweeping flights through the hot summer air, often returning to the very spray from which he set out.

"Family VII. SATYRIDÆ, or *Satyrs*, a large family of Eleven species, many of them differing much in appearance, but all ornamented on the upper or lower wings, or both, with rings of some size or other. The wings are rounded, mostly of a brown colour, while the whole family have straight knobbed antennæ, and four legs.

"Family VIII. LYCENIDÆ, or *Argus* Butterflies; so called because most of them have, on the under-surface of the wings, many shining eye-like dots, and remind us of Argus and his hundred eyes. This large family of Sixteen species includes all the Blues and Hairstreaks, some of the daintiest and most charming of our small butterflies. All of them are fond of creeping up the leaves and stems of plants and trees, and are therefore provided with six legs, which the Hairstreaks use very nimbly in passing from one side of a leaf to another. The antennæ are fine and straight, the wings mostly rounded; but the lower wings of the Hairstreak are distinguished by a small pointed tail, which adds much to their beauty.

"Family IX. ERYCINIDÆ, or *Dryads*. The Duke of Burgundy Fritillary has this family all to himself, and is chiefly remarkable for the male butterfly's having only four legs, while the female has six, as if the lady in this case spent her time more in creeping about the house, and the gentlemen in gadding abroad. Oddly enough, neither the caterpillar nor chrysalis of this butterfly has been found in England, though both Duke and Duchess have been taken plentifully every year.

"Family X. HESPERIDÆ, or *Skippers*. This small family of Seven species takes its name from the odd skipping manner in which they all flit rapidly over the ground, or from flower to flower when feeding. They are altogether an odd family indeed, having short, clubbed, or hooked antennæ, standing widely apart, while they differ from all other butterflies in not closing their wings in the usual fashion, but keeping the lower ones open and flat, while the upper ones are erect. Some few moths have, I think, the power of doing this, but not, as far as I know, any other British butterflies. They have small, stiff, rounded wings, and the full number of legs. Their odd, swift, skipping flight makes them difficult to catch, except when settled on a flower.

"These ten families include altogether Sixty-four species; but out of these, four—the Mazarine Blue, the Bath White, the Camberwell Beauty, and the Queen of Spain Fritillary—are so rare, that, however hard you may work, boys, for many years to come, you will never, I fear, get beyond sixty."

"But shall we get sixty here in Hampshire before we go back to school?" inquires Cecil, his eyes sparkling at the thought.

"Well," said I, "if we get thirty out of the sixty, I think we shall do well, considering that we have but seven weeks, and the May and June butterflies must now be all over. But it's just five o'clock, and here we are at the Vicarage. Our work is done for to-day; to-morrow we shall begin in real earnest."

The sunset clouds were rosy red that night, and everybody prophesied a good day for butterflying.

The next day was bright and sunny enough, but the wind was from the north-west, and rolling clouds every now and then came up and shut out the sun, so that we did not expect much sport; but at half-past ten

Henry and I set off with our nets to Starling Park, a fine old place, full of grassy lawns, among tall forest trees, patches of rough ground, and plenty of oak and hazel copse. Meadow Browns, Ringlets in abundance, and a few battered specimens of the Wood Argus flitted about us whenever we went down the long leafy avenues, but nothing worth keeping, until we got to the middle of a little open brake, among the tall trees, where I saw coming towards us a small butterfly, whose flight was totally different from that of any we had seen. It skimmed and floated along from bush to bush and flower to flower, with an easy, dainty flight, so that when it hovered for a moment over a spray of blackberry blossom, I caught it at once.

It was one of the small Fritillaries (Family IV.), *Argynnis selene*, a May or June butterfly, who, after his little life of six weeks, had nearly lost all his bright colours, and sadly battered his wings (see Fig. 1, p. 124), but still was worth keeping because of his untimely appearance. The colour of his wings on the upper side is of a fine rich chestnut brown, much like that of the Highbrown Fritillary, dotted with black; underneath, the lower wings are marked with small irregular spots of yellow, black, silver, and reddish brown, the lower edge being fringed with a row of silver dots. *Selene* is a very elegant little butterfly, and very much resembles *Argynnis euphrosyne*, except that it has more silver on the lower wings, and is of a darker brown. The caterpillar of a blackish colour, covered with dingy spines, feeds on the dog-violet. There are two broods of the butterfly, one in May or June; and a second late in August.

Scarcely had I packed *Selene* into the box than Henry set off full cry after a butterfly I pointed out to him, going away at a good pace over an open sward of grass.

"I believe," said I, "it's an Orange-tip, that has no more business to be out now than a Pearl-bordered Fritillary."

Away dashed Henry as hard as he could go, tumbling as he went headlong into a swampy pit among the trees, but catching his butterfly and bringing him back in triumph. It was the Orange-tip (*Anthocharis cardamines,*) and a perfect specimen of the male (see Fig. 2). When fully stretched, his wings expand from one inch and three-quarters to two inches, the upper wings of the male being of a soft creamy white, with a broad orange-coloured band (edged with black), which covers half the wing. In the wings of the female the orange patch is wanting; but the under-side of the lower wings in both male and female is marbled with pale green. This is a lovely little butterfly, and well deserves its name, the Lady of the Woods. The caterpillar, of a pale green with one narrow

stripe of yellow on the side, may be found in May and June feeding on the cuckoo-flower or wild mustard.

When we got back to the Vicarage, Mary had brought home a small thick web full of caterpillars which she found among the branches of two hawthorn and blackthorn-bushes, which grew up closely together. All the caterpillars were rough and hairy, but some of a fine rich brown, and others of a dark, tawny, black hue, so that they looked like two different species; but after keeping them a day or two in a flower-pot, we found that they were all of one family, but that the blackish ones had cast one skin more than their brown brothers; almost all caterpillars, you must know, casting their skins three or four times at least before they enter on the chrysalis state. This is a very curious process, and I shall have a word more to say to you about it in another chapter. Our brown and black friends turned out to be the caterpillars of a small Egger moth, and so we packed them all off into the hedge from which they came. Cecil had been out with his net along the hedges of the upper wood, and now came in with great triumph to say that he had seen several Purple Hairstreaks hovering over the oaks and ash-trees, and scores of Meadow Browns in the fields, where also he had caught a large Fritillary.

When we came to look at it, it proved to be a fine male specimen of Family IV., the *Argynnidæ* (*Argynnis adippe*), or Highbrown Fritillary (see Fig. 3); his antennæ both perfect, and the silver spots on his under wings uninjured. (I mention these points because many young collectors, when once a butterfly is in their net, are only in an awful hurry to kill him. If they do this roughly and in haste, the result will be that the specimen is not worth keeping.) The wings, both of male and female, are on the upper side of a fine rich sienna brown, spotted and barred with black, while the lower wings on the under-side are splendidly marked with large silver spots, and especially by a row of rusty-red spots with a dot of silver in the centre of each. It is this row of rusty-red spots which mainly distinguishes *Adippe* from *Aglaia*, the Dark-green Fritillary, both being often taken at the same time, July and August, and in the same feeding-places, along the edge of woods, or in any woodland paths where blackberry blossom and other flowers abound. *Adippe* is a swift flier, and not easily caught. The caterpillar, spiny, and of a dingy grey colour, feeds on the dog-violet, hanging to the leaves or stems of which the chrysalis, spotted with gold, may often be found in June.

That evening we devoted to stretching our butterflies, and hoping that it would be a fine day on Monday. The next day, Sunday, we had a glorious walk through the woods to a little rustic church at the edge of the copse, and, as we had of course no nets with us, saw many good butterflies. Monday came in due time, and with it a bright clear sky without a cloud, while the wind blew rather fiercely from the east; but, east or west wind, as long as there was a hot sun, the morning must not be lost, and so away we went, Cecil and I to the upper wood, the two younger boys to the lower wood, and some meadows abounding in long, rough, thick grass, and many flowers.

The path which we took brought us at once into a broad grass road, with thickets of green hazel, ash, and young oak on either side, the ground being actually carpeted with wild strawberry,* sweet basil and ground ivy, interspersed with clusters of marjoram. The air was intensely hot, and fragrant with wild flowers; while the commoner butterflies swarmed on every side of us. Presently we came to an open brake, where some clusters of blackberry blossom were just opening in the sun, and at that moment there came sweeping by us a large bright yellow Fritillary, *Argynnis paphia*, or Silver-washed, who is so well known by her under wings of silver green, as to need little more than a glance at Fig. 4, to distinguish her from all the other Fritillaries. She hovered for a moment over the blackberry blossom, then darted away through the leafy glade, presently skimmed quickly back, again hovered over the blossom, and settled. One swift sweep of the net, and she was ours.

The male differs from the female in being marked on the upper surface with broad black borders to the veins of the upper wings, while the ground colour of the lady's wings often inclines to a tint of greenish olive-brown. On the under side both sexes are closely alike. The caterpillar, thorny and blackish, with yellow lines on the sides, is found in June on the dog-violet. A spray of blackberry blossom in a woody glade is a sure feeding-place for this elegant butterfly.

We saw some Purple Hairstreaks high up on the ash-trees and hazels, but they all kept carefully out of reach.

To-day Henry and Willy came home in great triumph, with a box full of butterflies. They had been down in the open piece of rough grassy ground by the edge of the lower wood, and there among the long grass, marjoram, cinquefoil, and wild scabious, had actually caught a dozen Marbled Whites. So abundant were they in this one piece of ground, that the boys might have taken fifty if they pleased.

* So abundant were the wild strawberries, that a quart basin was easily filled with plain fruit by a couple of children in less than half an hour.

"You were quite right," said I, "not to take any more: we have now plenty for our own collection, and a couple or two to give away or exchange, when we meet with a friend in need."

The Marbled White (*Arge galathea*) is not a very rare butterfly, but a very local one, being taken only in certain districts, and only in certain small feeding-grounds in those districts. We, for example, had hunted for butterflies all round this very parish for six weeks in 1867, and not chanced to find a single specimen. In fact, this was the first time we had ever seen one alive, so that our dinner that day was well seasoned with Galatheas, how they flew, and how they puzzled a stranger with their strange dingy appearance. And then we thought of the odd butterfly Cecil had chased on the down near the Barrows.

"I knew it was a Marbled White," exclaimed Cecil, "he had such a dusky look about him."

And he was right, for within a week, in a strip of rough grassy copse near the Barrows, we saw a score of Galatheas in a single morning.

*Arge galathea* belongs to Family VII.—the *Satyrs*. The ground colour of the wings, which are slightly scalloped at the edge (see Fig. 5), is a yellowish-white, or pale creamy yellow marbled with black. The female is supposed to have wings of the yellower tint, and the male of paler white, the under-side (see Fig. 1) being even more elegantly chequered than the upper. It flies very near the ground, settling frequently, and is easily caught, except in a high wind. The caterpillar, of a greenish colour, with a yellow stripe down the sides, feeds on several kinds of grass, attached to which the chrysalis may often be found towards the end of June. Galathea is the only White which has four legs.

July 13th was a cloudy day, with wind from the north-west, after some thunder and heavy rain in the night, so that we did not reckon on many butterflies. But we went down for an hour to the Galathea meadows by the edge of the wood; and there saw a score or two flitting heavily about over the long grass. We allowed ourselves one good specimen each; and after watching the bright grassy field and busy Galatheas for a time, wandered pleasantly homewards; putting up, on our way, two fine coveys of partridges in a turnip-field that had suddenly grown green after last night's rain. As we passed through one of the narrow winding woodpaths I spied a Purple Hairstreak at the end of a blade of grass. He had just come out of the chrysalis, and could scarcely fly; and being very perfect, we caught him. The gleaming purple on his upper wings looked very lovely in the sunshine.

The next day, July 14th, after two hours' hard work at a pile of books waiting to be reviewed, we all set out again for the open down, to try our old hunting-ground for Skippers and Chalkhill Blues. But, though the wind was calm and the sun hot, not a single specimen did we get. We searched carefully all along the edge of a field of ripe oats, fringed with yellow bedstraw, clover, wild scabious, marjoram, and half a dozen other dainties; but not a Blue was to be seen where only two years ago we had caught dozens. On the edge of the down, however, perched on a tall thistle, I took one fine specimen of Galathea, clearly just out of the chrysalis; his wings being still slightly crumpled, like blotting-paper, in one or two places, but his plumage very bright and perfect.

On our way home through the woods, however, we took five very fine Silver-washed Fritillaries, and one of the smaller species of the same family, *Argynnis euphrosyne*, or Pearl-bordered Fritillary, which we still have in our cabinet; but, being a May butterfly, he was in so battered a condition that our artist declined to make a drawing of him. We must hope, therefore, for a better specimen at some future day.

Butterflies feed on the honey of flowers, and are said to love all sweet and dainty things; but on our way home through one of the narrow winding wood paths, we had a strong proof that they have a decided taste for much stronger meat. Up against an old oak the keeper had nailed a dead magpie—clearly some months ago, to judge by his rank odour; and upon this very high game was perched a large Fritillary,* evidently feeding to his heart's content. The Purple Emperor himself is said to have a mighty relish for strong dainties of this kind; and to have been caught more than once when drawn down from his airy flight to feed on some morsel of carrion.

July 19th.—Cecil came home this morning from Chalkhill Wood in a wonderful state of excitement; having seen, so he protested, a couple of Purple Emperors sailing round the tops of some lofty oak-trees. I am afraid that we put down this vision mainly to the power of a very strong imagination.

But he brought home with him a couple of those elegant little butterflies, the Azure or Holly Blue, belonging to Family VIII., the *Lycenidæ*, and about these there could be no doubt. *Lycæna argiolus*, or Azure Blue, is one of the purest and simplest of the Argus butterflies, and may easily be known from all the other Blues by the soft silvery grey of the whole of the under-side of the wings. This silver grey, which reminds one

---

* The very one copied in our plate after being caught by Henry.

of a pretty Quakeress's silk gown, is faintly sprinkled with very small black dots. The upper surface of the wings is of a fine purplish blue, the upper wings of the male (see Fig. 6) being edged with a narrow darkish border, and those of the female (Fig. 7) with a broader band of black; while on the lower wings of the female, close to the edge, is a row of black spots which greatly add to her beauty. When flying, *Argiolus* often looks like a pale butterfly with whitish-grey wings, and may thus at once be known from the other members of her family. The caterpillar, of a yellowish-green colour, is said to feed on the blossoms of the holly; and the chrysalis may be found in May and July. We took it in great abundance, both in the woods, fields, and lanes.

In still greater abundance, also, we took another small, very active, and charming little butterfly, of the same family, *Lycæna agestis*, or the Brown Argus (see Fig. 8). How he comes to be counted among the family of Blues it is hard to say, for he has not a speck of blue about him. The upper surface of all the wings is of a warm, rich, coppery brown; and at the edge of each wing is a row of dark orange spots, rather broader in the female than in the male. The under-side of the wings is exactly like that of the Common Blue, *Lycæna Alexis*, except that it wants two small spots, between the discoidal spot and the root of the wing, which Alexis always has. The small caterpillar, pale green with a brown line down the back, feeds on the *helianthemum*, or rock cistus, whose gay yellow flowers we found growing most abundantly along the edge of all the wood paths. This is one of the smallest and nimblest of British butterflies.

July 19th was one of the hottest days of the year; the wind being fierce from the east, with a blue cloudless sky, and a brassy sun.

"Just the very day," so said Cecil in his eager fashion, "just the very day for Purple Emperors!" as if he had been used to them all his life. As I had never seen one alive, I determined to give myself the best possible chance, and at 10.30 set off, in the blazing sunshine, for Chalkhill Wood, which lay on the sloping sides and summit of a low rising ground, covered with a thick coppice of oak, ash, and hazel, and crowned every here and there with lofty oaks. The lower slope of the hill was occupied by a field o scrabbly potatoes, which, I was told, from neglect, never bore anything worth eating. Next to the potatoes came a wide straggling patch of ground densely covered with one brilliant mass of scarlet poppies and snow-white camomile. Not one inch of ground was bare, and a more dazzling flower-show I never set eyes on. To us butterfliers such a field was impracticable, for the instant a butterfly crossed the area of dazzling crimson and white, we lost sight of him for ever. We therefore made our way into the wood at once, by a winding path among the copse; paying heed neither to small nor great, but devoted entirely to his Royal Highness alone. On every side of us were bushes of broad sallow, on which the caterpillar of the Emperor feeds, and on every side oak-trees of all shapes, sizes, and ages.

"It was over one of those splendid evergreen oaks," said Cecil, "that I saw the Emperor, sailing round, like a great swift."

"All right," said I; "all right, my boy. All I wish is that he would sail across one of them now."

The words were scarcely out of my mouth, when on a spray of hazel about six feet from the ground, perched on an outer leaf, with his head downwards, basking in the sunshine, right in front of me, I beheld a Purple Emperor in all his glory.

For a moment I was speechless,—fairly speechless with excitement, as I turned to Cecil, and pointed with my finger to the glowing prize still within my reach. My hand trembled as I raised my net, but nearer and nearer I drew it steadily on, taking care not to shut out the rays of the sun, and then suddenly, with one swift triumphant stroke, the Emperor was mine! My fingers trembled as I killed him, but in less than a minute he was safely in my box, and Cecil fairly beside himself for very joy. It was a perfect specimen, having plainly just emerged from the chrysalis, and crawled up to sun himself on the hazel-spray, before winging his lofty flight above the green oaks.

*Apatura* \* *iris*, belonging to Family VI., NYMPHALIDÆ, well deserves his name of Emperor; his dusky wings of black, shaded with most glorious purple, being of the largest, swiftest, strongest make, and well fitted for royal flights over his forest domain. On the upper wings are ten irregular white spots, and on the lower wings a bar of white, stretching down to the lower segment, at the end of which is a small ring of bright orange. The under surface of the wings both of Emperor and Empress is alike, and though very beautiful in its mixture of grey, orange, and black, it wants the royal purple of the upper. The wings of the female are of a dull brown, without a tinge of purple. The Empress is rarely taken, being content apparently to remain quietly at home, while her royal lord

---

\* This surname *Apatura*—a softer form of *Apodura*, which means *without feet*—is given to butterflies of this family because their caterpillars have *not* the two *usual* clasping feet at the end of the tail.

and master wings his way through the blue air, in search of fresh domains, or some rival to his throne.

On that very morning we watched two of these royal butterflies skimming on swift wings round the top of a lofty oak, and now and then darting aloft into the blue air, until the eye could scarcely detect them. But not one more ever ventured down to the perilous regions of earth near enough to be within reach of our net.

I had to go and see a sick man in one of the little cottages beyond the church, and so on we went through the copse, every now and then opening the collecting-box, and having a glance at our grand prize, to make sure that he was quite safe. We could, as yet, hardly believe that we had really got the Emperor safe and sound in our box; and though they received us with shouts of triumph at the Vicarage, it was some time before they would believe our incredible story—not, in fact, until we had opened the box and proved the treasure to be indeed ours.

At 7 P.M. that evening, when it began to grow cool, Mamma, Henry, and I walked down through Starling Park into a broad open glade of soft thymy grass, lined on each side with lofty forest trees, the beeches and oaks being the finest we had yet seen. The sunset was bright and clear, and every tiny branch and twig of the tall trees could be clearly made out against the soft blue sky; the wind had gone down, and everything was at perfect peace. As we strolled on, the last rays of the sun just caught the topmost boughs of a lofty beech-tree, and crowned them with rosy light, when to our great surprise we saw clouds of small butterflies hovering over the top of that tree, and an ash which grew next to it. After watching their gambols for some minutes, we clearly made them out to be Hairstreaks, many scores in number, and among them one large swift black butterfly, which from his peculiar flight was at once detected to be an Emperor, who had been probably disturbed on his airy throne by this cloud of saucy interlopers. They were fifty or sixty feet beyond the reach of our longest net.

The next morning at 10.30 Henry and I went down into the same glade, but not a single butterfly of any kind could be seen near either tree although the wind was from the west, and the sky without a cloud. Except one old battered Sulphur, surviving from the brood of last March, we did not in fact see one butterfly in the neighbourhood but Meadow Browns and Ringlets.

## II.

SATURDAY, July 24, was a blazing hot day, though the sky was well filled with great piles of rolling white clouds, and the wind very stormy. After breakfast Cecil and I took our nets and went away down by the edge of the woods to Starling Park, hoping that we might get a glimpse of an Emperor, or some of the Hairstreaks of which we had seen such clouds only a few evenings before. We came into the same beautiful glade, nearly a mile long, and about three hundred yards in width, with tall forest trees on the right bank, and on the left a low scattered copse of oak, ash, and hazel, with a superb row of lofty beeches for a background. But neither among the long grass, nor in the copse, nor anywhere near the trees, could we see more than a few butterflies, and all these were of the commoner sorts. On one of the grassy slopes there were several clusters of dark green shining holly, but not a single Blue or a caterpillar could we find anywhere near them.

And then, all at once, as we might have done long before, we found out that there were very few flowers in the park, either in the wood paths or on the grassy lawns, and therefore there were few butterflies.

"We have had our walk for nothing," said I, "so far as butterflies are concerned;" when, just as I spoke, up started a large female Sulphur from among the tall grass, and flew swiftly away before the wind. I set off as fast as I could go in pursuit, and Cecil with me, in case the butterfly should turn; but after two minutes' hard running, and several random strokes of the net, I was fairly winded, and obliged to give in. Then Cecil took up the chase, and for another minute went along at a great pace, zigzagging to and fro across the glade, until the poor Sulphur suddenly began to flag, and all at once actually fell down on the grass, unable to fly another yard. It is not an easy thing to run down a butterfly, nor one that often happens. As the specimen was a very perfect one, we kept it, and made our way rather wearily home, along a straight piece of turnpike road, which, unlike most turnpike roads, was fringed on one side by some noble walnut-trees and firs, and on the other by some capital brakes of bramble in full blossom, among which we took a couple of dainty little Holly Blues. As I strolled through the Vicarage garden, up started a greenish-white-looking butterfly, which turned out to be a female Orange-tip, one of the May brood, hardly able to fly, so thin and battered had the edges of her wings become.

What does a May butterfly mean by showing her face at the end of July?" asks Cecil.

"It reminds me," said I, "of that spray of hawthorn which Mary brought home the other day, with a cluster of flowers in full bloom, a bunch of buds and of berries on the same stem. A butterfly's life lasts, I suppose, about six or eight weeks in most cases; but here is a poor battered old lady who has weathered the storms of life till she has reached at least twice the usual age of the oldest native. It is, in fact, as if Miss Smith, Mary's governess, should live to be a hundred and fifty. But still, after all, in these burning hot July days it is pleasant to have only a taste of spring, once more, in a bunch of fragrant hawthorn, or in the lovely wing of an Orange-tip."

"Did you ever find a dead butterfly?" inquired Cecil; "for there must be thousands that die every year before they're half worn out, or have got to be old cripples like the Orange-tip."

"Well," said I, "I never found but one dead butterfly, and that was a Wood Ringlet, two years ago, in the wood beyond Loudwater Church. Henry and I were coming home one evening, after a long day at trout-fishing, and there before us in the path lay the poor Ringlet, quite dead and stiff, but quite perfect, and both his antennæ unbroken. What was the cause of death I can't say; and there was no coroner to hold an inquest."

"Heart disease, perhaps, papa?"

"No, not heart disease," I replied, "because butterflies are far wiser than men and women, and don't poison themselves with too much eating or drinking, or food which they can't digest: they breathe fresh air, and can't have their blood poisoned by drains. Nine hundred and ninety out of a thousand, depend on it, live out their full time. Here and there one gets snapped up by a hungry bird, a flycatcher, a water-wagtail, or a swallow; but most of them live to a good old age, and die in peace."

"But still," says insatiable Cecil, "they must die somewhere, and how is it we don't find them dead?"

"No doubt," I replied, "you *would* find some of them, if you looked in the right places, down among the thick leaves near the roots of the grass in the hedgerow, in the quietest secret crannies, wherever in wood and field and roadside there is most perfect shelter from cold and wind. There, the poor worn-out butterfly, after his little life of sunny pleasure and holiday-making, creeps in, and dies a lonely death. After which, perhaps, the ants eat him up, or a stray wasp or two, or some hungry insects in search of a morsel of food, set on him, and suck out of his body what little moisture there may be left in him. Then comes a touch of autumnal frost, cracks all his brittle legs and wings up into broken fragments, and perhaps half buries him with a sprinkle of earth from the next day's thaw, so no wonder that we find so few dead butterflies."

At 7 A.M. that evening, while strolling through the lower wood, we saw dozens of Hairstreaks hovering over the top of almost every oak and ash-tree in the copse; but none within reach of our nets. The next day, as we were all coming home from church through the Emperor wood, by a narrow winding path, fringed with flowers and thick bushes on either side, Mary suddenly spied a butterfly perched on a cluster of white flowers, to which he seemed to stick fast in a very odd fashion. The flower was

Wild Carrot (*Daucus carota*).

the wild carrot, which grew in great abundance along the path, in straggling clusters of white blossom, very much like hemlock. The butterfly, oddly enough, was a Wood Ringlet, who had settled down upon the cluster of tiny flowers, and was there held fast. We touched him with our fingers, and he never moved; we shook the flower, but still he remained fixed in his old quarters. At last Cecil's sharp eyes found out that a white spider (see Fig. 12, page 136) had lain in wait among the tiny flowers and pounced upon him as he fed. We looked, and then clearly saw that the spider had still got hold of the poor Ringlet, and held him fast by the throat.

He had sprung upon the butterfly, as a tiger springs upon his prey from behind an ambush of boughs, killed him with one fatal bite, and was now sucking his blood. It required several smart taps with a stick to make him relax his hold; but we forced him to give way at last, and carried him back to the Vicarage.

There were three other spiders of the same kind on that one cluster of flowers, and we afterwards found numbers of them in all the woodland paths, but rarely by the open roadside. The spiders were always white, and were rarely found on any but white flowers,\* while to judge by their plump bodies they lived on the fat of the land. They all had the same sort of round puffy body, and long thin legs, which they wrapped round their prey and held it fast. So vicious were they, that on another occasion, when I dropped a couple of them into a tiny bottle of spirits of wine, one instantly seized the other by the leg and bit it off. Twice I saw a Meadow Brown caught in this crafty fashion; once a bluebottle fly; and once a poor hapless wall butterfly (*Satyrus Megæra*), who had quietly settled to sip the honey from a tall thistle, been seized on, and been pulled into the very heart of the silky purple flower. His body, hind legs, and lower wings were sticking up in the air, in the most ungainly fashion, just as a duck turns up his tail in the air when diving.

Monday, the 27th, was a blazing hot day, with a fierce east wind and a bright brassy sky. But butterflies were out in abundance, and besides a number of fine Holly Blues and some Skippers from Mary's net, we had our first Clouded Sulphur (*Colias Edusa*) from Cecil. Close to the Vicarage gate was a large field of lucerne, just coming into flower, and after feeding there the butterfly took a flight across the garden, was at once seen, and chased. Away went the Clouded Sulphur at a great rate before the wind, down the turnpike road, in the blazing sunshine, and away went Cecil after it, as hard as he could go, for a full quarter of a mile. Then he got ahead of *Edusa*, and turning quickly round, with one quick sweep of the net against the wind had the game safe at last. It was one of the largest and most perfect specimens we had ever seen, a female, and, judging by her dainty plumage, just out of the chrysalis. The Clouded Sulphur is one of the swiftest fliers among butterflies, and Cecil came back from his chase in the noonday sun looking red hot and melting, but in great joy at his good luck.

\* They clearly chose clusters of *white* flowers for a hiding-place, so as better to escape the keen eye of their enemies, the wren and the titmouse, and at the same time to lurk unnoticed by their prey the butterfly; guided, no doubt, by that wise instinct which God has given to all His creatures in their search for food.

The Clouded Sulphur, *Colias Edusa*, belongs to Family III., RHODO-CERIDÆ, or *Red-horns*, and is one of the most favourite butterflies, both for beauty and rarity. The ground colour of all the wings is of a fine rich saffron hue, with a narrow band of rich dark brown, sometimes almost black, at the outer edge. This dark margin in the male is usually crossed by faint yellow lines (Fig. 3), and in the female dotted with a series of pale yellow spots (Fig. 2), which add greatly to her beauty. The female is considerably larger than the male, and apparently rarer, for though we afterwards caught six or eight more males, we took only two females. The caterpillar, of a pale green, whitish-yellow stripe on each side, may be found in June, or even in July, feeding on lucerne or common white clover.

Just after the capture of the *female*, in came Henry from the Barrows —where he had seen Chalkhill Blues and Harvest Blues in abundance— with an equally good specimen of the *male Edusa*, which he had caught in an old chalk-pit at the edge of a field of barley. This, with Cecil's, made a capital pair of specimens for our cabinet, being brighter and more perfect than those which we caught two years ago, when they were found here in great abundance.

The Clouded Sulphur is one of those few butterflies which every now and then are very plentiful, then for a season or two are scarcely seen, and again appear in great numbers. I had many long talks with Henry about this fact, and after searching far and wide for its cause, we came to the following conclusion. The fields of lucerne, on which the caterpillar feeds, in most places are allowed to lie fallow for a year or two, that the crop may grow richer and stronger for grazing on. When it so happens that many of the lucerne crops of a district or a county, or even a wider range of country, chance to be mown or grazed on about the same time in June or July, a great destruction of caterpillars and chrysalides takes place, and few Clouded Sulphurs appear. Then follows a year or two of rest, when the brood quietly increases, then a year of plenty, and next a time of scarcity. The question is a curious and interesting one, and, at all events, our solution of it, for want of a better, is worth considering.

For the next two days we had heavy thick rain and cold winds, which kept the boys busy indoors at their books and me with pens and ink, while the butterflies had to take refuge in the thickest hedges, at the roots of the long grass, and down among the ranks of waving corn. But the next day, when the sun shone brightly again, the flowers, yellow bedstraw, scabious, hardheads, yellow rock rose, and blackberry blossom, refreshed by the glorious rain, had come out in greater number and beauty than ever, and the butterflies were everywhere in clouds.

Cecil and Mary set off to the open down, near the Barrows, and came back with a box full of treasures, among which were a pair of Pearl Skippers and some very fine Graylings.

The Pearl Skipper (*Hesperia comma*, Figs. 4, 5), belonging to Family X., HESPERIDÆ, is a very elegant little butterfly, and perhaps the most beautiful of all the Skippers, if only for the silvery spots on the under surface of the wings. The upper wings of the male (Fig. 4) are of a rich tawny brown, with a dark outer margin, varied with spots of pale yellow, while the lower wings are of a still darker hue, and also spotted with pale yellow. In the wings of the female the colours are all rather dingier, and have not the dark stripe of black to be seen on the upper wings of the male. The under-side of the wings both in male and female is of a yellowish green hue, with clusters of whitish silvery spots, generally eight on the lower wing, and six on the upper (Fig. 5). The caterpillar, of a greyish-green colour, here and there touched with red, and with a narrow white band round the neck, may be found on one of the small trefoils in chalky ground in July.

Mary had caught both the Pearl Skippers together, hovering over a flower, and thinking that they were the common large Skippers, let them fly again.

"I don't believe they are common Skippers at all," said Cecil; whereupon Mary set off in pursuit again, and once more got them in her net. This time she luckily kept them.

If you look at Fig. 6, you will at once see that the common large Skipper (*Hesperia sylvanus*) is to be easily distinguished from *Hesperia comma*, and still more easily from the small Skipper (*Hesperia linea*, Fig. 7), by size, colour, and shape of wing. The general colour of the wings in both these species is a dull tawny orange; the male having a bar of black in the middle of the upper wing, from which some dark veins run towards the edge; while both wings grow darker towards the margin. Underneath, both the large and small Skippers are very much alike, the colour of the wings being a dull yellow-brown, tinged with green. The caterpillars of both these Skippers are of a dull green, and feed on various kinds of grass; the larger one being dotted with black, and the small one marked with whitish stripes. Both butterflies are common, and easily caught in woodland paths or meadows along the edge of woods, from May to August.

Among the butterflies of the soberer, quieter colours, are the Meadow Browns and Graylings; of both of which they had brought home good specimens from the down, and of which we saw great numbers all through our seven weeks' holiday, on the heath, and in the meadows and woods.

The Grayling (*Satyrus semele*), belonging to Family VII., SATYRIDÆ, may be found in good numbers on open heaths, stony fields, and chalky downs; though not nearly so common as the Meadow Brown. All the wings are of a dark tawny grey colour (Fig. 8); the upper ones having near the edge two black spots with white dots in the centre, like eyes; while the lower ones, which are beautifully scalloped at the edge, and marked with an irregular bar of dull orange yellow, have only one eye, much less bright than those on the upper wings. This orange bar is brighter and clearer in the female, which is also larger than the male. Underneath, the lower wings are shaded with a lovely mixture of brown and grey (Fig. 9), so exactly like the colour of broken, flinty, stony ground, where the Grayling loves to settle, that it often puzzled us to find *Semele*, even when we had just seen him drop down to the earth for a rest.

I have known this butterfly rest for more than five minutes on a bed of broken grey and white flints, where it was next to impossible to distinguish its brown and grey wings from the stones about it. The caterpillar, of a dull grey colour, striped with green, feeds on heathy grass, and is said to change into a chrysalis in the earth.

The Meadow Brown (*Satyrus janira*, Figs. 10, 11) is to be found nearly everywhere, flitting along with heavy zigzag flight, in her sober dull robes of russet brown. The female is larger and handsomer than the male, all her wings being of a dull reddish-brown, the upper ones having at the higher corner a patch of lighter yellow, in which is a black spot with a centre of *two* white dots; while the lower wings are of a pale dull brown, slightly scalloped at the edge. The male is smaller than the female, and still dingier in colour, though sometimes found with a flush of a redder hue across the upper wings, which are marked by one *smaller* black dot, with a speck of white in the centre. Underneath, the wings of the male and female are much alike. The caterpillar, green with a white stripe on the sides, feeds on many kinds of grasses, low down, near the roots, where it often remains in a half-torpid state all through the winter; waking up again, and feeding, when spring comes, and turning into a chrysalis in May.

But both these specimens (Figs. 10, 11) puzzled us very much by having stuck fast to them, close to the thorax, small round bits that looked like red coral. If you look at the plate, you will see that the male has one and the female two of these tiny red specks, each as big as a large pin's head. We searched all the books within our reach likely to tell us anything about this strange appearance, but could meet with no mention of it whatever. We afterwards found bits of the same red substance sticking fast to the bodies of Graylings, Harvest Blues, Meadow Browns, and small Heaths, but not, I think, of any other butterflies, with the exception of a single large White.

In the afternoon of the 28th the wind sprang up rather roughly from the south-west, bringing with it piles of rolling clouds which often shut out the sun, and at 3 P.M. Cecil and I set out for the Barrows, going along by the edge of the wood, through grassy fields abounding in flowers, through patches of open copse, and over the chalky down. But scarcely a butterfly was to be seen. In the strip of sunny copse by the roadside, near the Barrows, where last week we saw scores of Marbled Whites, not one was to be met with; the bank covered with scabious, wild thyme, and patches of tiny clover, where the Blues by the score were busily feeding this morning, were quite deserted; not a Grayling on the down, and scarcely a Meadow Brown by the thick grassy hedge. All had taken refuge in the bushes, among the stalks of barley or the thick grass, till to-morrow's sun should bring back light and warmth and joy to the dusky woods, fields, and springing corn, as well as to the birds, bees, and butterflies, and all the other creatures of God's hand, which He has made to praise Him by their beauty, their songs of joy, or their lives of patient harmless toil.

As we made our way homewards along by the edge of the wood, we came upon a little pool of water in the chalky ground from the previous day's rain; and near the water, where the ground was still muddy, was settled a large White butterfly.

"Look at him," said Cecil; "is he drinking?"

"Beyond a doubt," said I; "why shouldn't he drink, if he is thirsty?"

"I should have thought that he would have found enough moisture yesterday on the flowers and grass, without coming to a pool of water."

"That was yesterday; and he has perhaps made a long and weary journey to-day, through the hot sunshine, and now against this fierce wind,—a journey of many miles; so that he needs a good long draught. Two years ago, when Henry and I caught that famous dish of trout at Clatford, one burning day in August, in going down to the river we came to a very shallow patch of water which had leaked out from a spring in the meadow and crept on to the dusty road. When we got close to it, up started a whole cloud of White butterflies, who had been settled tippling at the edges of the crystal lake. We counted more than twenty, many of which, after hovering about, gossiping and flirting for a few minutes, quietly settled down again to their potations. This was towards the end of the butterfly season, when these Whites have a great trick of collecting together in great numbers; though one hardly knows why or for what purpose. But, Cecil, Purple Emperors are the fellows to tipple! Mr. Hewitson, a famous butterflier, tells us that at Kissengen in Bavaria, where he had many chances of watching the Emperors, after long and rapid flights in the neighbouring forest they would take refuge in its shady recesses to cool themselves, and sip the moisture from any puddle of water—the dirtier the better—with their long trunks. And so eager were they in tippling, that he once took seven under a flat net at one stroke!—he even caught one with his finger and thumb."

"Oh, papa!—you are not telling me something out of Baron Munchausen, are you? Fancy seven Emperors at one stroke! I only wish there were some pools of water in our Emperor wood."

"I only wish there were, my boy," said I, "for my story is a true one, every word. What happens among the Emperors in Kissengen would surely happen here in Hampshire, if our woods were not scattered over the stony hill-side, where a drop of water is hard to find."

"But you talked of butterflies being weary after a long journey; how ar do they fly in a day?"

"That is a hard question to answer. But, although butterflies are such idle truants, there is no doubt that they have favourite feeding-grounds, to which—even if driven away—they return again and again after a long flight, and that a gay young spark of a butterfly will often chase his ladylove a very long way, and that she delights in leading him a long chase; after which he will, perhaps, set off once more in pursuit of another charmer. Then, as his life cannot be all spent in love-making, and he must have something to eat and drink, he wanders from meadow to meadow, and from flower to flower, over and through woods, and along dusty roads, for many a long hour of sunshine, until, at last, quite worn-out, he slips down into some cozy quiet nook among the grassy flowers, or, if it is calm, among the sheltered leaves of some favourite tree, and there goes fast——"

"Asleep?" interrupts Cecil; "does he go to sleep? If he sleeps, he dreams."

"Well, my son, about dreams I can't say anything, though I *have* heard of a book called 'The Butterfly's Dream;' but with regard to sleep, all I know is that every butterfly has a good pair of eyes, though he may not be able to shut them. He may sleep with his eyes open, and if he does not sleep, he sinks into a state of deep rest; for I have with my finger and thumb, more than once, taken a butterfly with his wings folded closely up, when he was quietly resting on a flower or leaf, just as evening was closing in."

"And is everything you read in books true?"

"I am afraid not, Cecil; people who deal in fibs will write books, you see, as well as talk. But one thing you may depend on,—I think, that the people who really love the green fields and woods, the shining rivers, the lonely hills and valleys, with all the wondrous and beautiful creatures which they contain, and write as if they did love them, are the least likely to deal in fibs, or 'crackers,' as you schoolboys call them. And I will give you a good reason for this. The more a man truly studies these things, and the more knowledge he really gains of them, the nearer and nearer he draws to the wise and good God who made them all at the first, and in whom they live and have their being. The nearer a man draws to God, and learns His wisdom, the more and more he forgets himself and his own cleverness, his petty troubles, and foolish sins, the envy and ill-will of his neighbour, as well as his own; and the less he thinks of such things, the more likely he is to think of and delight in the truth, and care to write that alone."

By this time we had nearly reached home, when I heard Cecil, who was loitering behind, suddenly call out,—

"I say, what heaps of ladybirds! Where do they all come from?"

"That's more than I or anybody else can tell, but they never come until they are wanted, and when wanted they *do* come, in hundreds and thousands at a time, like a regular army of little red-coated soldiers."

"But they don't fight," inquires Cecil, "and of what good are they?"

"No, they do far better than fight with their enemies, they set to work and eat them up. The locusts that you read of in the Bible were sent as a curse to destroy the green crops and herbage; but these millions of ladybirds that just now are swarming all over England are a true blessing to the farmer and the gardener, because they eat up their greatest enemies, the aphides or 'plant-lice,' that infest the young trees here in the wood, the ripening corn, the hop-plants in Kent, the fruit-trees, and the flower, in such countless myriads as to be like a plague. Last week the rose-trees in the Vicarage garden were thickly covered with green aphides;

one morning, suddenly comes a shower of ladybirds, all hungry, and all having a relish for green insects; the next day all the rose-trees are clean. And the good work they have been doing in the garden has been going on all over England, among the pea crops, the hop-gardens, and the orchards. And the strange thing is that no sooner is their work done, than the little red-coated army all disappears. 'For weeks,' says one gentleman,* 'my best apple-trees were covered with American blight. A few days ago myriads of ladybirds swept down into my garden, and instantly set to work, and now after one week my trees are all clear, and as healthy as ever. The ladybirds are all gone.' The good which they do, therefore, is beyond all price; but I cannot tell you where they come from in such swarms, or whither they go when their work is done."

"They are very little things," replies Cecil, "to do such a grand lot of work."

"Yes," I replied, "they *are* little things; and if you watch the mightiest works that are done in this world, you will be surprised to find how many of them depend on what are called trifles, or little ways and little creatures. All the honey in the world comes from little five-sided cells, built up out of wax by a little insect, who, with a tiny trunk, sucks the sweet nectar out of flowers. The famous Plymouth Breakwater, a mile long and two hundred feet wide, stretching across the mouth of a harbour in the deep sea, was built by overturning thousands of boatloads of small stone into the wild waves, until a foundation was found for the mighty blocks of granite which now pave its upper surface. The coral islands in the South Sea were all built by myriads of insects; and the worst enemies which our old line-of-battle ships used to have were the Teredo or auger worms, which bored a tiny round hole through the hardest, toughest beam of oak.

"The ladybird always crawls up to the topmost leaves and branches until she finds her prey, the green aphis, the hop-fly, or the black blight, as the country people call them; and all we can say is, may they have a good appetite and a good digestion, especially as the aphides multiply at the most terrible rate, and the descendants of a single couple, in three weeks, would amount to more than a million. But though such a little creature, the ladybird is full of life, and very hard to kill, as you may fancy from a little story of Mr. Spence's. 'I caught a ladybird,' he says, 'in my study window one day, with twenty-two spots on her wings (most of the common ones have two, five, or seven), and thinking she was very

* Mr. H. Brook.

pretty, I determined to kill and keep her. Not knowing how to kill her I put her into a wine-glass of gin. There she stayed for a day and a night, when, fancying she must be dead, I laid her out in the sun to dry. No sooner, however, did she feel the warmth than she began to move, and presently flew away.'"

"Well," said Cecil, "after being drowned in gin for twenty-four hours, she deserved to get away, and tell her wonderful adventures, if she could."

"*If she could?*" said I; "but what is to hinder her? Ants talk to each other, and bees spread an alarm through the hive in less than half a minute. Is nobody to be able to say a word, unless we understand the language they talk in? I heard an old hen talking to her chickens, this very morning, down on the gravel yard near the stables, and she said as plainly as the *Cluck, cluck, keck, keck* language could say it, 'Now, you good-for-nothing children, mind what you're about, and don't get away too far from the coop, for there's that vile cat from the vicarage watching you from the corner behind the laurel hedge.' And the very moment she spoke they all came rushing in between the bars of the coop, as if they had seen an ogre. And sure enough, when I got to the corner, there was Grimalkin, ready to pounce out on the first little wicked chick who dared to disobey his mother——"

"But that doesn't prove that *ants* talk," interrupts Cecil.

"No," said I, "it does not. We will go home to our old smoky garden in London to prove that. Do you remember the tall ash-tree in the grass-plot?"

"Yes."

"Well, at the foot of that tree was an ants' nest, out of which they crawled up the dingy black trunk at all hours of the day; up and down again in two distinct, long, narrow, winding files, in and out among the crevices of the bark. One day I took off from a plant in the flower-bed a couple of large green aphides, and stuck them fast in small crannies just where the line of march of the ants came. Several ants went by without noticing the green dainty. At last one stopped, felt the little monster all over with his antennæ, and seemed to be trying to move him or taste him, I could not make out which; and then away he went. On his way up the trunk he met a friend, and stopped to talk to him with his feelers, and *back they* BOTH *came* and made another inspection. These two then waylaid half a dozen more passers-by, and presently eight or ten ants did what one could not. They seized on my friend in the crevice, and three or four at a time, with the others about them as a joyful crowd, carried the green aphis steadily down the tree and into the nest, where no doubt they made a kind of delicate preserve of him, as good, perhaps, as green-gage jam is to you and Willy."

"The ladybirds do a deal of good, I see now," replies Cecil; "and I suppose, as they eat the green aphides, so the sparrows eat them?"

"I hope not," said I; "and although Master Sparrow is a greedy chap in his way, yet poor little *Coccinella*, the ladybird, often escapes from his nimble beak. The very moment anything touches the twig on which she is crawling, she hides her feet under her body, and sticks fast to the leaf or stem; but if the danger comes nearer, she drops headlong down to the ground, as if dead."

For a few minutes after this long talk, we walked on in silence; Cecil still noticing from time to time the number of ladybirds to be seen on the flowers and bushes in the hedgerow, and now and then setting up a stray butterfly from among the long grass. But, though silent, he was, as usual, thinking over what he had heard about the green aphis; and presently began again:

"I wonder the rain didn't wash away most of these crawling things on the apple-trees and roses?"

"Well," said I, "you may wash away a good many dozens, or even hundreds, and not miss them after all, when thousands are born in a single night. And, besides, a good soaking of cold rain doesn't do them much harm, as Mr. Curtis will tell you. He once took four aphides and sunk them in water for sixteen hours. At the end of that time they were taken out and dried, and three out of the four crawled away, as usual, as if nothing had happened. So that, you see, a few hours' rain is a mere trifle to the clusters of green aphides on the plants and flowers."

## III.

WE were all coming home through the Lower Wood on a dull, heavy, hot morning, after a long ramble, and as we had seen very few butterflies, everybody was complaining of the heat.

"I should like," said Cecil, "to lie down in the shade under this great oak-tree, and sleep till to-morrow morning among the long grass. Perhaps the butterflies will be out by that time."

"They seem to be all asleep now, at all events. But it's growing hotter and hotter every ten minutes, so do not give up all hope yet."

Just as I said this, D—— called to us to come and look at a Purple Hairstreak which she had spied settled on the leaf of a nut-bush. He was clearly fast asleep, and she caught him easily with her fingers. He was a perfect specimen, but we all voted that he should not be killed, and so away he flew, zigzagging heavily up among the branches of a tall ash-tree.

Presently we came to a little open glade at the edge of the wood, and there to our great surprise we found Meadow Browns and large Skippers by the score, all as busy and gay as in the brightest sunshine. But not a Blue nor a Sulphur was to be seen.

"The fact is," said I, "that the Hairstreaks, Blues, and Sulphurs of this little district must have got drenched in yesterday's heavy rain, and are sleeping off their fatigue. Let us take our sticks and beat all the nut-bushes, buckthorn, and young ash-trees on both side of the next wood path, and stir up the sluggards."

We accordingly all set to work, laying about us right and left, and in less than five minutes had set up a score or two of Purple Hairstreaks, and a couple of the finest Sulphurs (male and female) that I ever saw. These two still flew very heavily, and we caught both.

The Purple Hairstreak, *Thecla quercus*, (Figs. 1, 2, p. 148), which belongs to Family VIII., LYCÆNIDÆ (*Argus* butterflies), is a very lovely little butterfly, much like a young Purple Emperor, if there were such a thing. But, as you know, butterflies are all full grown when born, as some other insects are, and so escape all the dangers of infancy. If it were not so, not one butterfly in a hundred would survive the touch of cold and rain, much less escape from the many enemies, birds, wasps, spiders, and boys, that are always on the watch against him.

The wings of *Thecla quercus* are all of a dull blackish-brown; the inner part nearest the body (especially in the upper wings) being beautifully tinted with royal purple, and daintily fringed with white. But the purple on the wings of the female is generally brighter than on those of the male. The lower wings have a short tail, which adds greatly to their beauty; while the under-side of all the wings is of a silvery grey tint (Fig. 2), touched here and there with bronze, with a pale irregular streak of silver white across them, and a spot of orange with a black central dot at the corner of the lower wing. All the Hairstreaks have this pale *streak* on the under-side of their wings, in some shape or other; except the Green Hairstreak, which has, instead, a faint line of white dots along the edge of the lower wing. The caterpillar of the Purple Hairstreak is small, of a reddish and yellow colour, and may be found on the oak towards the end of June. For the chrysalis, of a dull brown tint, you must search among the stalks of grass, near the roots, wherever oak-trees abound. The butterfly is to be caught from July to September.

The Green Hairstreak, *Thecla Rubi* (Fig. 3), is rather smaller than his cousin the Purple, but there is a strong family likeness between them as to the size, shape, and angle of the wings. The only pair we afterwards obtained were caught in this wood,* being among the last of the June brood; but though we hunted most carefully for them till the end of August, we never got another specimen.

All the wings are of a dull brown, slightly tinted with bronze, the colour being lighter at the outer edge. Underneath, both upper and lower wings are of a lovely pale green (Fig. 4), like bright verdigris, fringed with pale brown; and *some specimens* have a faint row of white dots across the middle of the wing. No such dots, however, are to be seen on the Green Hairstreaks in our engraving. The caterpillar, which is of a pale green colour, shaped like a wood-louse, feeds on the blossom of the blackberry, broom, and dyer's-weed, and may easily be found, by those who have sharp eyes, early in April and May.

As the two Sulphur butterflies were the largest and most perfect which we took during the entire holidays, we determined to keep them both. They were easily caught, in the hot still summer air of the woodland path with thick trees on either side of us, though in the open meadow at the edge of the wood, or on the windy down, either of them might have cost us a long chase.

The Brimstone or Sulphur butterfly belongs to Family III., RHODO-CERIDÆ, or *Red-horns*, and in spite of his being very common, is one of the purest and loveliest of English butterflies. In fact, he deserves a far softer and pleasanter name than *Gonepteryx rhamni*, and if he were only as rare as he is common, he would soon have one, say *Papilio primula*, or Primrose Butterfly. There is little to say about his personal appearance, except that he has six perfect legs and wings of a charming golden sulphury yellow, pointed with a short tail at the lower angle; and each wing adorned with a round spot of bright saffron (Fig. 5). The female differs from the male in being rather larger, flying more heavily, and having wings of a pale greenish-yellow. The under-side of the wings, both in male and female, is of a much fainter greener tint (Fig. 6), and

\* Given to us by that most generous of collectors, Mr. Pamplin, of Winchester, whose butterflies are worth going a long day's march to see.

the spot of bright orange on the upper side of the wing becomes a dot of reddish-brown, a colour which in some specimens (see Fig. 6) fringes the edges of the wings. The antennæ of the Brimstone are short and of a ruddy colour, but his legs are long, and he is much given to creeping up the stems of those plants which he likes. The caterpillar, green dotted with black, with a white stripe at the sides, may be found on the buckthorn in May and June; at the end of which time it turns into a chrysalis, pointed at both ends, at first of a pale green, dotted in part with red specks, then becoming yellow.

There are two broods of the butterfly, one as early as March and April, when the weather is mild, and a second in July and August; but the second brood is always the more plentiful, though in the case of other butterflies it is almost always the less in number, and occasionally in size. The Brimstone is found in fields, lanes, and woods, but especially in woodland paths where flowers abound.

On our way home, near the Vicarage gate, we met Henry, who had been up to the Barrows on the high ground. He brought us tidings of great numbers of Holly Blues, and of one Brown Hairstreak which he had chased along the edge of a corn-field, and then lost sight of among some thick nut-bushes. But, above all, he had seen, and chased, until he was dead beat, a good-sized blackish butterfly, as big as a Painted Lady, he said, which flew very swiftly after the chase began, over the tops of the bushes, and so got away. What this black swift gentleman was we could not quite agree. He asserted that it must have been the White Admiral, who, in spite of the white bar across his wing, really does look very black when flying swiftly. Perhaps he was right.

The next morning was very still and calm, and there had been a very heavy dew. But what struck us all most strongly, whether in the garden, road, or woodland path, was the abundance of cobwebs. They were everywhere. Skye, our nimble terrier, who had been scouring round the meadow and through the copse, came in with his nose and eyes covered with patches of them. There were webs across the window-panes; webs stretching from rose-tree to rose-tree, and across the creepers in the porch; webs on the grass; some of the finest gossamer streaming in long dainty threads on the summer air; and others that seemed like tough spiders' webs able to handcuff and fetter a stout bluebottle fly.

As usual, Cecil was one of the first to propound a question: "These webs are so tough and strong, why don't they make some use of them?"

"Well," said I, "one reason is that spiders are very difficult hands to manage: they have a disagreeable habit of biting and killing one another; and when compared to the dull, quiet, patient silkworm, are not pleasant things to handle or have to do with. But the attempt has been made, and not only have large numbers of spiders been brought together into a nursery, and there led to lay eggs, but the young have been brought up, made to spin, and the silk has been wound off and woven. The female spider is the spinner, and the utmost that can be got from a single insect is about 150 yards, weighing $\frac{1}{20}$th of a grain, while a large silkworm's cocoon will yield 300 yards, weighing 3 grains. Yet, fine as spider's silk is, its strength is something prodigious."

"Oh! I know that," says Cecil: "there were three jolly bluebottles in one thin tiny web this morning, all struggling to break through like so many mad bulls; but in spite of all their buzzing, roaring, and kicking, not one strand was broken; and there sat the little spider in her corner under the vine-leaf, afraid to venture out among three such great bullies, and yet knowing that they were all as safe as if a 'bobby' had handcuffed the three, and hung them up in her larder."

"Well, Cecil, I'll tell you another fact about the strength of spiders' silk. A bar of steel an inch thick will bear a weight of nearly sixty tons; but it is said—on good authority, too—that a rope of spiders' silk an inch thick would bear up a weight of seventy-four tons; that is to say, it is a quarter as strong again as the bar of steel. Whether this is positively true is not certain; but there can be no doubt whatever that a thread of the silk $\frac{1}{4000}$th part of an inch thick will bear up fifty-four grains, so that there is no reason why the rope should not bear up the seventy-four tons."

"What colour is the silk?" inquires Mary.

"It is of two colours, silver grey and golden, and both may be drawn from the same spider, at different points of her spinning organ, and of two different kinds also. The yellow is the strongest and most elastic, and after being stretched flies back again to its old length, like a thread of india-rubber; while the silver crinkles up, and is apt to snap if stretched too hard. But both kinds are wanted in building a web; and if you look at one carefully you will see with what skill and beauty every part is arranged—one kind of silk for the strong, straight, outer edges, and the other for the swaying, bending cross-beams."

"But, I say," interrupts Cecil, "what a terrific lot of spiders, and what miles of silk, one must have before a silk dress can be made!"

"Not so many spiders as you might think. You must remember that although each silkworm spins but one cocoon, and is then done for, a spider, after yielding 150 yards, has only to rest for a few days, and is then quite ready to have 150 yards more drawn off; and so on, a dozen or fifteen times in a month."

"What do you mean by drawn off?"

"Just what I say, Cecil. I can't stop now to give you a full description; but Dr. Wilder, a very wise man, who has been studying spiders for years past, and knows more about them than a dozen dictionaries, says that all his apparatus for winding off their silk consisted of 'two large corks, a bent hair-pin, two large common pins, a bit of card, and a bit of lead.' All I can tell you now is, that the Doctor catches the spider between his finger and thumb, so that two legs are turned back, out of the way, applies his machine in the right fashion, and winds away as easily and smoothly as if from a lifeless cocoon. The thread of a single spider is so fine that it cannot be wound off *alone from the reel*, so the cunning Doctor arranges a large number of spiders, and contrives to wind off all their silks together in one thread. The great difficulty is, as I told you, to prevent the bloodthirsty spinners from killing and devouring each other. Only a few out of every hundred young spiders, brought up together in one web, ever escape alive to marry and set up housekeeping and separate establishments for themselves; and a hungry wife has been known, first to kiss her husband, and then seize on him and eat him up, which, as she is one hundred times as big as her lord and master, she can easily manage.

"An ounce, says Dr. Wilder, is $437\frac{1}{2}$ grains, and as each spider yields one grain, it will take about 450 to produce a yard of silk, or 5,400 for a dress of twelve yards. Each silkworm yields about two and a quarter times as much as a spider of one season, so that we should want 200 worms for a yard of silk and 2,400 for a dress. This would make spiders' silk just two and a quarter times as dear as silkworms'; and so for the present, Mary, there is not much chance of your having a dress of spiders spinning."

In the afternoon Cecil, Willy, and I took our nets again and went off to the Emperor woods, up a steep hill of chalky, flinty road, along the edge of a copse. On each side of one part of the road was a high stony bank covered with rough grass and a few flowers, and here, as usual, we saw one or two Painted Ladies hovering over the grass and settling on the ground. But they were very shy and swift, and we caught none. Farther on, nearer the wood, we found Holly Blues, Purple Hairstreaks, and Silver-washed Fritillaries in abundance, and Willy took one very fine

Brown Argus. The Hairstreaks were flitting about over the tops of some young ash-trees, and we might have caught any number that we cared to have.

"Never mind the Purple Hairstreaks," said I to Cecil, "but go along the edge of the wood and look out for his cousin *Thecla betulæ*, for though his usual time of appearance is the middle of August, Henry declares that he saw one yesterday."

Away went Cecil in great spirits, and in less than five minutes I heard shouts of triumph as he came racing back to me, net in hand, and crying out, "I've got him, I've got him!" And so indeed he had. This was our first specimen, and was caught hovering over a hazel-bush at the edge of the wood.

The Brown Hairstreak, *Thecla betulæ*, though plentiful in some places, is in many others counted a rare fly, so that there were great rejoicings when the first was brought home to the Vicarage. The whole of the upper surface of the wings is of a rich, smoky, brown colour, the upper wings having near the edge a patch of cloudy yellow (Fig. 7), and the lower ones being tailed and scalloped into an outline of great beauty. The female is sometimes slightly larger, but always much gayer and brighter in colour than her sober husband. The patch of yellow becomes a bright orange on her upper wings (Fig. 8), while the tails on the lower wings shine with the same glowing colour. She, indeed, looks far more like a foreigner than an English butterfly, and the under-side (Fig. 9) of the wings in both male and female is still more striking. There the ground colour is a rich tawny brown, with an outer margin of bright orange running parallel with the edge of the wings, which are fringed with most delicate white. Across both wings wanders a clear streak of silvery white, which adds greatly to their beauty, and shines brightly amidst many dainty shades and touches of black and orange.

The caterpillar, of a pale green colour, striped with yellow on the back and white at the sides, feeds on the blackthorn, and is mostly found at the back of the leaves. In July it changes into a small, dull brown chrysalis, faintly spotted with black.

As we came home through the woods, we saw a tiny young rabbit scamper across the path, and hide away in a bunch of long grass. Cecil, with his usual quickness, pounced upon the little trembling ball of white and grey fur, and carried it home in his pocket, declaring that he knew all about the feeding and nursing of young rabbits, and would educate this one as a plaything for the children. After a long search in the stable, an old birdcage was found, and in it the poor little Bun was duly lodged, to his utter amazement and terror. They began to feed him the moment they got inside the Vicarage, on bread, apples, cabbage-leaves, grass, carrots, sops, milk, wild parsley, and fifty other dainties; but, at the end of the second day he insisted on dying, to the great mortification of his four nurses. I am afraid that he fairly died of stuffing.

On our way home, Willy caught so perfect a specimen of the Gate-keeper, that we resolved to kill it for our cabinet. The Gate-keeper (*Satyrus Tithonus*) belongs to Family VII., SATYRIDÆ, and is sometimes called the Small Meadow Brown. He is a very common little butterfly, of a sober brown hue, but still has a beauty of his own (Fig. 10), to be seen even in a woodcut. The upper surface of all the wings is of a rich dull brown colour, with a broad margin of a darker shade; and the male (see cut, p. 148) has across his upper wings a bar of blackish cloudy brown. Each upper wing has also near the tips a small black spot with two tiny white pupils, and each lower wing a black spot with one pupil of white. The lower wings are slightly scalloped, and he is a nimble flier when roused. The caterpillar, of a dull greenish colour, having a red head, may be easily found in May and June on the common meadow-grass, and there also the small grey chrysalis into which it changes. *Tithonus* is common throughout England, but, oddly enough, has not been found in Scotland. The female differs from the male in being without the patch of brown on the upper wings.

As we strolled homewards through the oak wood, we noticed one noble Emperor skimming swiftly over the lower trees, but he was far beyond the reach of our longest net, and soon out of sight among the green oaks. Of the Purple Hairstreaks, we might have caught fifty if we cared to do so, but we only took a few very large and choice specimens. We came upon several clusters of wild carrot, and upon them found many of our old friends the little white tiger-spiders (Fig. 12, page 136), all busily at work as usual. One big fellow was lurking on the head of a Scotch thistle among the long narrow purple petals, and into these he had dragged a small brown moth head first, the legs of the poor victim sticking up into the air in a very odd fashion, as we often found to be the case. After some little poking with a sharp stick we dislodged the tiger, who dropped nimbly down into the grass; then we released the prisoner. But he was dead, and sucked quite dry.

We got back to the Vicarage at last, before dinner-time, but very hot and weary, just as Henry came home from the Barrows, to which he had again gone off alone. We saw by the look of his face that he had got a prize of some kind, and so indeed he had, as you shall hear in the next chapter.

## IV.

WE were quite right about Henry. On his way up to the Barrows he had passed through a large field of sainfoin at the edge of the wood, and was strolling leisurely along, away from the path, where the rosy flowers were scanty, and the grass grew up thickly in rough patches. Among these patches of grass were many tall plants of the common dock, now covered with clusters of red seed; and on the stem of one of these docks he saw hanging, head downwards, a yellow butterfly, which he at first took to be a common Sulphur with its wings folded. On coming nearer, he saw it was a Clouded Sulphur, and apparently fast asleep. With one swift sweep of the net the prize was secured, and he went on his way to the Barrows with a joyful heart. It was, as he thought, a capital specimen of the female Pale Clouded Sulphur, *Colias hyale*, and, having just before come out of the chrysalis, as lovely and perfect a butterfly (Fig. 1, p. 156) as we had ever seen. The down on the wings, and the pink fringe on their outer edge, were both untouched, and Henry had managed to kill the butterfly without a single flaw on its perfect beauty.

But, when we came to examine it, we found that it was not a Pale Clouded Sulphur at all, but a much rarer butterfly, viz., a white variety of the female of the common Clouded Sulphur, named *Colias helice* (see cut, p. 156), in which the ground colour of all the wings is of a pale greenish-white, the margin of the upper wings intensely black, and broader than in *Edusa* (Fig. 3, p. 136), mentioned in Chap. II., p. 135.

Over the whole wings and body of *Colias helice* is a greenish-white tinge, which is wanting in all the other varieties of this butterfly; while the deep rose-coloured fringe at the edge of the wings gave it a still greater beauty.

Nor was this all Henry's good luck this day. Having secured the lovely specimen which I have just described, he went on mightily towards the Barrows in great spirits, and at the edge of the down saw two male Clouded Sulphurs chasing each other across a field of ripe barley. This was a prize that he could not resist, and so away he went galloping at full speed among the golden ears, at a rate which I am afraid did the future crop no good. He soon came up with the two butterflies, and, after many vain trials with his net, at last managed by one dexterous sweep to bring them both to the ground near the hedge, over which he climbed into the winding road on the down.

I gave him great credit for this double capture, because, in the first place, it was done under a blazing sun, across rough hilly ground, and while the two butterflies were going at full speed across a field of swaying yellow corn.

"As they flew," said Henry, "they were of the very exact colour of the brown barley, and if I had taken my eyes off them for a moment, as they flew, I should have lost them at once."

This grand capture excited us all so much that we set off with fresh vigour in the afternoon, once more for the Emperor woods. Away we went again up the steep chalky hill, putting up, as we went, dozens of Blues, Hairstreaks, and Painted Ladies, but able to catch none of them because of the fierce wind which had sprung up from the south-west, and threatened to blow our hats away. But still it was very hot, and along by the edge of the wood, and in the shelter of the winding paths, we found plenty of butterflies which were easily caught. Here I took a noble pair of High-brown Fritillaries (Figs. 2, 3), the silver spots on the under-side of the lady's wings (Fig. 3) the brightest and most perfect we had yet seen. The silver seemed as if laid thickly on with a brush. The boys took several other specimens, but all these were so battered, and their wings so faded and knocked to pieces, that not one was worth keeping. They had clearly been leading a very gay life of it for the previous four or five weeks; and the first heavy shower of rain would most likely be fatal to all further flight over these pleasant feeding-grounds. There was a very large field of sainfoin on one side of the road, as we went up the hill, and here we had constantly watched for Clouded Sulphurs, but, up to this time, in vain. We had all tried the ground carefully, but nobody had as yet seen a single specimen. To-day, however, we were luckier; for, just as we got to the brow of the hill, at the edge of the wood, up got a Clouded Sulphur, and flew very softly and gently across a bit of open brake to a patch of wild scabious and yellow bedstraw among some long grass. I crept cautiously after him, and presently found him perched on the purple flower of a hard head, swaying to and fro in the wind. He was so intent on getting a good sup of the sweet honey, that I came close up to the flower, and with my finger and thumb caught him with his wings folded, as you see him in Fig. 4. He was so perfect and bright a specimen that we all voted he must be kept.

"His trunk, or proboscis, is, as you see, slightly uncurled, as if it were

in the very act of getting ready for sipping the honey of a flower. If you watch a butterfly closely while hovering over a bed of flowers, you will see that he begins to uncurl his trunk before he settles, then straightens it out, and plunges it down into the flower. Presently, after drawing it out again, he gives it a little curve or twist, then straightens it out, and makes a second dip into the honey at the base of the petals. Having done this four or five times, away he flies in search of a fresh flower.

"If you cut the trunk into two pieces with a sharp penknife, and place it under a good microscope, you will see that it is divided into three separate canals, one of a square flat shape in the middle, with a rounder tube on either side. M. Réaumur, a wise and clever naturalist, was one of the first to observe this, and he managed to find out the use made of the three canals in the following manner. Having got a moth to settle  quietly on a lump of sugar, he held in one hand a powerful magnifying-glass, and brought it near to that part of the trunk he wished to examine. 'Sometimes,' he says, 'I was half a minute, or nearly a minute, without perceiving anything, but then I saw clearly a little column of liquid mounting quickly along the whole length of the trunk; and in the liquid some little balls, which seemed to be globules of air drawn up with the liquid from the flower through the two outer canals.'* Presently he saw some fluid rising straight up through the middle channel, and then, after watching some time, once or twice he saw some fluid *descending* from the root of the trunk to the point. Thus he found out how a butterfly or moth is able to nourish itself on honey, thick syrup, or even solid sugar, by sending down some liquid which falls against the sugar, moistens, and dissolves it. When this liquid has got charged with sweet, he sucks it up again through the two outer canals."

"I say," interrupted Cecil, "how splendidly the trunk must be made!"

"Yes, and when you remember that the whole three canals together are no bigger round than a common pin, you will see with what beauty and wisdom God has made even His tiniest creatures, and framed every part to do its own work with ease and perfection."

As we went down the hill, a small Tortoiseshell Butterfly sprang up from a bed of nettles near the hedge, and was immediately chased by a chaffinch. Away they went, down the road in front of us, their rate of

* "The Insect World," p. 175.

flying being pretty much about the same, the butterfly now and then darting ahead, and being overtaken by the bird.

"Two to one on the Tortoiseshell!" cried Henry; but nobody would take his bet.

"The butterfly is safe enough," said I; "for though the bird has stronger wings, and would soon tire out little golden-wings, the Tortoiseshell, you see, is continually zigzagging up and down, and thus changing his line of flight. Thus the bird is puzzled when to make a dart on him, and the butterfly escapes over the hedge, or sinks down among the grass or nettles in the hedge, before his enemy can seize on him."

At this moment came two sudden interruptions. Willy had suddenly found a patch of butterflies' eggs on a leaf in the hedge, and shouted out—"I say, Henry, what butterflies are these?" while Cecil had a fresh question to propound about the trunks of butterflies.

"First of all," said I, "we will deal with the eggs."

There were thirty or forty of them, of a dingy green colour, thickly scattered over two leaves of a nettle, in patches of five or six.

"Well, Willy, so many caterpillars feed on the nettle that it is difficult to say exactly to what butterfly these eggs belong; they may be those of the Red Admiral, or the Common Tortoiseshell—most likely the latter, as they seem to be rather small, and I see that there are many more of them on the nettles. If so, they are not worth keeping. The Red Admiral is a rarer and finer butterfly, and his caterpillar is always worth rearing."

"How is it," asks Willy, "that all these eggs didn't get washed off in the rain last night?"

"Because," said I, "the butterfly who laid them stuck them fast to the leaf with a strong sort of gum, which water cannot melt."

"Where does she get the gum from?"

"Ah," said I, "that I cannot tell you. But her eggs are covered with it when she lays them; and no rain can wash it away. If you look, too, you will see that most of the eggs are laid on the under-side of the leaf, so that the birds shall not spy them out. But some of them do get eaten, for all that; and some are trampled on and destroyed, or lost among the grass; so that not one-half ever become caterpillars. If it were not so with the eggs of butterflies and moths, the consequences would sometimes be terrible. Many of our trees, bushes, shrubs, and plants would be entirely stripped, very often of blossom as well as leaves; for some butterflies lay a hundred eggs, and a moth has been known to lay five hundred."

"All on one plant?"

"No, Willy; not all on one plant. She may lay her eggs in a dozen different places, or a score perhaps. But she *can* lay as many as five hundred eggs, because, when shut up in a box, she has been known to do so.'

"And do they ever really strip the trees of all their leaves?"

"I have seen many young ash-trees and nut-bushes in this very wood with a dozen bare branches on them—every single scrap of green stripped off, as if cut with a knife; and all done by caterpillars."

"What appetites they must have!" adds Cecil. "I hardly ever found a caterpillar but what he was eating!"

"Many caterpillars," said I, "in one day eat thrice their own weight in leaves! And if other larger creatures had such appetites, our green meadows would soon be bare and black enough; for an ox weighing sixty stone would in twenty-four hours eat three-fourths of a ton of grass!"

"But if these caterpillars eat so much," says Willy, "how is it that they do not grow fat and big?"

"Why," said I, "one reason is that the leaves they eat pass through them very quickly, without being digested. The caterpillar seems to live on the juices of the leaves which he eats; and these make him grow quickly, but not fat. One has been known to eat more than forty grains of leaves in a certain number of hours, and yet at the end of the time he was but one grain heavier than when he began."

"And do all little creatures eat and drink at this tremendous rate?"

"Oh, no!" said I. "The caterpillar of a moth or butterfly eats a great deal, and grows quickly, increasing rapidly in size more than in weight; while the little worm from which the flesh-fly comes grows in weight at a tremendous pace. Thirty of them, on one day, all together weighed one grain. The next day each worm weighed seven grains; having thus, in twenty-four hours, become two hundred times heavier than he was before. Keep a caterpillar without food for a day or two, and the chances are that you will then find him dead at the bottom of your breeding-box. He lives on plants, or leaves of trees, which are always to be found in plenty; and his law is to eat and drink as much as he can. But the little ant-lion, who makes a pitfall in sandy earth and traps all he eats, has been known to live without a scrap of food for six months; and a spider has been kept without food under a sealed tumbler for ten months, at the end of which time he caught a fly that was introduced to him as nimbly as ever. He deserved his fly, I am sure. While, to crown all, Mr. Baker tells us that he once kept a beetle (with the awful name of *Blaps mortisaga*) without food of any kind for three years. The food of all these creatures is scanty and hard to get when compared to the green leaves for the caterpillar; and so their law is, to be able to do without it for a long time.

"But here we are at the Vicarage gate, and I am sure you have heard enough about caterpillars to last you until to-morrow."

"I only want to know one thing more," says Cecil; "and that is, which sort of caterpillars birds like best?"

"Well," said I, "that is not an easy question to answer; but wrens and titmice and all that class of birds eat a great many green caterpillars, and the Robin, it is said, rarely touches the hairy ones."

"Those big ones that we found on the cabbages in the garden," cries out Willy, "look as if they were naked, and very clean. I am sure they are the nicest."

"I don't know about the nicest, Willy, but they are among the commonest, and some of them most easily found, though they are sometimes of the very exact colour of the leaf on which they feed, to our eyes. The brown hairy caterpillar is pretty safe so long as he is feeding on a brownish leaf, or holding fast to a brown stem; but the moment he shows himself above the edge of a green leaf, he is seen and gobbled up in a trice by some roving titmouse.

"And now," said I, "I have done with caterpillars for to-night, and I shall tell you nothing more about them till some day when I find a Looper in the wood."

Several days passed away after this before we took any new butterflies, but the August sun began to grow hotter and hotter, and the yellow corn to turn of a ruddy brown.

"Well," said I, one morning, as we all set out with our nets to Lower Wood, "if we can find none of the rarer butterfles to-day, we had better get a specimen or two of the common ones."

It was a still, sultry morning, and though heavy clouds almost covered the sky, now and then we had a gleam of sunshine, which the butterflies found out as soon as we did. Opposite the gate leading into the woods is the park lodge, and outside this is a little open bit of grassy waste with a stone wall at one side of it, under the shade of some tall trees.

Flitting over this plot of grass, and now and then settling on the stones of the old wall, we saw several pairs of the homely little Wall Butterfly, and of these we took a couple at once.

*Satyrus Megæra*, or the Wall Butterfly, belongs to Family VII., the SATYRIDÆ, and is one of the commonest of them all. He is to be found from June to August in almost every lane, zigzagging about at a slow pace, often settling on warm hedge-banks or walls, but starting up again as the shadow of your net falls on him. He is very easily caught, unless thoroughly roused or scared by your awkwardness in missing him with the net; but then often flies away at a great pace. More than once I have mistaken him for a small Fritillary, from his rapid flight.

All the wings are of a tawny golden yellow, irregularly barred with brown, and edged with the same colour. On the upper wings, in the middle of the yellow, a round black spot with a white dot in the centre, and sometimes a second black spot (Fig. 5) close beside it, also having a white speck in the centre. The lower wings, which are slightly scalloped at the edge, have three distinct spots of the same kind. The under-side of the wings is, both in colour and marking, much like the upper side, except that the brown bars are narrower, while the eye has round it a distinct circle of light brown. The female differs from the male in having no *cross* bar of brown on the upper wings. You may see this *cross* bar joining the bands on the upper wing of Fig. 5, a very fine male which we caught near the park gate.

The caterpillar of *Megæra*, of a light pale green, with a white line on each side, as well as stripes on its back, feeds on grass, and may easily be found by the roadside in May, when it turns into a green chrysalis that afterwards changes to a dingy brown colour.

When we got inside the wood, Hairstreaks as usual were flitting about from tree to tree, especially among the ash-trees and young oaks. But though we watched them narrowly, we could not detect a single Green or Brown Hairstreak; and of the Purple we already had some dozens. Even Alice (*ætat.* 4) had caught a pair among the long grass at the edge of the copse, for the express adornment of the nursery.

We set to work, therefore, on a little open piece of grassy meadow, lying between two pieces of copse, and soon found plenty to do among the commoner butterflies. Oddly enough, the first I caught were a pair of the Heath Butterfly, or, as he is sometimes called, the Least Meadow Brown.

The Heath Butterfly, *Chortobius Pamphilus*, like the Wall, also belongs to Family VII. SATYRIDÆ, and is one of the daintiest and most elegant little flies of the whole genus. If he were only a rare catch, instead of a common one, his beauty would bring him great praise. All the wings are of a pale tawny yellow, faintly shaded with delicate brown (Fig. 6) at the edges; the lower wings being more and darkly shaded with brown, and rounded at the edge. Near the tips of the upper wing is one faint round spot of brown, and this spot on the under-side of the wing appears almost black (see Fig. 7), with a white dot in the centre; while the lower wings are almost entirely shaded with brown, across which, parallel with the edge, is an irregular waved line of whitish yellow. The female (Fig. 7) is larger than the male, and on her wing this pale mark is clearly seen.

The caterpillar, of a green colour, with stripes of dingy white on the back and sides, is small, but may be often found on the dog's-tail grass in May or August, as there are two broods of the butterfly, in June and September. The tail of the caterpillar is slightly forked and tinted with red, so that it may be easily known.

The favourite feeding-ground of the butterfly is on heathy commons—hence its name; but, oddly enough, as I said, we found it most abundantly in little open clearings in the woods. Here, too, in the same patch of grassy waste, we took a couple of specimens of that curiously nimble little butterfly, the Grizzled Skipper. So oddly and quickly does he flit from flower to flower that Willy, who was the first to see him, declared that he was not a butterfly at all, but only, as he said, "a very *dodgy* little fly, with black wings."

The Grizzled Skipper, *Syricthus alveolus* (Fig. 8), belongs to Family X., the HESPERIDÆ, several members of which we have already noticed. When fully stretched his wings are barely an inch wide, all of a dull clouded black tint, marked with very small white spots. The upper wings have about fifteen of these tiny spots, while the lower ones have rather fewer; both being plainly fringed with black and white of the same tint. The under-side of the wings is of much the same colour, but rather paler, while the spots are fewer in number.

The caterpillar is said to be of a dark green colour, with a black head; feeding on the leaves of the common Potentilla, and turning into a chrysalis in April. The butterfly appears in May, so that we were very lucky in catching a couple of good specimens so late in the year as August.

Just after we had caught this little Skipper the sun came out strongly, and swarms of smaller butterflies began to start up from every cluster of grass and furze, or clump of bushes. Heaths, Blues, Common Skippers, and Meadow Browns might have been caught by the score, and the crowd yet have seemed no less. But, strangely enough, we scarcely saw a single

White Butterfly, or a Small Tortoiseshell, though the time for their second brood was fully come. Out of the meadow we passed on into a wide green road through the wood, from which there branched away several narrow winding paths.

"Now," said I to the elder boys, "Willy and I will take the lower path down to the open brake by the old sawpit, while you go away to the higher ground among the thicker copse, where you must look out for Wood Ringlets."

We soon reached the open glade at the edge of the wood; and as there was an abundance of wild thyme, wild scabious, and other flowers in full bloom, and the sun was out again, we saw swarms of butterflies. We strolled about for an hour, and got some very good Sulphurs, a pair of Brown Hairstreaks, and a Silver-washed Fritillary, and then set out on our way homewards. All at once Willy, who had crept through a great bed of fern in pursuit of a Sulphur, cried out,—

"A White Admiral! a White Admiral!"

I ran up to him as fast as I could, and soon got close to the hedge, where he stood pointing with his net to a thick rough bush of blackthorn.

"Where is he?" said I; "I can see nothing like a White Admiral; and there is not a scrap of blossom there of any kind for him to feed on."

"There he is; there he is! I can see him now, opening and shutting his wings at the tip of one of those branches."

I crept very cautiously up, and at last, after a long search, I saw him exactly where Willy's sharp eyes had seen him. The branch was five feet from the ground, and very rough and ragged. But there was no help for it; and so I made one quick sweep just above the bough, and by sheer good luck secured the prize.

"Well done, Willy," said I. "But how did you guess it was a White Admiral, for you have never seen one?"

"No, but I have seen his picture in the big book at home, and I was sure that he was like nothing else but a White Admiral."

It was rather a prize for us, because the White Admiral is a June or July butterfly, and it was now near the middle of August. So we packed him up securely, and went away home in great spirits.

The White Admiral, *Limenitis sybilla*, belongs to Family V., the VANESSIDÆ, or Angle Wings, and is one of the most graceful and beautiful of English butterflies. All the wings are of a dingy black tint, the upper ones being marked by a cluster of white spots, and the lower ones by a bar of the same colour across their centre (Fig. 9). The wings are also scalloped at the edge, but not angled so sharply as those of his cousin the Red Admiral. The under-side of the wings is far more beautiful than the upper, being of a rich tawny red with patches of silver grey, spotted with dark brown near the edge, and across the centre barred and spotted with white. The caterpillar, which feeds on the honeysuckle, is of a pale green colour, with a reddish head and legs, and short hairy spines along the back, turning at the beginning of June into a green chrysalis dotted with gold.

When we met the boys we found that they had taken some splendid Sulphurs and a couple of Wood Ringlets.

The Wood Ringlet, *Satyrus hyperanthus*, belongs to Family VII., the SATYRIDÆ, and is found plentifully in shady woods and bushy lanes all through the months of June, July, and August. The upper surface of all the wings is of a dull brownish-black colour; the upper ones, in some specimens, having a row of three small black spots, edged with pale brown, and a white speck in the centre; and the lower, three spots of the same kind more clearly marked. The under-side of the wings is paler in colour (see Fig. 10), the little *ringed eyes* on all the wings being much clearer and brighter. The caterpillar, of a brownish-grey colour, striped on the back, feeds on wood sorrel and several kinds of grass; and at the end of June creeps down to the roots of the grass, and changes into a small brownish chrysalis just below the surface of the earth.

They had also caught a very odd-looking variety of the same butterfly —a male, without any *ringed eyes* on the upper side of his wings (see Fig. 11), but having the black dots more strongly marked.

Mary, too, had been very busily on the watch for the second brood of the small Pearl-bordered Fritillary, already described at p. 128, and after a long chase by the edge of a corn-field, had caught one perfect specimen (Fig. 12), which, though very small, was the best in our whole collection. It had just come out of the chrysalis.

## V.

AFTER many sultry days, a cold north-east wind set in, and most of the butterflies in field and wood, except in very sheltered spots, had disappeared. But the large silver-spotted and silver-washed Fritillaries seemed to care nothing for the cold wind, and were still to be seen in abundance. Cecil, too, was not to be daunted by the cold, and set off as usual one morning to the broken copse at the edge of the lower wood.

For a time he saw nothing but the large Fritillaries, and a few Purple Hairstreaks which he routed out of the thick bushes with the handle of his net; but presently he came to the edge of an open brake, edged with low oak-trees, and over one of these, to his great delight, he saw a Purple Emperor sailing to and fro; then, suddenly sweeping down almost to the ground, and then hovering over a thistle, where at last it settled. Surprised and delighted at the sight, he made a hasty dash at the grand prize, missed it, and sent it off like a flash of light away across the top of a lofty oak. For a long hour he waited and watched, but all in vain; not a glimpse more did he get of his majesty that day; though a week afterwards Cecil and I both saw him again, sailing over the top of the same favourite oak-tree.

Meanwhile, Henry and I had gone away to a distant part of the Black Wood, which we had as yet never tried. But though there were multitudes of large Fritillaries on the wing all through the thickest part of the copse, we got not a single new butterfly, and were making our way slowly along by the edge of a corn-field, when, in the very thickest part of a thick hawthorn hedge, among some long grass, I suddenly spied a large yellowish-brown bird. We got close up to him, and at the first glance thought he was dead. It was a large brown owl, that had apparently been caught in the heavy storm of wind and rain during the previous night, and had crept into the thickest part of a warm hedge to dry his draggled feathers. He was standing bolt upright, with his eyes shut, and looked a great deal more like a stuffed bird than a living one; so we determined to catch him if we could. Henry, therefore, went round to the back of the hedge, to attack the enemy in the rear, while I kept watch in front. In two minutes, after a deal of scuffling and scratching, he cleverly managed to slip his long butterfly-net over the owl's head and body, and, *nolens volens*, drag him out into the long grass. The poor old half-drowned bird, winking and blinking in the bright sun, fought very hard for his liberty, and managed to inflict one sharp bite on Henry's finger, but at last we managed to get him thoroughly down into the bag-net; and after a long tramp of five or six miles, brought him in triumph home to the Vicarage. There we hunted up an old blackbird-cage in the stable, and after some trouble got our prize into it. He was a splendid fellow, and as evening came on soon began to set himself to rights and smooth his ruffled feathers. For a long time he would touch nothing that we offered to him in the shape of food; but at last a dead sparrow taken out of a trap, and partly picked, was pushed into the cage at the end of a stick. At this the owl opened his eyes, and making a sudden and fierce dash, seized it with his talons, and having torn off a few more feathers, swallowed the sparrow at a single gulp. But the next minute he disgorged his prey, and then with much wriggling about of his head, and many chokings, swallowed him once more, this time, as we thought, not to reappear. Next, bits of meat and bread went down like pills, when presently up came the sparrow again; and then, after two more gulps, was finally swallowed.

But that night few of us got any sound sleep. Fearful noises that sounded like the cries of people in distress, mixed with strange hissings and screams, were heard all through the Vicarage; and at a very early hour in the morning it was found that the owl had got out of his cage, and was ruling over the cat and kitten in the back kitchen like a tyrant. Then we held a council of war, and decided that the owl must go. But the difficulty was to get him out of his cage. Henry had brought home a dead mole from the corn-field, and popped it into the cage; and upon this the owl instantly pounced with outspread wings, and held it fast with his talons. Now and then he gave the unlucky mole a sharp dig with his beak; then he tried to tear off the feet before swallowing it; but finding this impossible, he had another try at it, feet and all. All in vain. Then we opened the door of the cage, and with a long sharp stick had a series of fights among the straw for the possession of the mole. We secured it at last, got it outside the cage on to the green grass, and then left it as a dainty for his majesty to come out for and make his own again; and at last he condescended to creep through the open door, made a sudden dash at the mole with outspread talons, missed his stroke, and then sailed softly away on his soft downy wings, over the Vicarage garden into the next meadow. It was a beautiful sight, and glad enough, no doubt, the owl was to be once more free outside the horrible bars of a cage.

The next day was bright and sultry, and we all went to work again with double vigour. Once in our day's ramble we came upon a little glade in the heart of the wood, where the ground was carpeted with flowers; and there, on a little patch of about twenty square yards, we counted more than a dozen of pale golden Sulphur Butterflies all hovering over the blossoms, or drowsily settled with wings close shut. We might easily have taken the whole of them. To-day we observed how strangely the colour and tint of butterflies' eyes changed after they became quite dead, light grey or green sometimes turning black, and yellow changing into opaque brown or red.

We picked up, too, some odd caterpillars to-day; one, that fed on the broad-leaved sallow, was of a pale whitish-green tint, exactly like the colour of the leaves. When tired of eating, he would curl himself up at the back of a leaf somewhat in this fashion:

though how he managed to stick fast to the leaf completely puzzled us. But, blow as hard as it might, the wind never shook him off; and unless we were quite close to him, it was impossible to make out the leaf from the caterpillar, so exactly alike were they in colour. This, no doubt, saved him from the sharp eye of hungry birds in search of a morsel for dinner.

Another very strange caterpillar of a pale brown colour, exactly that of the stems and branches of the sallow, belonged to the family of *Loopers*, so called because in walking they hunch up the middle part of the body thus:

into a sort of loop. Most caterpillars have sixteen legs, of which six are sharp and scaly near the head; but the Loopers have only ten altogether; four being thick fleshy ones, near the tail, called prolegs. These Loopers cannot shorten or lengthen their segments as they walk, but only bend their bodies. But the strangest thing about them is, that they have the power of stretching out their bodies in the air, at an angle of about 45, without any support, merely holding fast to the twig or bough by their hind feet (see engraving on next page.) Sometimes the Looper stretches himself out thus in the air, or across from one twig or bough to another, holding to the one side by his hind legs, and to the other by four of his sharp feet near the head. But in either case he is so exactly like the neighbouring branches in colour and shape, that it is impossible at the first glance to make out which is the Looper and which the stem of the sallow.

August the 17th was a blazing hot day, with a fierce east wind, and we worked very hard for many hours, but we got no butterflies of the rarer kinds, and not many of the commoner sorts. A noble pair of Peacocks was our best prize, and these we got just at the edge of the copse, on a sunny slope, just outside Emperor wood, where teazle and nettles grew in abundance.

The Peacock (*Vanessa Io*) belongs to Family V., VANESSIDÆ, or Angle Wings, and a glance at Fig. 1 (page 169), will show you how deeply and beautifully these angles serrate the edge of all the wings. He gets his name of Peacock from the large eye of orange, red, and other brilliant

Looper on Twig.

colours which he bears on each wing; those on the two upper ones being the largest and brightest. The base colour of the upper and the lower wings is a fine, tawny, reddish-brown, while a patch of black surrounds the eye, just below which also (Fig. 1) you will see two small whitish dots that add greatly to the beauty of the wings. The under-side of all the wings (see Fig. 2) is of a fine deep black, mingled with rich brown. The Peacock is a noble and gorgeous butterfly, and though a strong and swift flyer when disturbed, may be easily taken when settled on the blossoms of the thistle, on which he loves to alight and sun his splendid wings.

The caterpillar, which is common enough in all woody districts, feeds on the nettle. It is of a deep black colour, covered with sharp spines, and thickly sprinkled with white. The chrysalis, of a rich tawny brown colour, sometimes tinged with gold (see Fig. 3), is often to be found hanging head downwards to the under-side of the leaf—more rarely to the stem—of the nettle.

The August brood of these butterflies was just now beginning to come out, and we saw great numbers of them along the edge of all the woods for many weeks after this; but though we took many fine specimens, not

one of them was so large or so splendid as Fig. 1. He was fully three inches in extent from wing to wing. On this day, too, we took several very perfect specimens of the Small Tortoiseshell Butterfly, *Vanessa Urticæ*, the caterpillar of which is much like that of the Peacock, and also feeds on the nettle. The wings of this butterfly (see Fig. 4) are of much the same shape as those of *Vanessa Io*, but considerably smaller, and beautifully angled at the lower edge. The chief colour of both upper and lower wings is a fine tawny, rich orange brown, barred and spotted near the upper edge with deep black, and patches of lighter yellow between or near the spots of black. All the wings, too, have a lovely margin of brown and black at the outer edge, a row of faint blue spots dividing the two waved lines of brown and black. The body of this butterfly, like that of his relation *V. Io*, is thickly covered with rich down, the antennæ being long and fine. The under-side of the wings is marked much in the same way, but the orange is turned to stone-colour; and all the other colours are of a duller, dingier hue.

If the Tortoiseshell were as rare as he is common, he would be counted a very handsome, gay fellow. The young caterpillars, of a dull greyish or greenish-black, thickly spined, at first live together in a web; but after changing their first skin they separate, and wander freely about the nettles on which they feed. The chrysalis (see Fig. 5) is rather smaller than that of *Vanessa Io*, but much like it both in shape and colour, hanging by the tail from the back of a leaf or stem, the head also being divided into two sharp points or ears (see Figs. 4 and 5). There are two broods, in June and August, everywhere common in woods, lanes, and fields, wherever nettles abound.

The next day I set off alone to the Emperor wood, intending to watch for Emperors and Brown Hairstreaks.

It was just twelve o'clock when I reached the farther side of the wood, and stood watching under an immense oak-tree, over which I had often seen many butterflies hovering. The sky was without a cloud, not a breath of air was stirring, and I was very glad of the thick shade. But though I watched long and carefully, not a single glimpse did I catch of any Emperor. Peacocks, Fritillaries, and Purple Hairstreaks flitted about from flower to flower and bush to bush, in abundance. At last, however, I made out two or three small butterflies hovering over a spray of oak, within reach; and watching these very carefully, one by one, as they settled, I managed to capture three Brown Hairstreaks,—two females and a male, —all perfect, and most brilliant in colour. Why they haunted the oak so much I could not make out, as the caterpillar feeds on the blackthorn, birch, and wild plum.

Seeing no more Hairstreaks, I made my way into a thicker part of the wood, and at last found myself in a narrow path between two hedges of stunted oak. Everything was intensely still, and the heat very great; so I stood up in the shade and watched and listened. Suddenly something fell from the leafy boughs of the oak above my head into the grass and leaves; next it fell on my head, and then again among the brambles. And still I could see nothing moving either in the tree or on the ground. At last, however, just above my head, on the lower branch of the oak, I made out a cluster of large, yellow and brown, hairy caterpillars. In the middle of the bunch I saw one who was clinging fast to a twig by his hind feet, hanging head downwards, and working his body to and fro in short rapid jerks, as if in great pain. Every now and then, in the violence of his contortions, he managed to strike one of his companions, who instantly gave way and fell to the ground, I watched more than a dozen dislodged in this curious fashion; until at last the struggling tyrant was left alone in his glory. What the object of all this tumult was I could not discover. All I could make out was, that as soon as the other caterpillars reached the ground, they set off as hard as they could go to the hedge on the opposite side of the path, and there disappeared among the long grass. Whether the food on the branch of the oak was scarce, or whether the hairy family, after living together so long, had quarrelled and come to blows, or whether their instinct—for some wise purpose—led them to separate, it is hard to say; I only tell the plain facts. I caught and carried away one large hairy fellow, as big as my little finger, as a trophy.

August 19th was just such another day, with a fresh breeze; and we all set off at 11 A.M. to go over the same ground. Many Brown Hairstreaks were watched up into the same great oak-tree, all beyond our reach; but, oddly enough, we each caught one perfect specimen.

The next day, the wind having died away to a dead calm, and a white, hot mist stretching over the woods and hills, Mary and I set out once more to the Emperor wood, in search of the Small Pearl-bordered Fritillary (*Argynnis selene*), of which the second brood was now due. For some strange reason scarcely a butterfly of any kind was to be seen until nearly twelve o'clock; when among the thinner copse, where marjoram and other wild flowers grew in abundance, swarms of Peacocks, Sulphurs, Harvest Blues, Argus Blues, Coppers, and large Wood Fritillaries suddenly appeared in almost countless numbers. We might easily have taken a score of each, but we contented ourselves with a few Blues, which were very brilliant and perfect; but difficult to catch, because of their being constantly engaged in chasing and frightening their cousins, the little Coppers, who with shining golden wings flitted about in the sunshine; so swiftly at times as almost to escape notice.

The Common or Harvest Blue, *Lycæna Alexis*, belong to Family VIII., LYCÆNIDÆ,—a large family, all having six perfect legs, rather short knobbed antennæ, and wings mostly either tailed or scalloped at the edge. In the male, both upper and lower wings are of a fine lilac-blue colour, fringed with delicate white (see Figs. 6 and 7); but in the female (Fig 8) of a tawny brown, faintly shaded with blue, with a row of orange spots near the margin of each wing. The under-side both of male and female is of a soft silvery grey (see Fig. 9), dotted with many black spots, and a row of orange spots like those on the upper surface. The eyes of this butterfly turn to a deep black colour when he is dead. He varies much in size (Figs. 6 and 7), both being male specimens taken in the same wood and on the same day; though one is nearly twice as large as the other. In fact, No. 7 is scarcely larger than *Lycæna alsus*, the smallest of English butterflies (see Fig. 9), of which we took only one specimen all through our long holiday. The caterpillar of *Alexis*, of a dingy green colour, dotted with white at the sides, feeds on clover and trefoil.

The upper and lower wings of *Lycæna alsus*, the Small Blue (Fig. 9) are of a dark dingy brown, faintly touched with blue; rather brighter in the male than in the female, lightly fringed with white; the under-side of the wings of both sexes being of a pale silvery grey, veined and dotted with very fine black. The caterpillar, of a pale green colour, streaked with yellow lines, seems to have been rarely found. This, the smallest and daintiest of English butterflies, is not often taken; probably escaping many a sharp pair of eyes by his small size and nimble zigzagging flight. He looks, indeed, very much like a tiny moth when flying, and is generally found near chalky downs or old lime-pits and quarries.

On our way home from the wood, Mary found in a bed of nettles several caterpillars of the Small Tortoiseshell, which the next day turned to chrysalides of a rich tawny brown (Fig. 5), here and there touched with bright gold.

The next day Alice came running into my room to tell me that she had picked up a large hairy caterpillar in the passage, who could scarcely crawl. I saw at once what was the matter with him. He had been without food for three days, having escaped from the box in which I had placed him, and was starving; having spent his time, I suppose, in crawling up and down the long passage. I carried him out into the garden, and set him on a spray of green oak. He began to eat *instanter*—devouring leaf after leaf, fibre and all, from the top to the very bottom. We watched him at short intervals for about two hours and a half; and, as far as we could make out, he never ceased eating. No wonder, therefore, that caterpillars are said to eat in one day thrice their own weight of leaves (see p. 157), and lucky for us it is that our larger creatures do not cram themselves after the same fashion. Three-fourths of their time, in fact, seem to be spent in eating; and their strong horny mandibles are never wearying of work. One thing, however, must be said for Master Caterpillar, and that is, that what he eats passes very quickly through him, and that now and then he may be put to hard shifts for food, to which Master Ox is not liable. He may get knocked off his bough into the long grass, and spend hours or days in finding his way back to his proper feeding-grounds. This very one had been certainly more than sixty hours without any food whatever.

All the smaller insects, indeed, appear able to fast without being much the worse for it. M. Vaillant tells us that he once kept a spider without food for ten months, at the end of which time, "though much shrunk in size, it was as vigorous as ever."

"I wonder," said Willy, when I told him this, "how M. Vaillant would have liked to be shut up in a bottle for ten months without food?"

Presently came a question which showed me that he was still thinking of our friend feasting on the oak-leaves.

"I thought," said Willy, "that if a caterpillar fell into the grass, he could climb up again by his thread of silk?"

"So he can, Willy, *if he has let himself down by the thread*; but if one of his own relations gives him a box on the ear, and knocks him headlong down on the ground, or a sudden gust of wind and rain snaps off the leaf on which he is dining, he has no time or chance to begin spinning, and so is driven to wander about among the grass, or in the dusty road, till he happens to meet with some bush or plant which he can eat."

The next day we made a trip to the Chalky Down on purpose to hunt for the only one of the Blues which we wanted—*Lycæna Corydon*, the Chalkhill Blue; and this we knew was to be found in abundance all along the edge of the down, and specially near a corn-field beyond the Barrows, where wild scabious, marjoram, and other flowers covered a strip of heathy grass between it and the road. We got there in good time, passing on our way dozens of smaller Heath Butterflies, Coppers, and Sulphurs, and at last reached our favourite hunting-ground. There we found the lovely pale Corydon, in such numbers that we might easily have filled our boxes. But it was intensely hot, and we therefore only took a very few choice specimens, among which Figs. 11 and 15 are rare beauties.

The Chalkhill Blue, or *Lycæna Corydon*, is a very lovely butterfly; the male having wings of a pale silvery blue, shining like moonlight, with a broad fringe of delicate white barred with black, and a band also of black at the edge, within the silver fringe (see Fig. 11). At the edge of the lower wings, too, is a row of black dots, circled with white (Fig. 11), and the whole body is covered with shining feathers of the same colour as the wings; so that, altogether, he is a very handsome fellow. The female is of a much soberer, quieter appearance; her wings being of a rich, soft, brown hue (Fig. 12), faintly shot with blue; the fringe being of a dull white, with a row of black dots partly circled with orange at the edge of the lower wings. In some specimens, too, in the centre of each wing is a small blackish dot (Fig. 12), sometimes faintly edged with white; but in many which we caught that day no such dots could be seen. The under-side, both of male and female, is of a lovely, soft, greyish-brown; bluish at the base, near the body (Fig. 13), fading off into grey, and brown towards the upper part of the wings, which are adorned with small black spots, and a row of orange dots near the lower edge. The caterpillar, which is of a pale green striped with yellow, feeds on the small trefoil, so often found on chalky downs.

You may fancy what numbers of Blues were out that morning, when I tell you that Mary, with one sweep of her net, took four perfect specimens, and a Brown Argus. All the Fritillaries that we saw were very ragged and battered; but not a glimpse did we get of a Clouded Sulphur or Red Admiral, though we watched carefully among the brakes of bramble at the edge of the down. We were more lucky when we got into the thick of the wood, on our way home, as you shall soon hear.

## VI.

"I DO not wonder," said I to Willy one day, in reply to a remark of his, "that you cannot understand how a crawling caterpillar, perhaps covered with hair, having sixteen legs, and living on green leaves, can ever change into a gay butterfly, with bright and beautiful wings, flying about among the flowers, and living on honey. It is altogether a mystery, like so many other of the wise and good things God has done. But, though it is so wonderful, yet we know it is true. I cannot explain it to you fully, any more than I can tell you how an acorn becomes an oak-tree, or a little tiny egg, filled with white and yellow fluid, turns into a golden-crested wren.

"All I can tell you is this. When the caterpillar is full-grown, and his time for change is come, he leaves off eating, and becomes very dull and sluggish. Sometimes he spins himself a little cocoon of silk, or silk and grains of earth, or creeps down into the earth, or makes for himself a warm smooth nest on the bark of a tree, or hangs himself by a thread of silk to the back of a leaf or stem. There he remains perfectly quiet and still for a time, and gradually changes into a chrysalis or pupa.\* Little by little the wonderful change goes on inside the shining case, and the crawling hungry caterpillar is slowly transformed into a bright swift butterfly on golden wings. The change, in fact, begins in a caterpillar before he turns into a pupa; and if the chrysalis be opened only a few days after it is formed, you will find some traces of the legs, the wings, and the trunk of the future butterfly, all folded and packed away in the neatest fashion, but so as to be of no use to the chrysalis.

"The time during which the butterfly or moth remains in the chrysalis varies very much in the different species. It may be a week or two, or as much as several months, depending partly on the time of the year when the butterfly is to appear, or on the cold or heat of the season. The skin of the pupa is very thin, and when the butterfly is quite ready to come out, a very slight motion of its body and wings will crack that skin. The crack soon spreads; other cracks appear; next the head of the prisoner makes its way through, and presently its whole body is free. At first its wings seem short and small and wrinkled, but by degrees, in the light and the air, they spread to their full size, grow harder and straighter, and are fit for use. Then the happy golden butterfly opens and shuts its dainty golden wings in the light, and, drinking in new vigour from the sunny air, starts forth upon its new life of enjoyment, roving here and there among the flowers, green leaves, and woods, sipping honey and making love through its little holiday of six weeks with untiring zeal; and as unlike a crawling caterpillar as it can well be.

"And when its little season of pleasure is done, it will choose a proper place and a proper plant, and there lay a store of tiny eggs, out of which

---

\* *Pupa*, a Latin word, meaning a girl, a baby; and so a name given to the baby butterfly in its case or cradle.

in due time will creep a brood of crawling, hungry caterpillars, each in due time to turn into the quiet chrysalis in its case of gold, or grey, or brown, or black; and each as duly at last into a bright and nimble butterfly. And so, Willy, year after year the mystery goes on, and these strange, ugly, crawling caterpillars are changed into creatures who spend half their time in the air, whose wings are covered with feathers that shine with all the colours of the rainbow."

"*Feathers?*" repeats Willy, as if in doubt whether he had heard aright. "I thought their wings were covered with dust."

"Yes, feathers. And on the first wet evening that comes this way we will get out the microscope, and you shall see the feathers plainly enough for yourself. But here we are at the Vicarage gate."

Now, about ten days before this, Henry and I had been hunting the broad sallows in the upper wood for a caterpillar or egg of the Emperor. For many days we searched in vain, but at last Henry found at the back of a leaf one large green clear egg, which we at once pronounced to be an Imperial one. It was, therefore, taken immense care of, covered up (leaf and all) under a wine-glass, and only brought out on rare occasions into the garden for air, when it could be watched. After a deal of watching and waiting, all that we could make out was that the egg changed colour, got darker, and had in the middle one tiny black speck, which was looked at every morning with profound interest, but looked at in vain.

One morning, however, there were great shouts heard in the drawing-room, "He's out! he's out!"

And so he was. There, on the withered leaf of sallow, lay what looked like half of a tiny frozen bubble, and close beside it a little thread of bright green, with a horn on his tail almost as big as his own body, creeping along at a great pace along the edge of the leaf. He looked like this:

Messengers were sent off at once for food of every kind likely to suit his Royal Highness: fresh sallow-leaves, willow, silver birch, and lettuce but, alas! all in vain. He did nothing but crawl about incessantly over the leaves, trying now and then to nibble them, and then giving it up in despair. This went on, to our great horror, for the whole day. The royal infant could eat nothing, and by the next morning had died of hunger, and shrivelled up into a mere bit of skin.

The terrible truth dawned upon us afterwards, that we had kept him too constantly under the hot air of the glass, and thus forced him to come out of the egg before his proper time, when he was unable to bite the leaves so carefully got ready for him, though he did try very hard at the lettuce.  After all, too, he was not a young Emperor, but a Poplar Hawk Moth caterpillar.

August the 25th was a blazing day, the thermometer at 106 in the sun and 85 in the shade.  Not a breath of air was stirring, not a cloud in the sky, as Mary, Willy, and I slowly crawled up the white dusty road towards the Emperor woods, stopping on our way at the Manor House, where one of our kind friends took us into the fruit garden at the back of the house.  There were few flowers, but almost at once I spied a butterfly.  It was a Red Admiral, which settled on the bough of an apple-tree, and was soon caught.  I only notice him here because he was the smallest specimen we ever saw, though perfect in every respect.  When stretched to the very widest extent, his wings, from tip to tip, only measured two inches, nearly an inch less than several specimens which we took during the following week.

After a rest in the cool pleasant garden, we made our way out into the open sun again, along by the edge of some ragged fields, by a huge pond, now almost dry, and through a noble avenue of elm-trees, under which we saw several large Wood Tortoiseshell butterflies (*Vanessa polychoros*) flying very swiftly from tree to tree; but we could not get near one of them.  Very slowly and wearily we made our way on across one or two corn-fields, and at last into the shady woods, scarcely seeing a butterfly the whole way.  The green winding wood paths, which a day or two ago were all alive with gay wings, now seemed deserted.

At last, however, we had traversed the whole wood in that direction, and came out into the turnpike road beyond it, from which we were not sorry to see the Vicarage just a mile off on the opposite hill.

Luckily we had a crust of bread and a flask of sherry with us; and so we sat down under a huge oak-tree at the corner of the road, and had a good rest.  It was a noble old tree, partly covered with ivy, and on the outer bark, about a foot from the ground, grew the largest fungus we had ever seen.  It was of a dingy yellow colour, spotted like a tiger-skin, as large as a dinner-plate, fringed at the edge with scales like small oyster-shells, and thickly studded with large drops of sticky moisture, which glittered in the sun and streamed down the bark to the ground.

"Just the very place," said I, "for a butterfly to come and drink."

The words were scarcely out of my mouth when a noble Red Admiral skimmed swiftly over some low bushes, and after circling swiftly round the stem of the tree, settled quietly on the fungus. In a trice he was safely in Mary's net, and hardly secured when a second splendid specimen flew swiftly to the very same spot, to sip of the same dainty spring. This one I tried for, and stupidly missed; but a third, which came about five minutes later, I caught; and both specimens are now in our collection.

The Red Admiral (*Vanessa Atalanta*) belongs to Family V., the VANESSIDÆ, or Angle Wings, and is one of the swiftest and most brilliant of English butterflies. The base-colour of all the wings is a rich, soft, shining black, the upper wings having an irregular band of scarlet running across them, and between this band (Fig. 1, page 180), and their outer tip five or six irregular spots of snowy white. The base of the lower wings is also barred with a scarlet band at the margin, through which runs a row of black spots, ending in a larger blue one at the inner edge. The upper wings are slightly angled and the lower ones scalloped at the edge, which in most specimens is tipped with white. On the under-side (Fig. 2) the scarlet band and white spots reappear on the upper wings, while the lower ones (Fig. 2) are marbled in a most lovely manner with tints of soft grey, brown, or bronze, edged with waves of the same colours of a lighter hue and faintly tipped with white.

The caterpillar, of a dingy greyish colour, covered with short spines, with a yellow line along each side, feeds on the nettle, where it may be found in June and July, and sometimes later. The chrysalis, hanging head downwards from the stem of the nettle inside a web, may often be found in July and August.

As we went lazily homewards down a dry stony road, with high grassy banks on either side of us, we managed to catch a fine pair of Painted Ladies, as graceful and elegant a butterfly as the Red Admiral, but far more difficult to catch. This is also a *Vanessa*, and in size, shape, and flight much like her brilliant cousin, and as she feeds on the thistle gets the name of *Vanessa Cardui*.* She is more difficult to catch because she far more rarely settles, and when she does so is just as likely to choose the ground as a flower. *Vanessa Cardui* belongs to the family of Angle Wings, though its wings are scarcely angled at all, but scalloped and fringed with great beauty. The tip of the upper wings (Fig. 3) is of a dark black colour, with five or six white spots near the upper edge, the middle and lower part of the wings being of a delicate rosy orange hue, spotted and

---

* *Carduus*, the botanical name of one family among the thistles.

barred with black, as indeed are the whole of the lower wings (Fig. 3), with a treble row of black dots of different shapes near the outer margin. The down on the wings of some specimens is of the very richest orange colour on the under-side of the upper wings (Fig. 4), the black colour is paler, and the tawny pink hue more widely spread, while the lower wings are mottled in the loveliest manner with grey, brown, yellow, and buff-colour, all of the most delicate tint. This *Vanessa* well deserves its name of Painted Lady. It is a very local butterfly, and a single specimen will often haunt one little piece of stony road for weeks together, when driven away coming back again in the course of a short time without fail. The caterpillar, of a dark brown striped with yellow, and covered with spines, feeds on the thistle, the nettle (?), and the mallow, and may be found in a little web of its own spinning among the leaves in June and July.

The Painted Lady is one of those butterflies which are abundant one year and very scarce the next.

In the shady lane leading out of the wood, too, we caught a couple of those quiet little, yet nimble butterflies, the Wood Argus, of which we had seen many specimens, but had as yet taken none. The Wood Argus, or Speckled Wood (*Satyrus Ægeria*) belongs to Family VII., the SATYRIDÆ, and has not much (see Fig. 5) to recommend it in the way of beauty.

The wings are of a dull dingy brown, the upper ones spotted with irregular dots and patches of yellowish-white, with one round black dot having a white speck in the centre near the higher edge, the lower ones having a row of similar dots (see Fig. 5) edged with white along the lower edge of the wing, which is slightly scalloped. Oddly enough, the eyes of this butterfly are said to be hairy. The male is smaller and darker than the female, and less good-looking.

The caterpillar, of which there are several broods in the year, of a palish green colour striped on the side with white, feeds on several sorts of grass, and may be found in March, May, and June. The butterfly is a common one in all woody districts.

As we went home that afternoon we saw, as we had done on many other sunny days, darting along in a buzzing, irregular flight from flower to flower, a very nimble little brown-winged creature (Fig. 6), which at first we could not clearly make out to be either moth or butterfly. He was feeding very busily on honeysuckle and other flowers full of honey, and, swift of flight, was very hard to catch. But we secured one at last, and as he was a very bright and perfect specimen, kept him among our

butterflies. He turned out to be a little moth of the kind that feed by day.

By this time we had once more reached the Vicarage, and glad enough of a good long rest after our weary tramp in the dust. The next day, and the next, the heat and the dust were more intolerable than ever; but as our holiday had now all but come to an end, we determined to have one more good ramble through the woods, and try for another sight of the Emperor. Our first day's walk was through Loudwater Wood, three or four miles in extent, full of green winding paths, beds of fragrant flowers, and noble trees of beech and oak. It was a famous place for butterflies, but on that day we saw nothing rare or good, though we searched far and wide. The only thing we brought home, in fact, was a couple of Green-veined Whites, which Willy's sharp eyes spied among the long grass by the roadside, where he caught both. They were as perfect as if only just out of the chrysalis, so we kept them, though both were females.

*Pieris napi*, or the Green-veined White, which is found in every green lane and roadside that is fairly out of the reach of the smoky town, belongs to Family II., PIERIDÆ; but, though common, is an elegant little butterfly. The upper wings, of a soft greenish-white, are tipped at the upper edge and spotted with black, and veined with streaks of greenish-grey (see Fig. 7). The male is easily known from the female by having no spots on the lower wings, and only one on the upper. On the under-side both (Fig. 8) male and female are much alike, the upper wings being of greenish-white, through which both nerves and spots show from the upper side; the lower wings being of a soft creamy yellow, veined (Fig. 8) with green. The caterpillar, of a pale dingy green, dotted faintly with red and yellow at the side, feeds on the cabbage and the leaves of the radish.

But though we got few butterflies that day, we got a good glimpse of an Emperor, though he was far beyond the reach of our nets; as well as of another gentleman, to Willy a still greater stranger. We had just crossed a wide open brake, where a colony of young pollard oaks and ashes had been thinned and stripped of underwood, when on suddenly turning round, I saw what looked like a good-sized reddish-brown dog creeping among the bushes, and jumping now and then over little clumps of bramble.

"There goes a fox, Willy," I cried. "Look! just beyond the hurdle on the ground, by the stack of wood."

He had just time to catch a glimpse of Master Reynard, who, startled at the sound of voices, went away at a great pace into the copse. He was looking after the young pheasants. Presently we came to a grove of lofty green oaks and beeches, and all these we watched with longing, eager eyes, hoping for a glimpse of an Emperor; but for a long time in vain. All at once, however, midway between two tall beeches, but fifty or sixty feet above the taller of them, sailing round in wide circles, or darting upward with a sudden spring through the bright sunshine, we made out against the clear blue sky a large butterfly that we first took to be a Swift. But after watching him for some minutes, we made out quite clearly that it was a Purple Emperor, roving to and fro over his sunny kingdom in the heart of the summer woods. And that was our farewell glimpse of his Imperial Majesty.

The next day, after a long ramble through the lower wood, where Blues, Hairstreaks, and common Skippers were still to be seen in abundance, oddly enough, we took our first perfect pair of Copper Butterflies, which were flitting about in the sun like sparks of burnished gold.

The small Copper, *Polyommatus phleas*, belongs to Family VIII., LYCÆNIDÆ, or Argus Butterflies, all famous for their beauty, and mostly for some touch of brilliant colour. The upper wings are of a bright golden copper colour (Fig. 11), especially in the male, edged with dark brownish-black, with eight or ten black dots scattered across the centre. The lower wings, scalloped and tailed at the lower edge, are of a rich dark brown, with a margin of bright copper. The female (see Fig. 10) differs from her husband in being rather smaller, and wearing much darker, soberer colours on her upper wings, and on the lower a row of four blue dots just inside the copper margin. Underneath both are much alike, the upper wings being of a pale coppery brown, spotted with black, and the lower ones of a tawny buff, speckled with rows of golden brown dots, with five small ruddy crescents close to the lower edge.

The Copper is a most pugnacious little butterfly, and seems to be always making fierce love or fierce battle among friends or enemies, whom he chases far and wide with nimble wings. Nearly every specimen we took was caught in full chase of some wandering Blue, Argus, or Skipper.

The caterpillar, small and of a pale green, feeds on the wood-sorrel, and as there are two or three broods, may be found from March to July wherever such plants abound.

This was, in truth, the last of our Hampshire butterflies; but yet I have one more specimen to add (the Small Meadow Brown), which deserves special mention. He was quite perfect in every respect—his antennæ unbroken, his wings without a flaw (see Fig. 9), but almost every particle of colour seemed to have been washed away, or dried up with the heat of the sun. His wings (Fig. 9) were almost transparent, nearly white, edged with a band of pale dusky brown.

"He looks," said Cecil, "as if he was just recovering from the influenza."

And so he did; but whether it was a low fever or a stroke of the sun that had befallen him, he was altogether the oddest little butterfly we had ever met with, as I think you would say if you now saw him in our collection, among all his gaily-dressed companions.

Out of the sixty butterflies which we had counted up (see p. 122) at the beginning of the holidays, as possibly to be met with in England, we had actually caught two-and-forty distinct species, making altogether a grand total of just three hundred specimens; so that we now have plenty to spare for a friend, or to exchange.

And now, before we say good bye to all our kind young friends who may have followed us in our butterfly rambles thus far, we must notice a few points about which we talked during the few wet days that kept us indoors, or that the microscope taught us in the evening.

All butterflies have four wings, which are covered with very tiny scales or feathers, which to the naked eye look like fine dust, and when touched come off upon the finger. Scales is the true name for them, because they are arranged on the thin membrane of the wing like scales on the skin of a fish, so that those of one row partly overlap the next. They all look alike to us when shaken off from the butterfly's wing, but under the microscope are found to be of very different shapes and even sizes; and from these tiny scales shine out all the lovely colours of the azure Blue, the royal Emperor, the golden Sulphur, the purple splendour of the Peacock, and the fiery gold of the Copper, as well as all the sober greys and silvery browns which help to make up the beauty of the butterfly world.

Here is some dust, which we took from various butterflies, as it looked under the microscope, when magnified about 300 times (see next page).

Group No. 1 are from the under-side of the large white butterfly, and semed to be pretty much alike in size and shape; Group No. 2 are from the upper side of the Peacock; No. 3 from the wing of the Chalkhill Blue, which seemed to be almost all of a rounder, shorter form; while Group No. 4 were taken from various wings, and mostly of the finer dust. Some we found to be notched with three or four teeth at the upper edge, but the commoner number was five, and rarely six; while a very few, like one or two of Group 4, ran off into single or double sharp points.

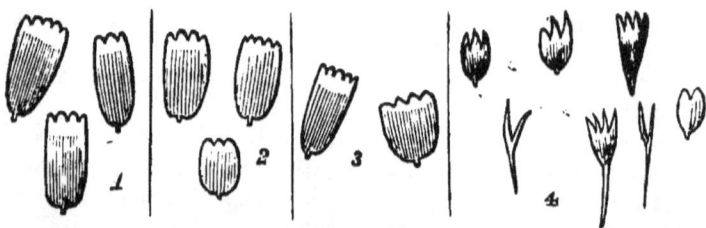

By a very ingenious contrivance* with the microscope, we were able to cast a shadow of the scales, when magnified, on to a sheet of paper, and so to sketch them in their true exact size.

The eyes of the butterfly, which we were not clever enough to prepare for the microscope, are as wonderfully formed as the scales on the wings. Each eye is made up of about 17,000 facets, each of which is so contrived as to catch the rays of light on its centre, and thus serve all the purpose of a separate eye, though they are all joined together in one orb; just as the little facets cut on the surface of a diamond are all on one shining stone. The face of the whole eye being globular, the butterfly is able to see in almost all directions; and thus it is that he is so wary and so difficult to catch, even when settled on a flower.

The colour of the eyes in different butterflies differs very much. Those of the Meadow Brown seemed to be of a dingy grey, but when dead of a deep brown; in the silver-washed Fritillary of a pale clear green when living, of a nut-brown when dead, as they were in the large White; while those of the Sulphur were brown both when living and dead; and of the Harvest Blue, black in both cases.

But look where we would throughout the butterfly's whole framework —legs, wings, antennæ, and trunk—everywhere were found traces of the same Divine wisdom, skill, goodness, and beauty; every part exactly and perfectly fitted for its exact work, and no part without some special use. The more deeply we looked into any part, the more perfect we saw its construction to be, the more full of grace and beauty, the more lovely and wondrous in the eyes of man, the more worthy of the great and all-wise God who has made everything beautiful in its season, and able to praise and magnify its Maker by a life of joy and obedience, and perhaps

---

* The neutral tint reflector.

with sounds and songs of praise which none but the ear of the Almighty can hear.

Our six weeks' holiday among the woods and green fields had taught us many things which we hope that all our young readers will soon learn for themselves. It taught us how little we really knew, and how much there was to learn, about the living creatures that haunt the fields and woods; that of all books, next to the Bible itself, the Book of Nature is by far the grandest, fairest, and wisest; that every page in it is full of beauty, joy, and truth—truth so simple and so clear that the child may read it, and wisdom so deep and noble and lofty that the wisest man cannot learn it all. That, turn where we would, everything was still "very good," as Our Father first made it in the glorious garden of Eden. That the more we learned, the more eager we were to get fresh knowledge; and all the lessons we learned were a pleasure; the fresh air and exercise bringing us new health and strength and vigour. And so in the country lanes, by sunny banks, on the open downs, and in green woodland paths, we rambled happily on, finding pleasant work both for mind and body; learning to be patient and to persevere, to be accurate in marking the difference between things nearly alike, to use our memories, to think, and to judge, and yet to have a jolly holiday.

And so, at last, our six weeks "among the Butterflies" came to an end, and we went back to smoky London and our regular work all the better, stronger, and happier than we had been for many a long day.

## POND LIFE.

IN the depths ot a pond, whose still waters are ruffled by no tides or currents, thousands of little lives are brought into being, nursed through infancy, and supplied with a dwelling-place for their brief existence. Weeping willow-trees wave their long pliant arms above this quiet insect home, graceful water-lilies cradle their waxen cups upon its surface, the pretty duckweed and starwort strew the waters with their petals; and on the banks, among the reeds and rushes, grows the gentle blue-eyed forget-me-not, and there also the wild marsh-marigold, which

"Shines like fire in swamps and hollows grey,"

blazes resplendent mid the soberly-coloured horsetails; everywhere flags, osiers, and tall feather-grasses are mingled in nature's beautiful confusion with the gorgeous purple iris and the willow-weed.

On sunny days what merry gambols take place on the surface of this pond, what wonderful changes and transformations occur beneath the shadow of those lily-cups! Gnat boats float about with their precious burden of embryo life, little gnat larvæ wriggle and dart among the stems of the water-plants below. The Great Water-Beetle spreads dismay into the heart of the small fry of the pond, and tyrannizes over another aquatic giant, the Black Water-Beetle, who in spite of his great size is of a mild amiable disposition. This beetle, like most other aquatic species, is very fond of still ponds, where it may be found in the egg, larval, and perfect states. The female encloses her eggs in a beautiful silken covering. For this purpose she possesses organs most unusual in beetles, and quite peculiar to this species; namely, abdominal glands secreting a glutinous fluid which can be spun out into silken threads. From these spinnerets she draws forth delicate white silk, and weaves it into a cocoon shaped very much like a turnip, with an upright projection on one side: the exterior soon hardens, and being covered with a gummy surface, is perfectly water-tight. When about to construct this aquatic nest, the beetle fastens herself, head downwards, upon some water-plant, and expelling her eggs, encircles them with the silken threads, which she attaches to the stem or leaf on which she has taken up her position. The number of eggs thus enclosed varies from fifty to sixty. There the larvæ are hatched, which is said to take place in warm weather in little more than a fortnight, although as a rule six weeks elapse ere they are excluded from the eggs. They forsake their common nursery within a few hours, and begin to grow with great rapidity. They are long grubs with a horny head; the end of the body having two little tail-like appendages, which form part of the breathing apparatus of the insect, and through which it inhales its supply of air. To obtain this the larvæ constantly rise to the surface of the water, where they also pursue small molluscs and insects. For the most part they creep about among the roots and stems of aquatic plants, seizing their prey by suddenly poking back the head, which places the curved jaws in a highly favourable position for entrapping and crushing the tiny shell creatures that form their chief food.

The Black Water-Beetle remains in the larval state about sixty days, during which period it casts its skin three times, increasing in size after each moult, till it finally attains the length of three inches. When it is full fed it quits the water, but does not stray far from its native element;

it usually selects a soft spot in the adjacent bank, into which it burrows by means of an organ situated at the end of its body, and, forming an oval cell in the earth, there assumes its pupa shape.

In its perfect state the Black Water-Beetle is a handsome insect of a blackish-olive hue, the elytra (wing-cases) incline more to blue, and the breast is covered with a yellow down. It is a truly aquatic species, residing entirely in the water, although it may now and then be surprised while seated on the leaf of a water-plant surveying the world around in solemn grandeur. In the evening, however, like the Great Water-Beetle, it flies about on land. The full-grown beetles seem no longer to prey on other insects, but satisfy themselves with a light vegetable repast, and in spite of their large dimensions they are said to fall victims to the Great Water-Beetle, which is equally active and voracious in its larval and perfect condition.

In tranquil spots near the banks the merry Whirlwig Beetles congregate in little companies and spin their ceaseless rounds, glancing like crystals on the surface of the water. The structure of the most familiar of these beetles (the Common Whirlwig) is marvellously interesting. Destined to live in and on the water, it is at the same time dependent on atmospheric air. The exigencies of its existence, therefore, demand that it should be adapted for an aquatic life, and that it should also be sustained by an abundant supply of oxygen. To enable it to perform its gambols on the water it is furnished with limbs peculiarly serviceable for the purpose. The Whirlwig (in common with all true insects) has six legs. The fore pair are of considerable length in proportion to the others, and are consequently of little use in swimming, although they are employed most effectually in the seizure of prey; but the middle and hind pairs of legs, which are short, broad, and flat, and fringed with stout hairs, form admirable oars, and assist the little creature to progress through the water and revolve upon it with a velocity which is all the more astonishing as the locomotive apparatus is entirely hidden from sight by the horny elytra.

The eyes of insects are usually compound, and consist of numerous facets, reflecting objects at various angles and distances, but immovably fixed and incapable of direction. The eyes of the Whirlwig are perhaps the most curious feature in its anatomy. Each eye is highly compound and divided horizontally into two parts by a horny partition, so that the insect has practically four eyes, two situated on the upper portion of the head, giving timely warning of aërial dangers in the shape of birds and entomologists, and two on the under surface, whose vision is directed to

the depths below. The extraordinary efficacy of these organs and the quickness of sight of the Whirlwigs may be attested by attempting to catch them with the naked hand; it is almost impossible to perform the feat, as the tiny creatures dart hither and thither, dive, and disappear before the hand has been approached within six inches of them. When the Whirlwigs take refuge thus below water, they carry with them a store of air which may be seen adhering in the form of a tiny glistening air globule to the hind part of the body. A little tuft of hairs is disposed at the extremity of the abdomen, in which the atmospheric air becomes entangled while the insect is above water, and among which it remains when it retires below until absorbed by the air-tubes. These tubes are the respiratory apparatus and convey the air through the whole system of the insect, the openings by which the oxygen is admitted into them being situated between the segments of the abdomen.

The life history of this species does not differ materially from that of other aquatic beetles; the eggs are deposited on some water-plant; in the course of a week or so the larvæ—long grubs bearing a marked resemblance to centipedes—are excluded. When the moment arrives for assuming the pupal form, the larva crawls up the stem of a plant for some distance out of the water, and there spins itself a small silken cocoon, in which it awaits the critical hour of its new birth.

Looking down to the bottom of the pond we may behold little objects, apparently strips of leaf, dried straws, and shell-covered tubes, moving about with every evidence of life. These quaint things are the larvæ of Caddis-flies bearing about their portable homes. In these strange dwellings they remain concealed from view, with the exception of the head and the three first rings of the body, which protrude from the entrance of the cylindrical abode. Réaumur, who devoted much time to the study of the habits of the Caddis-worm, gives the following account of their mode of constructing their cases:—

"The body of these grubs is lodged in a silken tube, the inside of which is smooth and shining. On the exterior of this case are fastened fragments of different materials, serviceable in strengthening it, and necessary to make the garment complete and to give it useful qualities. The elegance of the external form seems to be a matter of small moment to the Caddis-worms: the appearance of some of these tubes is most grotesque,—the outside is often bristling and full of inequalities. Others, again, make themselves garments which have a neater air, the pieces which compose them being symmetrically disposed one upon the other.

They change their garments whenever occasion demands it—that is to say, when the tube they occupy becomes too narrow and short, they make themselves one of more suitable size."

For this work they employ every variety of material, and with the happy faculty of "making the best of everything," adapt it to their requirements with wonderful ingenuity. Leaves, scraps of withered foliage, stalks of plants, small sticks, slips of straw, roots, tiny pebbles, shells of aquatic molluscs, and grains of sand are all stuck together without the slightest semblance of arrangement. When the little abode is as yet incomplete, should the Caddis-worm alight upon an object likely to add to the solidity of the tube, it is appropriated with all speed.

"A Caddis-worm occasionally finds two scraps of the stem of a reed bruised and split lengthways, and if it has as yet only adorned its sheath with small materials, so that it is neither strong enough nor sufficiently large, the Caddis-worm makes a very nice little overcoat with the two slips of reed it has had the good luck to discover, and which it can adjust without much difficulty. It places the tube in the two reed fragments, which it then draws together as closely as possible."\*

Then, again, there are Caddis-worms who decorate the outside of their tubes with the shells of fresh-water mussels and snails; and sometimes, with complete disregard for the feelings of the owners, these shells are attached to the sheath while the occupant is still alive, and so firmly are they cemented that no effort of the creatures enables them to remove themselves or to regain their freedom, and in this ignominious manner are they compelled to move about at the will of the insect into whose power they have fallen. Though the Caddis-worm seems so indifferent to the external beauty of its residence, it has sense enough to attend to a far more important particular. It is endowed with a wonderful instinct, which teaches it to avoid making the specific gravity of its dwelling greater than that of the surrounding element, as otherwise the freedom of its movements might be materially hindered. As it is even, it progresses at no great rate in the water.

When at rest, it retires completely into the tube; but when it wishes to creep about among the stems of water-plants, out pops the little head and the fore part of the body, and, clinging by the six fore legs, it drags its little home along with it. The portion of the body enclosed by the movable tent is soft, like that of a caterpillar, but the head and shoulders, which are exposed to possible collisions with hard substances, are pro-

---

\* Réaumur, *Memoires pour servir à la Histoire des Insectes*, tom. iii., p. 157.

tected by a horny integument. When the larva is about to metamorphose into a pupa, as if conscious that it is about to pass into a defenceless condition, it weaves a stiff stout web across the open ends of its little habitation, and in this quaint cylindrical tube it enters on a dreamless repose and patiently awaits its entrance on a brighter and merrier life. In due time it emerges from the case, and assumes the perfect form of a beautiful four-winged fly. There are different modes of performing the critical feat of bursting from the pupa-case. The larger species, such as the Great Caddis-fly, completely forsake the water, and climbing up the stems of aquatic-plants, fix themselves firmly thereto, and on this support await the splitting of the pupa-case; whereas the smaller species seem to undergo the final transformation upon the surface of the water, and, after the manner of gnats under similar circumstances, use the cast-off skin as a boat on which they can steady themselves till their wings have grown firm and steady.

These pretty Caddis-flies, with their beautiful hairy wings, flit about over the surface of the pond. The female carries her eggs about at the extremity of the abdomen; they are covered with a glutinous fluid, by means of which they are fastened to the stem of some water-plant. The mother-fly is very daring in her anxiety to deposit her egg-cluster in an advantageous locality: she will descend under the water, no longer her native element, and if disturbed in the act of ovipositing, will swim from one stalk to another. The most popular of these insects are the Great Caddis-fly and the Lesser Caddis-fly.

On fine summer days the quaint little Water-boatmen may be observed sitting motionless for an hour at a time, sunning themselves on the surface of the water. The form of these insects resembles that of a boat, the head being almost as broad as the rest of the body; the fore legs are strong and curved, and well fitted for entrapping aquatic insects, on which the boatmen chiefly feed; the hind legs are greatly elongated and feathered at the tips, and act as oars, which propel the body along at a high rate of speed. If the little creatures are frightened while basking in the sun, they give one stroke of their long paddles, and disappear into the depths of the water. Their exquisite wings are also very strong, and enable them to fly about on land with perfect ease, but their attempts to walk on the ground are exceedingly ludicrous.

They always swim on their back, a position which offers great facility for the capture of small insects that may fall into the water; they do not eat their prey, but, seizing it with their fore legs, drive their powerful beaks into its body, and after sucking its juices, leave it floating on the water an empty husk. The eggs of this insect are generally affixed by the parent to the stem of a water-plant. They are soon hatched, and the young larvæ launch forth into the water. The Water-boatmen pass through the usual stages of insect life, but these stages are not distinctly marked as in the beetle and butterfly, for the larvæ and pupæ in form strongly resemble the perfect insect, except that they are without wings, which appendages characterize the full-grown insect.

The Water-scorpions are very crocodiles of the water kingdom. They are inactive creatures, with a flat leaf-shaped body, having the two fore legs strongly incurved and adapted to their predatory instincts. Two hair-like protuberances are situated at the end of the body, through which the supply of oxygen is drawn in; and as they breathe atmospheric air, they are constantly obliged to resort to the surface to obtain it. They crawl along the edges of streams and ponds, seizing the insects that pass by. They also lurk among the stems of aquatic plants, and, clutching the larvæ of other insects in their strong claws, drain them of their juices, after the manner of the Water-boatmen. The Common Water-scorpion is the most familiar of this family.

An allied species, which has no familiar name, but whose scientific title is *Ranatra linearis*, is an exceedingly quaint insect, and very active in its motions on the water. It is highly amusing to watch one of these creatures skimming about by means of four spider-like legs, holding the two fore legs, developed into powerful claws, in readiness to snap up its terrified victim, who is scuttling away in the most precipitate manner. It feeds for the most part on the larvæ of the May-fly, and little beetles and flies that fall into the water. The *Ranatra* has wings fitting closely to its side, which serve to bear it on overland expeditions.

Numbers of insects frequent the surface of the water which are yet incapable of living below it, although they may occasionally dive for a moment. Among these are the Water-gnats, so called from their gnat-like bodies and long slender legs. The most common of these is the Water-measurer, who glides gracefully along, appearing to measure the water with its legs as it proceeds. It is often found in company with a more active species, the little *Gerris locustris*. This lively insect skims upon the water with great speed, darting hither and thither and turning about with much ease. The Water-gnats feed on aquatic insects, which they grasp in their fore legs, and holding them tightly, plunge their long suckers into their bodies, and thus extract their juices, but they in their turn constantly fall victims to the Water-boatmen, who treat them in a manner no less barbarous.

The best time to become acquainted with the infinite richness and variety of Pond Life is towards the middle of fine summer days. Somewhere about noon, Dragon-flies, May-flies, Gnats, and Chameleon-flies rise to the surface of the water and undergo their final metamorphoses into perfect creatures.

The Chameleon-fly is a beautifully-coloured insect with a soft velvety black body adorned with bright yellow patches. It belongs, like the gnats, to the order *Diptera*, one that comprises two-winged flies of various shapes and sizes. The Chameleon-fly arranges its eggs on the under-side of the leaves of water-plants, on the upper surface of which it greatly enjoys to remain basking in the sun's rays. When the larvæ are excluded they swim about in the water, after the manner of the larvæ of the gnat, to which they bear a striking resemblance. The body is elongated and worm-like, having the last segment or tail lengthened into a tube, the extremity of which is fringed with thirty hairs disposed in a circular form and looking like a feathery star.

Through this tube the insect inhales its supply of oxygen, and the hairs are capable of being folded up to retain the globule of air thus drawn in. It remains for the most part, like the larva of the gnat, poised head downwards near the surface of the water. The head is horny, and the mouth is bearded with many small hairs, which are kept in constant motion for the purpose of entangling minute insects, which are then seized and crushed by two small hooks with which the mouth is furnished. The larva of the Chameleon-fly does not shed its skin to pass into the perfect state, but its own skin hardens into a case, inside which the pupa is developed, and as the insect in this condition is much smaller than it was as a larva, there exists round it a vacant space occupied by air. This causes the puparium to be specifically lighter than the water, consequently it floats at liberty on or near the surface. In this tiny boat the pupa reposes, borne along among lily-cups and forget-me-nots by gentle summer breezes,

"From morn till eve, from eve till dewy morn,"

till the term of its slumbers has expired, when, if its juices have not been sucked by Water-boatmen, or it has not been bodily devoured by the Great Water-beetle, it rises in beauty a gorgeous fly, shining with metallic lustre, to bask in the cups of flowers and suck their honeyed sweets.

* *Metamorphoses des Insects*, p. 648.

Blanchard, the French entomologist, says, "the Chameleon-fly is the most common in our country. In its perfect state it haunts flowers, where it finds insects whose juices it loves to drain." But most English entomologists seem to think that it is satisfied with the honey of flowers.

Under and above the water, Pond Life abounds with insects in every stage of development; eggs in soft cocoons, or agglutinated to the stems of water-plants, larvæ eating, growing, gambolling, or trying to escape from the rapacious and hungry attacks of their foes; pupæ patiently waiting in their little cases till their perfect form has been attained and they can burst forth to play their part in the aquatic republic. And then the perfect insects in all the happy exuberance of life; gnats and May-flies humming, Chameleon-flies buzzing, beetles booming as they rise on the wing, Water-boatmen paddling, Water-gnats washing and rubbing themselves with their long legs, Whirlwigs gyrating, all predacious insects darting, hunting, and leaping with eager haste and speed; and those more interesting species who busy themselves in the construction of wonderful dwellings, afford unending amusement for long summer hours. By means of the water-telescope, the proceedings of these submarine toilers may be watched with great satisfaction. The Water-spider can be seen spinning its cocoon, attaching it by slender filaments to some plant, and then with untiring perseverance making journeys to and from the surface to obtain air-bubbles with which it fills its tiny home. On the margin of the pond, among many other semi-aquatic creatures, the Raft-spider may be seen collecting dried leaves, twigs, and herbage, and weaving them into a little craft, on which it sails about with sinister intentions towards the welfare of the pond community.

Thus so many lives are perfected among lily-cups and blue-eyed forget-me-nots,—so many minute hopes, fears, and instincts called forth within the narrow boundaries of a tranquil pond.

## INSECT PETS.

I AM usually a good-tempered person: romping girls and frolicsome boys never ruffle me. I can endure to have my furniture scratched and my old china endangered without being cross, and although an old maid, I can see my dogs teased and my cats insulted with only mild protests that the torment shall cease before it becomes cruelty. But (there always is a *but* or an *if* to everything) one kind of folly and cruelty completely destroys in me every remnant of patience—it is the way insects are treated by boys and girls, and I am sorry to add by grown-up people too.

In attempting to introduce my pets to little folks, and to promote their future friendship with beetles and their allies, I could wish for a fairy pen and pencil to do justice to the subject, so that every unfounded prejudice might be swept away. Of course, I do not expect any one to pet wasps or pick up bees, but I cannot see why a fly should be screamed at, buffeted, and injured because its unwieldy destroyer does not know what the harmless creature is. I suppose that almost every one knows a moth from a butterfly, and the bees from the wasps; but that there are many flies mistaken for these insects is perhaps not so well known. Many books have been written about butterflies, wasps, and spiders, but I have not seen any notice of the beauty and intelligence of the largest insect tribe known as Beetles. If the days of spinning cockchafers were over, I need not have made this little effort on behalf of my pets; but my experience of boys and girls prompts me to attempt such a slight sketch of the insect world as will enable any ordinary person to know a fly from a wasp or bee, and a beetle from a cockroach — mistakes which are as common as the little creatures are familiar. The first fact to recognize is rather startling, Spiders, wood-lice, and centipedes are not insects. The division *Insecta* contains only living beings who develop six legs and two or four wings, a distinction which will be readily recognized by any one who calls to mind the small round body and legs of the spider, and the heavy armadillo-kind of covering that clothes the creeping wood-lice. It is so difficult to contend with the general dislike to these little beings, that many people have no idea a spider has more legs than a bee or a daddy-long-legs, and perhaps I may not succeed in tempting any child to take more interest in a fly than can be evinced by pulling off its legs. This is, however, only a boy's trick, and is less trying to me (whatever it may be to the fly) than the senseless screaming of young ladies at a harmless cockchafer or an active beetle who may unwittingly disturb their gardening operations. Who but an entomologist ever stops to consider whether the intruder has two wings or four, six legs or eight, and whether it be injurious or useful? Few people, I imagine, or this simple knowledge would long since have been insisted upon in our educational system, as more practically useful than Algebraic Equations or the Latin Grammar.

It is really ridiculous to any one who is willing to take the insect side of the battle, to hear how the farmers and gardeners grumble because the wire-worm eats the barley, or the grub spoils the roses or eats off the China asters. One can only say to the tiny enemies, "Go in and win, my dears, because although you injure them, they despise you too much to take common precautions to prevent your living upon them." Nothing is more exciting than contempt, and I dare say this urges the little creatures to obtain their living where they can. In Paris there was once an exhibition of useful and injurious insects, and, as well as we can by the help of pictures, we intend to reproduce it here. To be of any use the broad distinctions between great divisions must be first recognized by comparing insects that are easily caught with the drawings and descriptions given, and then special notices of the familiar species of each Order will teach a fair amount of useful insect knowledge.

There are nine distinct Orders of insects, which are divided into the Biting Insects and the Sucking Insects.

Into these two divisions the most opposite habits and characters are introduced. Amongst the biting insects there are many which feed only on vegetable food, but whose preference is for leaves and petals rather than the nectar which is sought by sucking insects like butterflies and moths. In the second division there are many insects who have an intense craving for animal food, and there are many of us who have experienced the torture of the so-called "bite" of the bug or flea.

The habit of the insect in taking its food has caused great importance to be attached to the mouth parts, and although we shall not attempt to notice minute differences, we will describe simply the construction of the two kinds of mouth. Any one who has watched a silkworm feed has seen that its two jaws or mandibles move in an opposite direction from those of animals. With us the lower jaw moves up and down against the fixed upper one. But the insects move both jaws, and send them outwards and inwards, or from outside to inside, instead of up and down.

In beetles there are other jaws, an inner pair (the maxillæ) to which are attached small instruments for softening or grinding the food, a lower lip with two more instruments attached to it, and an upper lip which generally conceals the mouth when looked upon from above. This form which I have mentioned, the most familiar to illustrate, is entirely altered when the caterpillars develop into moths and butterflies. Perfect beetles (not including the weevils), earwigs, grasshoppers, and cockroaches have these

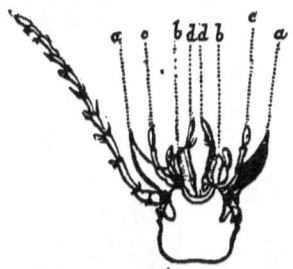

*a*, Mandibles, or large jaws ; *b*, maxillæ, a smaller pair of jaws ; *c*, palpi, two small attachments on the maxillæ ; *d*, palpi of lower lip.

biting mouths. In sucking insects the difference is something like this. The lower lip is drawn out to a great length, and turns up at the sides to form a groove for the other parts of the mouth to rest in. The two mandibles and the two maxillæ are drawn out into delicate lancet-shaped organs, and rest in this long lip. The four palpi are generally wanting, which may be accounted for by the fact that liquid food requires less preparation before being transmitted to the stomach, and these extra organs are not needed.

In the long and beautiful proboscis of the butterflies the mouth is enormously prolonged, but by a most delicate arrangement is spirally folded up when they are not feeding. It is easily uncurled and examined by the help of any small pointed instrument.

Some beetles have enormous mandibles, which seem to be adapted for catching and holding their prey rather than for mastication. In my collection there is a fine Longicorne who was killed just as he had captured a large fly; he still holds it in his mandibles, and the inner jaws or maxillæ

198  THE BOYS' AND GIRLS' BOOK OF SCIENCE.

are closing on the body of the victim. Even in death this strong creature did not release his hold, and his end was a benzine bath.

The English stag-beetle uses his mandibles generally to bite off leaves and bore the wood of trees for the purpose of depositing its eggs. It is supposed to be a vegetable feeder, but I once saw one of these insects with a caterpillar between the jaws, so I think they will eat anything which it is possible to grasp.

As I have said, there are nine Orders of insects, separated into two sections—those who have a mouth with jaws, and those who have a mouth with a sucker or proboscis.

| Biting Insects. | | Sucking Insects. | |
|---|---|---|---|
| I. Beetles | Coleoptera. | VI. Butterflies and Moths | Ledroptera. |
| II. Earwigs | Forficulidæ. | VII. Flies | Diptera. |
| III. Cockroaches, Grasshoppers, Locusts | Orthoptera. | VIII. Bugs | Hemiptera. |
| IV. Bees and Wasps | Hymenoptera. | IX. Lantern-flies, Cicadas, and the tiny green insects of the roses and plants, Aphis | Homoptera. |
| V. Dragon and May-flies | Neuroptera. | | |

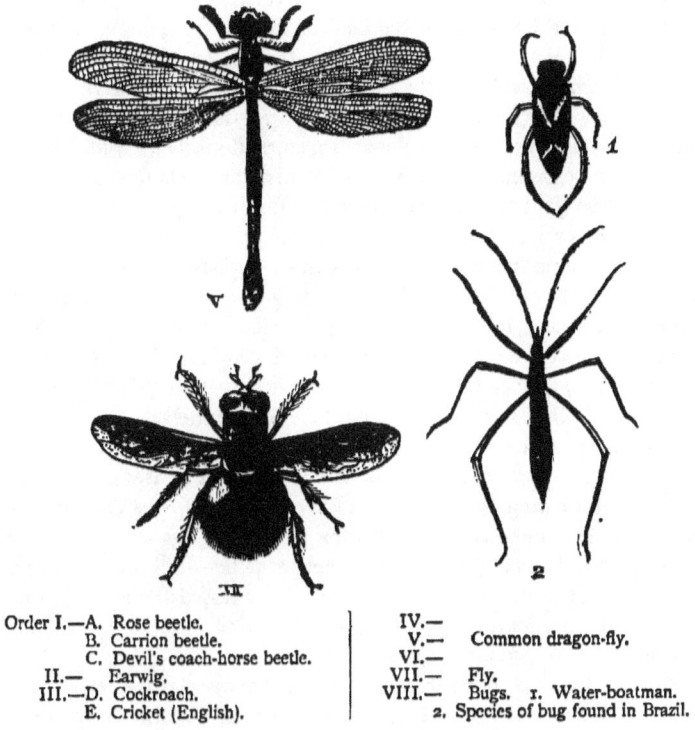

Order I.—A. Rose beetle.
  B. Carrion beetle.
  C. Devil's coach-horse beetle.
II.— Earwig.
III.—D. Cockroach.
  E. Cricket (English).
IV.—
V.— Common dragon-fly.
VI.—
VII.— Fly.
VIII.— Bugs. 1. Water-boatman.
  2. Species of bug found in Brazil.

That the beetles come first in rank is a proof that something more than mere beauty governed the uncertain minds of entomologists when they put lovely jaws in the place of lovely wings, and drove the butterflies to a position behind the beetles and earwigs. The motives for this arrangement are too unintereresting to enter upon in a new introduction to the whole division, but it is principally founded upon the changes which the insect undergoes between its first appearance as an egg, and its final perfection as a being with wings and legs. To explain this more clearly it is necessary to refer again to the familiar silkworm. The changes which intervene between the egg and the moth are not peculiar to this species.

Every insect undergoes metamorphoses which resemble those of the silkworm. The egg state is constantly seen upon the leaves of plants or meat which has been visited by a blow-fly. The caterpillar or maggot state is also known: the presence of these creatures in nuts and fruit, and upon cauliflowers and lime-trees, is too familiar to need description. Between those on the trees and those in the nuts there is an interesting distinction. The maggots in apples and nuts have no feet. The thoughtful mother knew her children could not walk after their food, so she placed them in the very heart of the flower, which, as it grew into fruit, entirely enclosed the egg, and when this was hatched there was food enough to enable the maggot to live, and eventually find its way *out* of the nut or apple. Wherever a hole is in the outside of an apple or nut, it is quite certain there is no insect within, unless an earwig has taken advantage of the hole and crept in for the sake of a hiding-place. But the maggot has worked its way out, having grown too big for its first home. These footless grubs are very common, and are great destroyers; but it certainly is not their fault, as they have no choice of a home. Caterpillars have feet like the silkworms, and can easily find food for themselves. In the chrysalis state insects differ considerably from each other. Some are quite like silkworms, others are almost perfect insects, and have sometimes the use of two little wings, which are lost when the perfect ones are grown at the last metamorphosis. In others the wings and legs are perfect, but enclosed in a thin skin, which binds them together until the creature is grown enough to cast it off. Every insect has passed from an egg to a caterpillar or maggot (the larva state), then to the chrysalis (or pupa state), and then to a winged and perfect insect.

In these pages it is only with the fully-developed insect that I can attempt to deal, and it is when they are grown up that their different forms are so little known, although many of them are frequent but unintentional visitors to our houses. The fathers and mothers of insects have always wings and legs, and are full grown. No caterpillar has children, having quite enough to do to look after its own food; and to expect a chrysalis to look after a family would be too stupid. But in using the expression "look after a family," I may lead you into some slight error. Most insects die when they have laid their eggs, and long before the little maggots come to life. There are some remarkable exceptions, as the earwigs and bees, who show great affection and intelligence in providing for the wants of their families; and almost all have an instinct or memory of their own early days, which prompts them to place their eggs where they may come to life in the midst of food suited to them,—the blow-fly, whose eggs are placed upon meat, and the moth, whose caterpillars live in our carpets and muffs. Indeed, it is always the children who do the mischief. Who does not know a midsummer daw? but when one is killed it is not considered a useful thing done, but merely an end put to a present nuisance. Nevertheless, so voracious are the cockchafer grubs, that they can destroy acres of planted fields, and are a great torment to almost all agriculturists. At the same time I have been often laughed at when I have suggested that boys should be rewarded for every score of midsummer daws or cockchafers which they could catch in the season. Indeed, it is impossible to imagine what would become of the crops if the birds were not so swift on the wing. I saw one evening last summer dozens of swallows and martins catching the flying insects as they buzzed in swarms over the beeches and lilacs in my garden, and they destroyed as many as the two cats playing on the lawn permitted to escape. Cats are very fond of having what my cousin calls "a lobster supper," and they will spend hours on a lawn in seizing the cockchafers as their wings harden and enable them to fly. But it is only just at one moment that these delicacies seem to suit the cat palate. When I put the cat close to an insect which had just wriggled from its chrysalis case in the ground, she would not touch it, but no sooner did it feel strong enough to fly than pussy's paws settled the attempt directly, and her jaws soon crunched the cockchafer to death. Whether cats dislike the stickiness of the new arrivals, or whether their sporting instincts tempt them to give their prey a chance of escape, I cannot tell, but I am quite sure of the fact.

So destructive are the cockchafers, that they have really roused some few people in France to enter into a perpetual war against them, and combined action has certainly saved many a district from their ravages. One of the most interesting accounts is furnished by the Mayor of La Sarthe in M. Figuier's "Book of Insects." The neighbouring country had been so much injured, that when the fields were ploughed the pale-looking grub had to be carefully picked from the furrows. In a field of twenty-nine acres, which was ploughed into seventy-two furrows, the number of grubs averaged 350 per furrow at the first ploughing; at the second, 250 per furrow; and at the last turning over of the ground there were found fifty in each: a total of over 40,000 in one field, which so effectually raised an alarm that the mayor, who had more sense than is usually displayed where the enemy is *only* an insect, offered a reward for the destruction of the perfect cockchafer, and the district was saved. In Switzerland, a few years since, 150,000,000 were caught in a cockchafer hunt, and this number might have been equalled in England in 1865, when the country was infested with them.

The principal enemy to the crops in England is the caterpillar form of a beetle known as the wireworm.

This little creature is a pale pink or red-looking worm, extremely attenuated, and whose preference for the very young wheat and barley is most trying to the farmer. At present he has found no remedy for this apparently insignificant enemy, and I have never heard of any attempt being made to collect the fully-developed beetle, which is familiar to most

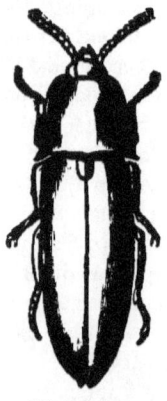 

*a* Skipjack (enlarged).      *b* Wireworm.

of us as the common skipjack.* The crows are frightened away from the fields because they pull up the plants when searching for these favourite morsels, and the wireworm gains time to live and develop and add yearly to its numbers. That is, *yearly* when the seasons agree with it; but if not, it remains a wireworm for two or three years, and grows to a beetle only under favourable conditions.

It will be easy to learn the chief features by which a beetle may be recognized in the common midsummer daw and lady-bird. The hard shiny skeleton which encloses the whole of the muscular, nervous, and digestive systems of an insect is nearly as complicated and quite as worthy of study as our own. It is marked off into three great divisions—the head, thorax, and abdomen. Divide insects transversely where you may, and you cut through blood-vessels, a digestive canal, and a string of nerves and breathing-tubes. It is a point to remember that by these tubes respiration is carried on at the sides, and not at the mouth of an insect,

---

* *Skipjack*, "Packard's Transformation of Insects." Westwood says it is a species of *Staphylinus*, or *beetle*, like Fig. C ( p. 198), but smaller.

## INSECT PETS.

and the buzzing of their flight is produced by the quick movement of the wings, which accelerates the power of respiration and drives the air out from the sides of the body with great rapidity against the wings; these

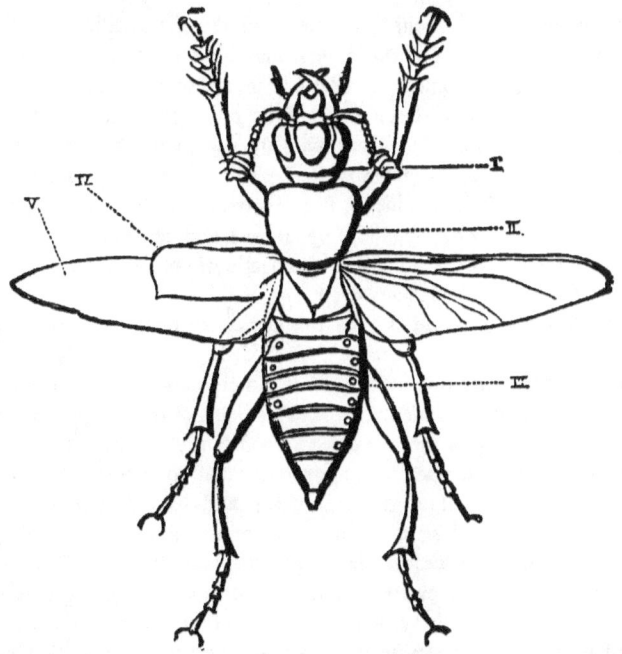

I. Head.   II. Thorax.   III. Abdomen.   IV. Wing-cases.   V. Wings.

are extended to their utmost tension in flight, and cause a vibration of the air which is familiar to us as a hum or a buzz, and is totally distinct from the whistle of a cricket or chirrup of a grasshopper.

### I.—BEETLES.

THE head of a beetle is furnished with a biting mouth, eyes, and antennæ; these last are sometimes inserted between the eyes, and sometimes below or above them. The eyes themselves are as hard as the body, but are faceted in manner to be explained when the bees and dragon-flies come to be noticed. It is said that each one of these facets is an organ of vision, and if this be true, it would hardly be possible to catch an insect without its seeing the hand put forth to capture it. How this may be I cannot say, but you will observe that the hard structure of the eye renders the protection of an eyelid unnecessary, and therefore there is not the rudiment of such an organ throughout the division.

To the thorax are attached the front and also the middle pair of legs, and from this a plate of chitine is sent under the abdomen, to which the hind pair of legs are joined. A beetle is thus shown to be a hard insect, with head, thorax, and abdomen; but its chief distinction from other insects are the wings. The front pair are not so much flying wings as cases to enclose the soft membranous pair when at rest. These cases (or elytra) are also hard chitine, firmer in many beetles than the body; and the second pair of wings fold up transversely beneath them. These cases lie straight upon the back of the beetle when it is at rest, and it is not easy to force some species to fly.

Ladybirds will sometimes take the advice so freely offered to them by children, and "fly away home;" and if they are suddenly arrested, will crawl about with one or both of the membranous wings projecting from the pretty red and black cases. This habit of the Ladybirds is very trying to collectors, as the insects will constantly die with their elytra and wings open, and spoil the appearance of a collection. The membranous wings are longer than the cases, and have a transverse joint which folds them across the abdomen. Some beetles have no wings at all, but the cases are almost always present, and although of no service to protect the wings, they are capital shield-armour for the upper surface of the body, which is always softer than any other part of the insect.

And now it is necessary to impress upon the timid reader that every British beetle is so harmless that the most ferocious can neither sting nor bite. I suppose a Stag Beetle could nip tightly if one put one's finger between the mandibles; but I have picked up many dozens with perfect impunity. The unpleasant habit which some species have of sending out an irritating fluid when caught, generally leaves a disagreeable scent on the fingers, but this would certainly be no reason for fright, as it is merely unpleasant. The power of emitting a disagreeable odour is certainly objectionable, but it is of great importance to the beetles who are likely to be eaten by larger species. No sooner are they aware of the proximity of an enemy, than they eject a fluid which covers their own body and renders them unpalatable. This habit is extremely annoying, because the insect makes no distinction between a beetle who pursues it as a sensual luxury, and a collector who designs to give it decent interment and immortality in a cabinet. The fluid is ejected under any circumstances, and dries upon the insect, and strong benzine and boiling water fail to make certain species of the genus *Tenebrio* pleasing or even clean objects in a collection. This is the more remarkable from the fact that if the beetle had escaped its enemy, the fluid would have disappeared, and the insect have regained its shiny black surface. Their cleanliness is remarkable, and they may be often picked up from the dirtiest-looking heap with perfect safety either to delicate gloves or fingers. But the means of defence used by the Bombardier Beetle must be mentioned. This insect is very often pursued by a beautiful beetle who desires to refresh himself with a dainty morsel. No sooner is the Bombardier aware of its danger than it uses the artillery with which it is provided. A loud explosion, a blue smoke, and a disagreeable scent proceeds from the anus, which immediately stops the pursuer. If he be courageous or hungry enough to follow his prey, he is as constantly repelled, the ammunition lasting long enough to enable the hunted creature to fire twenty times if necessary. This peculiar property of emitting a fluid is common to beetles and a few other insects as means of defence, but in all the British species its effect upon human beings is too slight to cause apprehension.

The size of these insects varies considerably: the largest English beetle would not exceed two and a half inches, and the largest I have seen did not reach two. But who shall describe the length of the smallest? for I have in my possession many Ladybirds (some British) who are too tiny for their markings to be seen without a microscope. They are perfectly formed and beautifully marked; some are shiny and dark, others pale, yellow, and brown, and others covered all over with long silky down, which is so long in proportion to the size of the insect, that it must be as cosy a covering as the fur of a cat or the feathers of a bird.

Whatever the colour or covering of a beetle, the presence of wing-cases is an easily recognized character, and no English species are liable to be mistaken, except the female Glowworm and the *Staphylinidæ*, or Devil's Coach-horse tribe, of which a drawing is given (Fig. C, p. 198). These look at first like Earwigs, but are easily distinguished from them. The membranous wings fold up under the extremely small cases, and the body is uncovered for some distance, but is of firm chitine, so that it is not easily injured. The points of contrast are easily distinguished between the Earwig and a beetle of this family. The abdomen of the latter has no hooks at the end, has not so many rings or joints, and is generally (except on the thorax and elytra) of a dull black.

It is hardly possible to gather a chrysanthemum or aster without seeing tiny beetles, and the shiny wing-cases will always proclaim the Order to which the insect belongs. The cups of the convolvulus and bells of the hyacinth tempt them to enter, and thus the pollen-grain is carried from flower to flower to perfect the seed for the next year. It is a matter of fact that if there were no insects there would be no flowers.

### II.—EARWIGS.

THESE insects (*Forficulidæ*) have a stiff hard body, which is formed of chitine like that of a beetle, and at a first glance there appears some resemblance to Fig. C (p. 198). But mistakes, which are common enough in other insects, seem never to have befallen the Earwigs: their terrible hooked bodies, and the superstition about them, making every nursemaid impress the danger of touching them upon children, so that we all grow up familiar with the peculiarly stupid and unmeaning name.

These insects have leathery and stiff wing-*covers*, which are distinct from the *cases* of the beetles. They are covers, because the posterior wing of the Earwig is never entirely closed by them. The description which I give of this wing is taken from a dead insect, and I have never yet succeeded in making an Earwig fly so that I might see its motion. This wing I disturbed myself, and the end of it at once explained the uselessness of a perfect case. It is a tough leathery patch, from which there are radiating veins and a membranous expansion to form a flying wing. The veins are distributed in graceful curves, which render this an infinitely prettier object than the posterior wing of beetles. It is folded up by a most exquisite arrangement. It is first closed like the bars of a fan, and then the wings lie like two delicate feathers over the body of the insect; at about two-thirds of its length the transparent part doubles upon itself; and to draw the wing to its proper position another joint is close to the leathery patch, and this turns the whole of the wings under the front pair, with the exception of the thickened part. The trouble of folding their wings must be a source of fatigue to Earwigs, and is probably a reason why they are so seldom used, and the abdomen with its hooked end is occasionally brought into service to fold and unfold them. It would appear that they fly when food is scarce, and there is no doubt but they effect an entrance into every garden where dahlias grow. I was going to say that these flowers are the natural food of Earwigs, but I am not at all sure that they have such an exclusive taste; my experience certainly is that they destroy these flowers when they can, in spite of the elegant traps which are set for them. Flowerpots, and bowls of tobacco-pipes stuck upon sticks, disfigure dahlia grounds, and every morning dozens of Earwigs come to sudden death in a pail of boiling water into which these traps are turned.

When there are no flowers to destroy, these ruthless creatures seek the fruit. How often when a ripe plum has tempted you to pluck and eat, has the tempting morsel been suddenly dropped because some crawling insect has fallen out of it! Startled, you have looked on the ground to discover the intruder, who constantly proves to be an Earwig; but here arises a difficult question: "Is he the intruder, or are you?" He had the right of possession, which may be nine points of Earwig as well as human law. If neither fruit nor flowers are to be had, these voracious creatures commit other enormities. The Earwig mother does not die, like so many insects, as soon as her eggs are laid. She lives five or six weeks watching her children develop, and evincing the greatest affection for them. Soon after they reach the maggot state she dies, and her undutiful children demolish her body, and that of any sickly brother or sister who may fail to reach perfection. Although Earwigs are so numerous in England, there are but few distinct species, only seven varieties being found here.

### III.—COCKROACHES AND GRASSHOPPERS.

BITING insects *(Orthoptera)* have a mouth with transversely movable jaws, four wings, the front pair being tough and leathery, but these are the wing-covers, which overlap at the points; six legs, armed with exceedingly stiff and sharp spines.

The whole Order is again subdivided into four groups:—

1st. The Cockroaches, with legs formed for running.

2nd. The Praying Crickets (not found in England); front legs much longer than the other four, and held up when the insect is at rest.

3rd. A group also unknown in England.

4th. A group, which comprises the Crickets, Grasshoppers, and Locusts. Legs formed for leaping.

The first section is unpleasantly well known in the familiar insects who have rendered themselves so thoroughly at home with us as "the Black Beetle." This troublesome individual is not a beetle at all, but, as any one may prove upon examining by the rule given, a truly Orthopterous species. Beetles are never found in the kitchens of town houses, although I have seen a black species in the damp cellars of country places. But the destroyer of the maids' comfort in the kitchens of London is a Cockroach, and even this British settler is a most disagreeable pest. Independently of the alarm caused by discovering one in the folds of a dress or in a tub of flour, there is the fact of their unpleasant odour, which remains in boxes and cupboards long after the insects have been destroyed. Fortunately the Cockroach is six months before it arrives at its full perfection, and they cannot multiply so rapidly as other insects from this cause. The most interesting feature about them is the beautiful adaptation of the mandibles for their work. The left is considerably longer than the right, and furnished with sharp teeth: these fit into the grooves of the right mandible, which is irregularly notched to receive them; the outer surface of both mandibles is smooth and rounded.

It is amusing to consider what these insects did with themselves if they found their way to England before kitchen stoves and bakers' ovens were invented. They have been brought here from the Levant along with the boxes of plums and currants, and finding that we have conquered the difficult climate by artificial warmth, our Eastern friends continue to abide with us. An unwary Cockroach in olden times must have perished in an English winter. But troublesome as they are to us, they are an infinitely worse nuisance on board the ships trading to India and the Cape of Good Hope.

The following extract from a letter is only one of many that could be produced about the punishment these insects are to the passengers:—

"I did not open the boxes of biscuits until we changed into this ship at Suez, but now I wish I had thought more of them, for everything, from the tea to the soup, tastes of cockroaches. They eat our boots and gloves, and ladies who have made the journey several times tell me that they have found them eating the nails and hair of sleeping children."

This is a tempting prospect for any traveller! but, by dint of setting traps baited with bread soaked in beer, persevering people sleep without fear.

In this rapid sketch of such large Orders I must not stop to tell any very interesting stories about these well-known insects, but pass on at once to a species not known in England, called the Praying Crickets, distinguished from the preceding family by the extravagant length of their front legs and their long thin bodies. They are also fond of daylight, and may be often seen sitting about the ground in the south of Europe, perfectly at ease, with their two front legs held up in the air. This curious attitude is the cause of their name, and in all countries they are called religious and holy. Sometimes another interpretation is put upon the attitude, and mendicant and pauper is applied to them. However, St. Francis Xavier (according to tradition, for I believe he was too sensible ever to have uttered the story), seeing a Mantis or Praying Cricket sitting as a devout creature holding up its legs in an attitude of supplication, tested its motive for doing so by ordering the insect to sing the praises of God, and it immediately carolled forth a fine canticle. We are also informed "that if a child loses its way and asks the Mantis for direction, the intelligent creature immediately points out the right way, and will seldom or never miss."* It is very hard to destroy such pretty illusions, but truth must be told at any cost. The true reason is that the Mantis is a very hungry creature, and lives by catching insects to devour. Naturally she does this with as little work as possible, and moving softly upon four legs, the two long front ones are ready to devour the victim before the Mantis is close enough to frighten it away. Although this fact must be clear to any one who will take the trouble to watch for it, the superstition has not yet died out, and certain nations hold these creatures in as much reverence as the ancient Egyptians did the Sacred Beetle.

In the last of the our divisions we find the creatures with the jumping legs, *i.e.*, Crickets, Grasshoppers, Locusts; these are slightly different from each other and from the preceding groups.

The first is given in Fig. E (p. 198), where it may be closely inspected, a privilege the living specimen distinctly refused. It is necessary to notice the colour of the Cricket, which is much lighter than the Cockroach, and is a pale brown. The antennæ or horns are extremely long. The wings, when folded, are too long to be covered by the front and thicker pair, and project in the points marked *a a*. The song of the Cricket is not made by any movement of the mouth, but by the front wings. One of these has on its upper side a thickened vein which reaches from the base to the tip, and its surface is roughened like the edges of a rasp. The under-side of the other wing is supplied with a similar instrument, and the friction of these when the insect is at rest produces the sound. I say when at rest, because the ordinary movements of respiration (they breathe by the pores or spiracles at the side of the body) cause a tremulous motion of the wing-covers, and this is communicated to certain spaces of the wings, which are thus converted into musical wind instruments, and produce the chirruping sound with which every one is familiar.

* Taken from Westwood's "Classification of Insects."

The poet Cowper was probably more ot a poet than a naturalist when he ascribed to the

"Little cricket on the hearth,
Chirping gaily, full of mirth,"

a tendency to merriment; because it is quite evident that our noisy friend is dull when he is singing loudest, and is calling for cricket society. The Crickets like dark holes and corners, and avoid the light; but although in this respect like the Cockroaches, they are not mischievous. Indeed, it is said that they destroy these pests, but this requires proof, for a Cockroach could certainly match an enemy his own size, and probably is never stupid enough to be caught by a noisy Cricket. Besides, these little things prefer moist food, such as yeast, soaked bread, milk, or broth, &c., and their fondness for liquids often causes their death. For the sake of the supposed good which they may effect in destroying other insects I never kill a Cricket, and it is difficult to make a country servant do so; the superstition that "crickets bring luck" being still prevalent with the poorer country people.

Grasshoppers differ from Crickets in the following points. They are generally of a green colour, and have their wing-covers arranged when at rest like a slanting roof. These project beyond the body of the insects, and this is the first species yet mentioned whose covers are longer than the abdomen.

The song of the Grasshopper is produced by the front pair of wings. Two round plates are found at the base of each cover, surrounded by strong ridge-like veins, which serve as a bar to produce sound when overlapped, the left wing lying over the right.

They feed upon the leaves of trees, but are often seen in damp grass. They are not found in such enormous swarms as the Locusts, but are fairly voracious when pressed for food. Mr. Westwood relates in his book on the "Classification of Insects" that he once shut up the green Grasshopper in a box, and in doing so its leg was accidentally jerked off. The next morning it was found to have devoured its own leg. The baby Grasshoppers when first hatched are as active as their parents, but have neither wings nor wing-covers.

Locusts differ from Crickets and Grasshoppers in the arrangement ot the wing-covers, in the organ which produces noise, and in the short antennæ. The wing-covers lie flat upon the body, and this being compressed at the sides causes the wings to slope from the middle of the back, like the roof of a house, and brings the outer edges in close contact with the first long joint of the posterior pair of legs. This joint has a rough surface on its inner side, and the insect when it makes the sound rubs this against the wing-covers, standing firmly as it sings on the four front legs.

But how is it possible to explain what food is preferred by these insects? A flight of Locusts soon destroys every green thing in a district, and, in one of the most exquisite books ever written, there are many accounts of the fearful destruction caused by these creatures, who are as much dreaded as the plague in some countries. That they travel in societies renders them dangerous, and they have been seen flying together in swarms three miles broad, and darkening the air so that it was impossible for people to see each other at a few yards' distance. To turn such a swarm from their path of destruction was once attempted by the Bashaw of Tripoli, who raised a force of 4,000 men to fight the Locusts, and ordered all those who thought it beneath them to fight such pigmy foes to be immediately hanged.

In 1650 a cloud of Locusts was seen to enter Russia in three different places, which from thence passed over into Poland and Lithuania, where the air was darkened with their numbers. In some places they were seen lying dead heaped one upon another to the depth of four feet; in others they covered the surface like a black cloth, the trees bent with their weight, and the damage exceeded all computation.

In 1841 the province of Ciudad Real in Spain was visited by so many Locusts that three hundred persons were constantly employed to collect them, and they destroyed seventy or eighty sacks a day.

The whole chapter of the book* from which these accounts are taken is so tempting to transcribe, that I can with difficulty refrain from doing so, but probably many will seek it for themselves, and be tempted to follow the example of the dear old gentleman who wrote it, and watch insects for their own instruction.

The next tribe, of Biting Insects, will require a chapter to itself, Bees and Wasps being too important to be glanced over with the Earwigs and Grasshoppers. I must also apologize for the extremely rapid manner in which I am treating a subject whose importance is too great for me to grasp. The patience and work of a life, and that life a long one, would fail to include all that might be written: to endeavour to remove the absurd terrors of young ladies, and to tempt boys to show human love rather than human cruelty, is the only aim I have proposed to myself

* Kirby and Spence: "Introduction to Entomology."

in attempting a slight sketch of the whole division. It may be well to recapitulate and notice the chief points mentioned in the three sections:—

Order I. *Beetles.*—Distinguished by their wing-*cases*, and the familiar species are Ladybirds, Midsummer Daws, Devil's Coach-horse, and common Dung Beetle. They are generally useful to man by living on other insects when perfectly developed. There are many vegetable feeders which will be described when the chief Orders have been noticed.

Order II. *Earwigs.*—Straight short wing-covers and long hooks at the end of the body. No foundation for the idea that they cause deafness, although they are called Earpiercers and Earwigs in all countries where they are known.

Order III. *Cockroaches, Grasshoppers, and Locusts.*—Long tough wing-covers, overlapping at the tips, and in the Cockroaches and Grasshoppers very long horns to the head. Wing-covers like a sloping roof in Locusts and Grasshoppers.

## IV.—MAY-FLIES—DRAGON-FLIES.

IT is not an easy task on a dreary winter's day to write (as I am now doing) of these graceful insects, whose very existence depends on the sun's warmth and the earth's brightness. At this season of the year there are no perfect May-flies living: they rest quietly enough as larvæ under the stones and in the mud of rivers. Not one of them would be so imprudent as to venture forth in the midst of snow and ice, like three unwise bees which I picked up while skating this winter, and warmed back to life in my muff. Poor little things! they had been nearly starved to death, because no food was left near the hive, and ventured forth in search of provisions, and were all but killed in the attempt. No May-fly could have done this for three excellent reasons. In the first place, there are not any of them in the winter; in the second, they would not be hungry even if they existed; and thirdly, they could not eat, for they have no mouths. Why, then, describe them at the same time with biting insects? This question certainly deserves to be answered. Insects are arranged according to a definite plan, as I have already said, and this involves a careful consideration of their habits and structure from the time they leave the egg; and it is the May-fly's children which bring it into the same Order with the biting insects. These young people, like the children of the Dragon and Caddis-flies, are true water-babies, and have very good mouths, and, so far as I know, very fine appetites also, seeing that mud has been found in their digestive organs. I must ask my readers to put out of their heads any idea that these insects are true flies or ants: the insects we know by these names belong to two Orders we have not yet mentioned, and our House-flies differ from this Order by such a visible difference in their limbs that I shall not point it out, but give the characteristics of *Neuroptera* as arranged in our list. Some of the Order have such an inclination to look like Crickets or the Praying Mantis that they have their front wings tough and leathery, or at least much thicker than the back ones; some of them are just as voracious as the most violent and ill-tempered Cockroach in existence, and the modification of the mouth parts demands a place for them before the Bees and true Ants, and after the Locusts and Crickets. The Order is called *Neuroptera*, or nerve-winged, because the tendency to thickening of the front wings belongs only to one or two species.

A Neuropterous insect is known by four transparent wings, generally of the same size; mouth with transversely movable jaws; long thin bodies of brilliant colours; and soft, instead of hard chitine, like the Beetles and Earwigs. At the end of the body there is a sharp-pointed arrangement that looks like a sting.

Now, this is the distressing part of the business, for not one of the whole Order could hurt us if it tried. This appendage is a complicated apparatus to lay eggs with, and causes many of these harmless creatures to be put to death. I have found it difficult to persuade people that this useful instrument was not a poisonous weapon even in such well-known insects as Midsummer Daws, and it is certainly unfortunate that, this being the case with Wasps and Bees, every other winged creature is made the scapegoat of their offences.

I intend to mention only four families of this Order, and these differ so much in appearance and habits that we cannot wonder entomologists have had very grave doubts about the propriety of grouping them all under the name *Neuroptera*.

These four families are—1. May-flies—*Ephemera*. 2. Dragon-flies—*Libellulidæ*. 3. Caddis-flies—*Phryganea*. 4. White Ants—*Termites*. Because the May-flies are quite sure to be seen on the fine and early spring days, they are put first on the list, but anything more inconvenient than this arrangement it would hard to find. First of all, I give four rules for knowing *Neuroptera*, and the very first example does not fit any of them. A May-fly is an exception, when a perfect insect, to all but the first rule. It has transparent nerve-veined wings, but the front pair are very much larger than the others. It has no mouth, but merely masses of membrane, which look like an attempt at a biting mouth, if the creature meant to live long enough to make it worth while to grow one. Last, but not least, there are at the end of the body three extraordinary tails, which any sensible person must consider a most extravagant supply for an insect. Why a May-fly should have more tails than its neighbours, except it be to make up for the loss of a mouth, I cannot tell. It has been said that they guide themselves in their downward flight by these

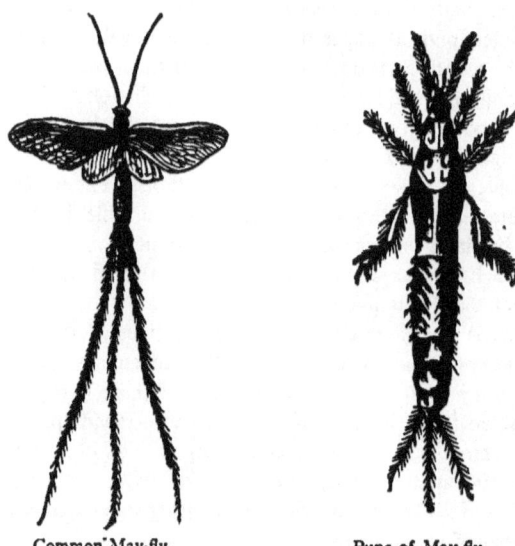

Common May-fly.  Pupa of May-fly.

appendages, but I am much inclined to doubt the truth of this, for then Dragon-flies, who live longer and fly more, would surely have developed some such means for balancing themselves in the air.

These three tails give a very easy way of recognizing a May-fly, and it can be referred to the Order by its soft body, transparent wings, its unceasing flight during its short existence, and by the fact that it may be seen always with a large number of its own species.

This is the first instance we have met in our descriptions of large societies of the same insect coming to perfection at the same time, and sporting together for a few hours, and then the whole of them ceasing to exist almost at the same moment. This is one of the curious facts

about May-flies. They never have anything to do with the land or its
nhabitants when alive, but when they die they are swept up in large
numbers and used for manure. All their childhood and much of their
old age is spent under the water, for it is said they live two or three years
as swimming caterpillars or larvæ, and moult or change their skins,
though not their form, twenty-one times. Thus they spend all their
youth under the water, until their wings arrive, and then their grown-up
life is passed in playing over its surface. In this last stage their only
object is to lay their eggs: these are dropped in two yellow packets in
the water, and then sink (if the fish permit) to the bottom. Here they
change to a curious sort of larva, with feet, gills, and a tail. Not being
of a very active disposition, the first business of the young larva is to find
a comfortable house, and without much delay it burrows two passages in
the mud. These two openings are made to unite at the bottom, and as
the greatest space is here, the individual who is builder, landlord, and
tenant enters, and rests at the point where the two passages meet. These
circular burrows have been a puzzle to many people who know the lazy
habits of the larva, and it has been suggested that the entrance and exit
are formed thus, to save the creature from the awkwardness of turning
round. I do not think this reason will answer, because, if saving trouble
was the object, why should it not make one large hole, and then there
would be no turn either? There is another reason from which we may
all take a lesson, I think. By the arrangement of the burrows the water
enters at one hole, reaches the insect, who inhales the fresh supply,
derives a certain benefit from it, and then, exhaling it, drives it behind
him, and out the other way. Thus the water, which is of the same use
to the gill-breathing creature that air is to us, is constantly supplied from
a fresh source, and is not mixed up with the impure water that has been
once used, as it would be if there was but one entrance and no exit.

My opinion is that these creatures teach a lesson in ventilation which
might be followed by some others who have fewer legs and no gills.
The little tenant changes its skin many times, as I have said, but retains
its gills always; and when it slightly changes its form to the pupa state,
the gills are not cast off. When this change takes place the wings can
be seen through the outer skin, and on the first warm day it comes to the
top of the water, and throws off the thick coat of the chrysalis. Then,
with some difficulty, it uses the half-liberated wings and reaches the edge
of the stream, where it settles on the leaves or stalks of a plant. Here
comes the final change: the sun dries up the membrane left round the

insect, and it slips off in two pieces at the sides close to the wings, and the liberated insect starts off at once to seek some new friend who is already disporting itself in the air.

The gills, which have hitherto remained through every change undergone by the insect, have now disappeared, and our little friend, who is now an air-breather, is supplied with wings, which have been sometimes called aërial gills. It must be realized that these external air-providers act in both instances as locomotive organs, for they are in constant motion whenever the insect guides itself through the air or the water; but gills are of more service than wings, because they last two or three years, and the wings and the perfect insect live only for four or five hours. Of what use are the May-flies? To tell a very solemn truth, I do not know of any pretty service that they are as May-flies, but artificial imitations of them tempt many fish into danger. There are many girls who read this book who know the difficulties of making a Dun-fly well enough to satisfy the critical eye of a grown-up brother. For my own part, I do not believe hungry fish are nearly so difficult to deceive as the anglers try to make us believe, or they would never be caught by dead flies, much less by artificial ones. To copy natural conditions is the golden rule of the amateur fishermen, and as May-flies seem to appear and disappear in the most sudden manner, there is not much doubt about the way their final departure is managed. Those which fall dead on the land are swept up for manure, but those who die after depositing their eggs on the water are immediately snatched by the fish, who may be seen waiting near the surface for their favourite delicacy. A ravenous fish sometimes swallows a hook with the morsel, and these pretty insects thus serve the double purpose of food for fish and aids to the angler's sport. From the first moment to the last of their lives there seems a constant succession of change in their bodies, and the final one from a water to an air-breathing creature must be the most wonderful of all. Just imagine what a wonderful change is undergone at the moment it emerges from the water. For two years at least it has been a mud-feeding, grovelling insect, with but little beauty or grace, and no movement but a jerking wriggle that cannot be really called swimming, and then with the smallest effort it finds itself poised on the golden Vallisneria or lovely white water-lily, with wings more delicate than the petals of either, and the power to move through the air in any direction. One can almost suppose such a mere upstart giving itself airs, and saying to the flowers, "Really I lived at the bottom of the water with you for a long time, and thought you important indeviduals, but now we have both grown out of it I seem to be a very superior creature, for I go where I please, while you remain always in one place." There would certainly be some truth in this, but I do not believe flowers and insects have divided interests, and it is more probable that our little friend is at first frightened at his new appendages, and clings to the lily until his strength increases, when they chant together the praise of the glorious sunshine which has kissed them into life.

Leaving the May-flies, we must turn to their more beautiful but less amiable relations, the Dragon-flies, and the rules given for distinguishing Neuropterous insects apply here very well. This insect has a biting mouth as well as the Beetles, Earwigs, and Cockroaches, and it would be worth while to consider how we might know a Dragon-fly's head from a Beetle's if we found them separated from their respective bodies.

The enormous eyes, covering nearly the whole of the head, would be the first remarkable contrast; and if these were closely examined, the shape of the facets would be found different in both insects. This we will consider when insects' eyes are to be studied; but we will search at present only for broad distinctions, and these are best seen in the mouth. The beetle's mouth, with which we are already familiar (p. 197), shows all the appliances for mastication externally—mandibles, maxillæ, and their palpi. The lower lip and palpi and the upper lip are in many species visible without dissection. This is not the case with the Dragon-fly. At first we notice only two large lips, and the lower one looks a most powerful weapon. Upon closer observation two square-looking mandibles show between them; but this is all the external mouth exhibits. As we are not very learned entomologists, we will be satisfied with the outside, only adding that within the mouth the organs are similar to those of the beetle, with the addition of an instrument, which, for want of a better term, is called the tongue. The rudiments of the ligula exist in *Coleoptera*, but are not sufficiently defined for us to talk about. Why should not the two insects have their mouths exactly alike? This important question wants thinking about; and to give a fair judgment we must take a beetle that lives upon insects, for these are the only food of the Dragon-fly. We have already determined that beetles walk more than they fly; and Dragon-flies never walk at all if they can help it. Here, then, comes a reasonable inference—one will find its food on the ground, the other in the air; the beetle must work its way through the ground or upon it, and when its prey is found, it must be held by the force of the hard external mandibles and maxillæ. But the Dragon-fly is quick on the wing, larger very often than its prey, who are soft-bodied, tender morsels, that may be transferred at once within the mouth and swallowed without danger to digestion. This well-developed mouth is well used; and it is a great mistake to kill Dragon-flies, who do good work for us in clearing the air of many thousands of smaller creatures, who would be a great nuisance if some quick-sighted creature did not destroy them.

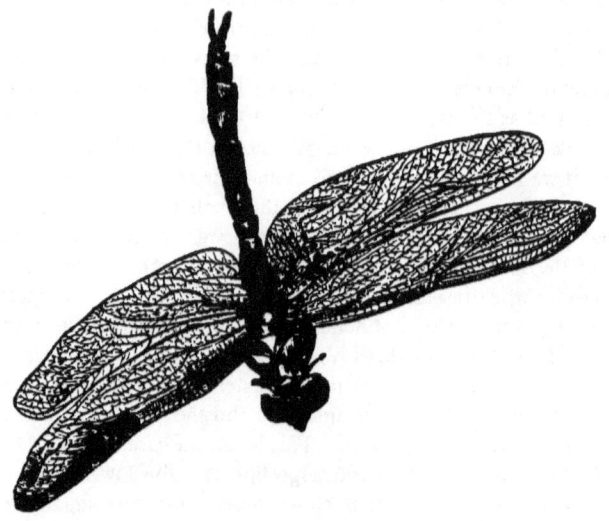

Dragon-fly—Perfect Insect.

Beautiful blue Damsel-flies, "Demoiselles," are amongst the names for these insect monsters, but "Devil's-darning-needles," or the old English term of "Dragon," is certainly more suited to their nature, if not to their appearance. Not that we know exactly what dragons are, but then these insects equal in savageness and bloodthirstiness any fable ever invented, and this name may be much nearer the truth than any of us have imagined.

Their brilliant colours and gauzy wings cause them to be much admired; but one cannot exactly conjecture the use of such gay clothing to an insect whose food depends upon its success in hunting down its prey. One cannot suppose the colour a means of defence, because birds will capture both duller and brighter objects. Is it possible that their beauty tempts some inquisitive insect near them, who pays the penalty of the fascination by death? Evidently the colours serve some useful end to the insect, for their beauty is universal, and some foreign species have nearly six inches of this rainbow brilliancy. Our own species are very large; it is not unusual for a Dragon-fly to be found who will be three and a half inches long. There are no very small insects throughout the *Neuroptera*, and in England there are not many different families.

Of the good character of Dragon-flies I have very little hope of finding any example, and I know nothing about them which will introduce any useful moral into their history. Probably there is more fun in watching them than the sober-minded respectable insects who crawl sluggishly over the leaves, and run only from the first enemy who attacks them. "Their courtship is a scene of general bad behaviour and violence," writes an eminent entomologist, and he finds nothing but treachery and spitefulness in their whole conduct. Fortunately, the life of a perfect insect is not a very long one, although it lives to a good old age when compared with the May-fly, whose existence can be reckoned by minutes. It is said Dragons live several weeks, but my own experience would certainly limit their existence to eight or nine days, but I have not seen enough of them to give a positive opinion, for it is easy to be deceived when judging from a specimen that has lived under conditions necessarily artificial.

The female dies soon after she has deposited her eggs on the stalk of some water-plant, and the future history of her unprotected children is extremely curious. Placed close to a running stream of water, there seems but a slight hope that each egg will have the good fortune to go through all stages of its existence, and the chances are so much against them that many thousands perish. In the mud at the bottom of the water the eggs change into larvæ, and they resemble in this state their young cousins, the May-fly larvæ, who are already reposing in their burrows. But there is an important difference in structure: the Dragon-fly larva has no external gills, and converts five plates at the end of its body into a breathing apparatus. The larva draws the water into its body by moving in a direction which sets free the edges of these five plates; a valve within transfers the water into a position which puts it in close connection with the blood in the creature's body. Something is given from the water to the blood, which is called oxygen, and the blood gives to the water a gas which it no longer needs, and this exchange once effected, the used water is expelled a considerable distance, and the insect forced through the water. Their larvæ are very active, and need no burrows to live in; but one can easily trace the motive which prompts the larva of the May-fly to build its home so that the used-up water can escape when we see these insects using a powerful apparatus to drive it away from them. Their ravenous disposition soon develops, and the lower lip of the larva in some species is a powerful instrument of attack. This is hardly to be doubted when we are told on good authority that they attack and eat small fish.

When the pupa (or chrysalis)—this term is only used to make clear that the third state of the insect is meant—is ready to make the final change, it creeps up the stalk of a plant, and remains there perfectly still until the skin splits on the thorax; the head is then drawn back, and the insect sets itself free, after the manner shown in the drawing, leaving the

Dragon-fly—Perfect Insect changing.

case of its whole body on the stem. This is the prettiest part of its history; but it goes on flying and eating all it can find, with just as much voracity as it previously did when it went swimming through the water. In this last metamorphosis the poets and moralists have found many pegs to hang pretty stories and useful lectures upon; but my business must not touch on imaginary beauties. I am only a special pleader in defence of these much-abused insects, and claim for them only a fair share of politeness, and as much consideration as would be awarded to the birds. I have said that the May-fly may pity the fixed condition of the water-lily; but how very sorry the Dragon-fly must be for its companion insect, if it has the power to reflect upon the fearful deprivation of a mouth which the other suffers. One can imagine no greater sorrow for our quick-flying *Darning-needle* than a loss of appetite or the absence of prey.

### V.—CAD-BAIT—CADDIS-FLIES.

EVERY one knows the odd little creatures who are so useful to the anglers, so that a few words will tell the story of their absurdities. If there be anything more comical in nature than the proceedings of a Caddis-worm, I certainly do not know where to find it. They are amusing without being immoral, and this is more than one can say for some other creatures who are also great fun to watch. Hermit-crabs look ridiculous enough when walking about with whelk-shells on their backs, but one can never help the uncomfortable sensation that they have been burglars and murderers; one has a conscientious objection to laugh at an individual who has obtained a freehold property by eating the original owner. A Caddis-worm does not fall to such a depth of degradation, for he only appropriates the house of the water-snail, and carries its living tenant about with him. This seems to be attended with discomfort to the struggling victim, and I have been often tempted to take the cad from the water and release the unhappy mollusk; upon second thoughts I have decided not to interfere, and as I should probably injure the Caddis-worm more than help the snail, I neglect the opportunity of aiding one animal at the expense of another. These worms, which are commonly called Cad-bait, find themselves inhabitants of the water when they leave the egg, and so far resemble the Dragon and May-fly larvæ. But from this point of its life it behaves differently from either of the other water-babies. Without hesitation it sets to work to protect itself from any natural enemies who may be prowling about with good appetites. "Ah," sighs the poor unprotected larva, "a soft worm like me is just the thing such creatures as trout and perch eat. I will just make myself an unpleasant morsel." And then she sets to work, and carries out her resolution with such goodwill that any creature attempting to eat her would have good courage and better digestion; even then such game would be hardly worth its capture.

Different species of Caddis have different tastes in architecture, and may be known by the materials and fashion of their houses, just as well as the old Greeks and Romans. One species prefer leaves of grass, and taking four or five of them, will fasten them together with delicate threads to form a long case open at both ends. Another takes rushes, and unites the pieces with a kind of glue in such a pretty fashion that its house belongs to a very regular order of architecture; another has fine sand curiously packed together; and a fourth, old pieces of stick. But those that I know best are utterly unscrupulous, and take sticks, leaves, pebbles, and snails to carry out their defensive works. The ridiculous part of the proceeding is that the occupants of the shells are alive, and have a great wish to move one way, while the Caddis, whose property they are, persistently takes them in an opposite direction. The worst of

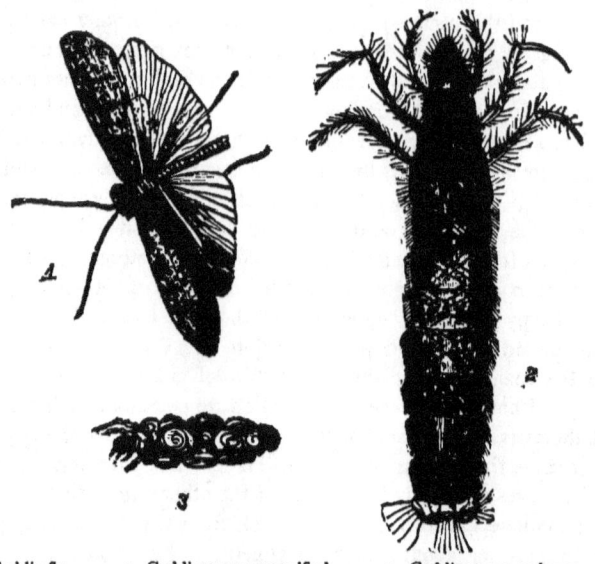

1. Caddis-fly.   2. Caddis-worm magnified.   3. Caddis-worm and case.

it is they can never get free, unless it be true that each time the Caddis grows too big for the old case, she makes a new one from fresh materials. This has been proved to be her plan, but not in a way that helps the snails much. Caddises who were originally content with homes of leaves and rushes, as they grow older, seize animals and sticks, that they may have stronger defences. The bottom of a clear pond on a fine day, it there are Cad-bait in it, is as much fun as an aquarium at the Crystal Palace. At the bottom of the water are seen little masses resembling wood, pieces of straw, or moving stones: catch some of these in a fine hemp-gauze net, and when you first look at them, you will believe that mud and stones are the only prizes obtained. Hold the net quietly just below the surface of the water, and watch anything that appears like a symmetrical mass.  Presently, a little head peeps out at one end of it; not seeing or smelling any danger, six little feet soon follow, and then part of a long thin body.  Then the case clears itself of the surrounding mud, because it is fastened to the end of the worm's body by two hooks, and begins to move solemnly about in the net.  Shake the net, or drop something into the water, and in a moment the little being draws back into the case, and looks a lifeless object, nor does it again venture forth until fairly sure that peace is restored.

This kind of amusement is harmless enough, if there be no fishermen about, but there is little chance of the worms again touching the bottom of the pond if any anglers know where they are to be found.  A wise old sportsman who thought himself a match in craftiness for any salmon, trout, pike, or perch, in any river within the four seas, wrote a book to inform other fishermen of the wonderful things he discovered.  Amongst these is an account of the use of Caddis-worms, which he says are very tempting to fish; if he had not called them pipers, cockspurs, strawworms, ruffcoats, sand-flies, grannums, cinnamon flies, and silverhorns, his directions would be clearer to the non-fishing part of the community, although I daresay the other part understand him quite well.  However, it is easy to imagine that the fish are soon caught when they see the tempting worm unprotected by any vegetable or animal abominations n the way of cases, and for this reason Cads are very popular bait for anglers to use.

It may be asked what becomes of the old cases?  Now, we are not quite sure about what happens just at this time, but we can guess, because there is a large limestone rock where the Caddises who lived long ago have left their story.  Some thousands, possible millions, of years ago, a great deal of water was forced from the country of Auvergne in France.  How this happened does not matter to us just now, but it left behind it the sediment or mud which is under all rivers and brooks.  This has hardened into a mass of rock called limestone, and when pieces of it are broken away, it is found to contain numbers of these Caddis-worm cases such as we find now.  It is so long ago since this rock was under water, that there are no stories written in the world of what happened then, and only these unwritten ones of the Caddis-worm cases help us to guess how things went on.  Fancy what clever little atoms they are! all that long time since they knew exactly the best sort of houses to live in, and we have never discovered this important fact for ourselves; even the right way to warm our homes is amongst the mysteries of the future. But if a Caddis were really as intelligent as it ought to be, surely it would have found out something after all these years; but no, it is the most conservative creature, and declines the privilege of letting the old order give place to new, either in its life or its architecture. This, you know, is very provoking, and not at all what Mr. Darwin, Prof. Huxley, and all the sensible part of the world expect. Everything ought to be capable of improvement, even Caddises, and so we must hope they may some day find themselves in circumstances that will cause them to add new conveniences to their houses, *if* they want them. One thing is certain: they helped to make that limestone rock, but only because they could not avoid it; and just the same thing may be said of the old cases of the Caddises now: a really industrious creature is a rock-builder like the corals, but these things leave their cases behind them, and the rock makes itself, which is a very different thing.

One thing Caddises are very careful about: their houses must be fairly easy to carry; to make sure of this they have many contrivances to regulate the weight of them. They manage to adapt themselves to the quality of the element they are in; the same species behaving differently in the same pond when any change occurs in the water. In a "Fairy Tale for a Land Baby" there is a fact which illustrates this, and we may be sure that the author saw it himself, although he did not tell us the reason for it in a fairy tale. Of course we can find a cause now for the curious proceedings he relates:—

"Tom went into a still corner, and watched the Caddises eating dead sticks as greedily as you would plum-pudding, and building houses with silk and glue. Very fanciful ladies they were too: none of them would keep to the same materials even for a day. One would begin with some pebbles, then she would stick on a piece of green wood, then she found a shell and stuck it on too; and the poor shell was alive and did not like at all being taken to build houses with; but the Caddis did not let him have any voice in the matter, being rude and selfish, as vain people are apt to be; then she stuck on a piece of rotten wood, then a very sweet pink stone, and so on until she was patched all over like an Irishman's coat. Then she found a straw as long as herself, and said, 'Hurrah! my sister has a tail, and I'll have one too;' and she stuck it on her back and marched about with it quite proud, though it was very inconvenient indeed. And then tails became all the fashion among the Caddis-worms of that pool, and they all toddled about with long straws sticking out behind, and getting between each other's legs, and tumbling over each other, and looking so ridiculous that Tom laughed till he cried."

Now we can see that this straw was stuck on as a sail to help the Caddis along; something in the water or in the materials was uncomfortable, and by way of making the best of it, the Caddis stuck on a straw. A leaf or part of a rush is sometimes used instead, but some object of the kind is an absolute necessity, as her home must be light enough to carry about. So much for the home, but now for the food. Truly, I think the word "cad," as boys commonly use it, to express some mean and grasping companion whose motto is, "Get what you can, never mind by what means," must have been taken from these worms. Anything vegetable or animal that comes in their way they will eat with as little scruple as they would take a living snail to put on the roof of their house. They are most certainly "cads," and I can find no excuse for their selfish conduct; and if the word has not been adapted from them, it might be.

After a time this roving, piratical life ends, and the Caddis comes to its *pupa* state. This is interesting to us who watch it, but I doubt if it be so pleasant to the insect. When she is sensible that a really serious life ought to commence, the Caddis moors her case to some heavy pebble or firmly-rooted plant; then she puts a grating over each end of her case before she finally loses the use of her jaws and legs. She is now determined to make no more additions either of snails, straws, or leaves, but has a most firm intention to make her toilette, and appear in the upper world as a modest, neat, and well-dressed insect. To accomplish this, intruders must be kept out, and water let in. The two gratings effect this, and are at the same time defences and ventilators. The water comes in one way, is carried by the imitation gills in and out of the creature's body, and finally sent out at the other end. This precaution, not to breathe the same water twice, is a common instinct in all the insects we have studied. Having thus prepared her home, the little worm folds up her legs and draws her head close to her breast, and soon covers herself over with a soft pink skin. She remains quiet and helpless for some time, until she finds herself provided with all the appliances of a perfect insect under the soft skin. Upon the outside of this, at the head, are two powerful jaws, and these she uses to break down the grating at one end of the house; then she comes out and swims about for a while with great energy. See how she has improved. Before she could only walk at the bottom of the water, now she can swim in it, and presently she will be able to walk, swim, and fly. When the weather suits her, she comes to the surface, and lying on some grassy bank, the sun dries up the delicate skin, which then soon splits by the movements of the being within it. Poor thing, she feels very delicate and sickly for a moment when the air first rushes into the gills, which are now inside her: it is really as terrible an alteration for a moment as we should feel if plunged suddenly beneath the water. Mrs. Caddis soon gets accustomed to the new sensation, and pausing only for a few moments to get a breath of air, instead of a breath of water, she spreads her soft wings, and flutters, rather awkwardly at first, over the clear surface of the pond. Now is the time for us to find out which Order to put her in; and we cannot wonder that these insects puzzled the clever patient people who thought it of the greatest importance to put them in the right place.

The Caddis-fly has four wings; so far she is like the other insects we have studied; but the two front ones are different from the two back ones, and this makes a difficulty at once. Has the creature wanted to copy beetles, cockroaches, and grasshoppers? These two front wings are covered with soft brown fur, but the others are transparent membrane, like those of Dragon and May-flies. She has six legs, by whose long tipped spurs she can walk; and by means of the fringes on them she can also swim a little; but this privilege of swimming I do not think she uses until she is looking for a safe place to lay her eggs. This is the last act of her life; and the memory of her own early days prompts the good little mother to put her children where they may get plenty of food and building materials. Think how much nicer she is than the careless May-flies and Dragon-flies, who just drop their eggs on the water, not caring if they sink, or are dried up by the sun, or are eaten by the fish. Our brave Caddis risks all dangers—and there are many for her in the water —by going quite a foot below the surface and swimming from leaf to leaf until she finds one suited for the little green gelatinous packet she intends to put on it. Soon after this she dies, for, like the May-flies, her whole business as a flying insect has been to carry her eggs to some place where provisions are plentiful. This is a necessary precaution, because if they always grew up in the same place they would cause the supplies to run short. So much occupied has she been that from the time of losing her pink skin she has never eaten anything, for Caddis-flies have no mouths and no sensation of hunger. If she does not die a natural death, she is quite sure to be seized in her weak helpless condition by a hungry fish. Pretty much disgusted he is too when he has eaten her, for as she has never had any food and has laid all her eggs, she is only a bony skeleton, and is very much more a nut-shell than the soft sweet kernel within it.

There are some learned people in the world who would say that Caddis-flies ought not to be put in our Order *Neuroptera* at all, but have a fresh one all to themselves. Two very wise people, named Linnæus and Fabricius, decided that Caddis-flies, May-flies, and Dragon-flies were all closely related, and I follow their example, because they knew as much about insects as could be found out when microscopes, lenses, gas-lamps, railroads, and electric telegraphs were not invented. Mr. MacLeay, who came many years afterwards and had all these things to help him, said they were both wrong. As we want to make use of our own eyes, with a true love for the insects themselves, instead of the science, we follow the old fashion instead of the new, as it is more easily recognised.

### VI.—WHITE ANTS.

We must now take a long step in advance of our former position, and have a peep at some foreigners. I would not bring them in at all, only their conduct causes our own good little ants to be scandalised. These new creatures are called White Ants, and live in warm countries, but have nothing to do with real ants, who are closely connected with the bees and wasps. White Ants are *Neuroptera*, but do not live separate lives like those we have studied. No sooner do the insects we have seen leave the eggs than each one sets up a business for itself, and has nothing to do with its own brothers or sisters. The child of a White Ant is differently brought up: a nurse, a nursery, and food are all prepared for it, and the greatest care is taken to keep enormous colonies in proper order, They are the most destructive and the most interesting insects in the world; but may they always keep a respectful distance from us! Their homes are as large as small cottages, and they are governed with a precision that would be a model to any human institution. They live together in these hills in countless numbers, and are united by interest and affection in preserving their community from attack, and in defending their queen if any misfortune destroy the dwelling, which is a marvellous piece of industry, and shall be described before the tenants.

This erection is as large as a haystack, when built by the tropical species. The ground floor is the most important part of it, as in the centre of the hill at this part is the royal chamber. Into this the workers carry an insect who is in a perfect condition, having legs, wings, and all the

15—2

proper attributes of a Queen Ant. Then she brings with her a husband much smaller than herself, but with better wings. He is not of much importance, in fact, is only king-consort, and the queen would be in a

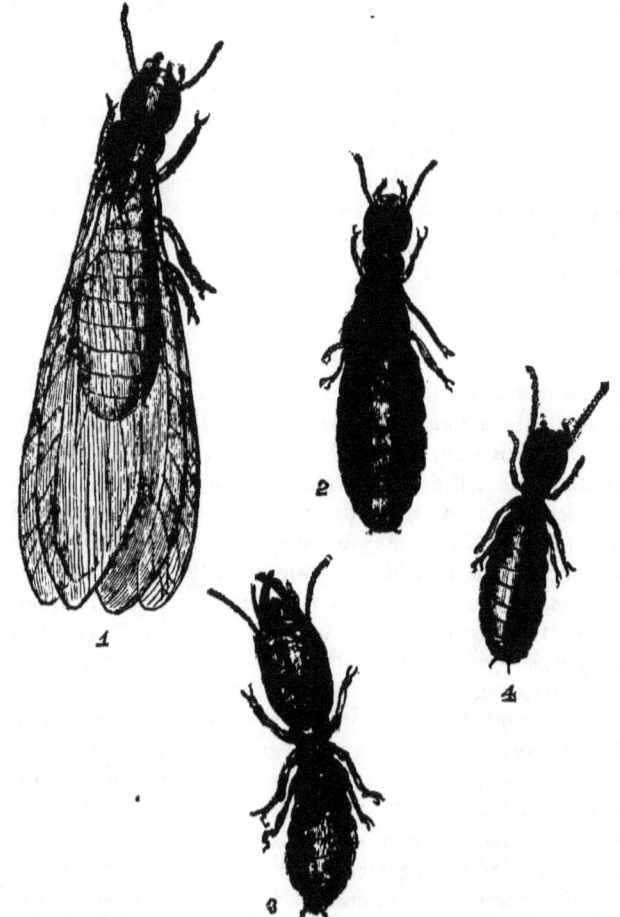

1. Male White Ant. 2. Female who has lost her wings. 3. Soldier. 4. Labourer.
(The queen is not drawn, as she is only No. 2 enlarged.)

strange colony alone, if he were left out. The new subjects take them both prisoners, shut them up in this centre chamber, and order the queen to lay eggs. Round this apartment are arranged numerous arched chambers, which serve for magazines and nurseries. As the cares of the workers increase, new rooms are added to the building, and connected always by passages, galleries, and staircases. This goes on until the ant-hill, or rather ant-mountain, is as high as an English haystack. We certainly do not want the African species to visit us, for their houses would take up too much room; there are, however, better reasons for keeping them at a distance than the space their colonies would occupy.

There are five kinds of individuals in each colony, and it has been almost decided that they are thus marked off in social rank:—The king, who is only a male with wings, is kept a prisoner with the queen. This unfortunate lady is a perfect insect, who once had four wings like her husband, but finding them inconvenient, she lost them as soon as she commenced her reign. Now the poor thing lays eggs incessantly, sometimes as many as sixty a minute, and eventually has 80,000 children. If she is not as badly off as the "old woman who lived in a shoe," I do not know where her equal exists. After the king and queen, the military force is of the greatest importance. This is represented by a set of creatures very much like the queen, only they never have wings, and their heads are enormously large in proportion to their bodies. These heads have powerful jaws, which come very energetically into action when anything goes wrong in the colony. Next in rank are the labourers, who do everything that is needed for everybody. These industrious creatures are nurses, woodcutters, purveyors, builders, and everything else, but they are so ugly! Indeed, there is nothing pretty about any of them, and those insects who are fifth on our list are the labourers changed into pupæ, who are waiting for their final metamorphosis into perfect insects, when they will have wings and fly by myriads from their old colony. When this happens, misfortunes follow to the inhabitants of the district near them; thousands enter by doors and windows, extinguishing the lights, smothering the table-cloth, sticking to the butter, falling into the sugar, and all the other miseries which attend a visit from them quickly follow. A few hours finds these insects in a most pitiable condition: their wings have fallen off like the petals from a flower, and with them their courage has gone also. True ants, who are not half their size, soon kill them: the frogs, the birds, the beasts, and human beings capture them by thousands, and one and all for the same purpose. Men, beasts, birds, and fish, have mouths and probably some sense of taste. Who then can wonder that the White Ants are eaten when they have the flavour of sugared cream and almond paste? I wonder an enterprising French confectioner has not found them out and sent us some *Crême des Termites* (this is what scientific people call them), or *White Ants à la Vanille* for bonbons. A traveller, who had eaten many of them, tells us he found them of a most delicate flavour and very nourishing, and that the natives mixed them with flour to make pastry. The Africans parch them over a fire and eat them as we do comfits.

Now I am sorry to say that I must leave my character of apologist general for insects, and tell the plain unvarnished truth about White Ants. They are destructive, vicious, thievish, ferocious, useless (apparently), and impudent. If I could think of anything worse to say I would, because they have done such serious damage to us poor humans. The only excuse for them is that they always work in the dark, and so anxious are they to do this, that if they feel under a moral obligation to eat the whole of a chair or a book, they cover it with a kind of cement which hardens. Inside this they commit their ravages, and you have a cast in white ant-clay of your favourite story-book or easy chair. It really is not fiction, but an unfortunate fact, that they mistake timbers of houses and furniture for their native hills, and carve out passages in them in a few hours. It is known that the owner of a room went out for a few days, leaving it comfortably furnished with good solid furniture, and returned to find it apparently as he left it. But the difference, though unseen, was soon felt. The first chair he touched crumbled to pieces in his hand, the back parted from the seat, the legs fell on the ground, breaking like rotten wood; the tables and sofas were all in the same condition; the looking-glass was plastered to the wall with cement, and the frame eaten up. A great traveller lost a pipe of fine old Madeira because these creatures took a fancy to the cask. The writer of the most interesting account of them that we have lost his microscope because the White Ants had a turn for scientific work, which they displayed by joining the brass-work and glasses with cement and then eating the wood. Enormous trees are reduced to the finest dust by their perseverance, and retain their branches and twigs like solid trees until the first wind scatters them with as much ease as it would smoke. Some one suggests that this is the use of their wood-eating propensities; these trees are so large that they would block up streams if drifted into them in a solid state, and White Ants thus serve the natives by making their wood rotten. This seems to me a very unsatisfactory apology, but as a poor excuse is better than none, we will give these foreigners the benefit of it and leave them, merely altering the title of our article to *Insect Pests* so far as the White Ants *(Termites)* are concerned.

## BEES AND BEEHIVES.

*"Where the bee sucks, there suck I."*
*"The bee is small among the fowles, yet doth its fruit pass in sweetness."*

HOSE who live in glass houses must not throw stones;" but as there is no rule without exception, we throw it out as a suggestive inquiry to any of our captious young friends whether the little winged dwellers in glass hives, exemplary in all the relations of life and faultless in their social and moral qualities, may not be privileged to have a fling in any direction.

"Not at me, if you please," tartly replies our wiry friend, whose indefatigable industry bears testimony to the fact that he learned to some purpose when a child the infantine ditty, "How doth the little busy bee," &c. "Not at me, if you please. I mind my own business. I am up early and late, never trouble myself with other people's concerns, and no one can accuse me of idleness; and the hoards I have laid up against hard times are plain proof that I gather honey all the day, and sometimes half the night too."

None of the sweetest, we fear, if it smacks of the tone and temper wherewith it is proclaimed. It may sometimes be found to be but lost labour that we haste to rise up early, so late take rest, if our worldly store lack the mellow sweetness of an abundance culled from earthly flowers under the sunshine of a heavenly blessing among the unselfish fellowships and countless charities of life, which are as the pleasant hum of bees in the sultry air of a summer's day of toil. But not being ourselves the denizens of a glass house, we will leave that stone for our winged friends to fling.

"Not at me, surely not at me," cries a second, with careless confidence. "I was never out of temper in my life; I take things easy. 'Live and let live,' 'Care killed the cat;' and if I cannot get things to my mind, I never fret; I e'en let them pass. 'It will be all the same a hundred years hence,' if things *are* all at sixes and sevens now."

Ah, good friend free-and-easy, bathe as you may in the waters of self-approval, I think we shall yet find a vulnerable point, even in that happy-go-lucky style of yours. What work will you ever accomplish, what edifice rear, that will not bear the marks of such careless ease? In the hive of glass it is all sixes and no sevens. Very particular gentlemen are these builders of a suitable residence and storehouse for their colony and their queen. Destitute of any apparent guide of measurement, what need have they of inch rule, of compass, or artificial hexagon? The eye, the mind —whatever that may be in the insect world—the hand, all work in faultless union, and produce a result as marvellous in its exact proportions as in its adaptability to the several needs of the architect; and if, in the material fabric of their city walls, there be a strength and adhesiveness inseparable from a degree of bitterness to the taste, the sweetness and abundance of the store within those walls invite us to leave to its winged workers the task of flinging this stone at the slovenly and the careless.

"Not at me," sternly ejaculates a third. "I am particularity itself. I see to it with my own eyes, and if the work done by my orders is the thousandth part of an inch awry, I have it all undone. Those that work for me must make straight work of it, and look sharp, or I come down upon them short and sharp."

Sharper truly than that sting for which the colonists of the hive find no such use. No task-master is theirs, no idle hands, no careless workers; each works for all, and all enjoy the fruit of the labour of each. Excepting the royal lady, who, engrossed by cares maternal, sits apart in the spacious chamber constructed for her by her loving subjects, all work in harmony. Each one knows his task, and though man knows not to what tribunal of conscience their fidelity may be referred, or by what meed of self-approbation rewarded, we have yet to hear of one among the Bees who will bind heavy burdens on others which he refuses to touch himself. So, in some interval of leisure from their more immediate duties, some speculator among them on human systems may fling this stone also.

"Not at me," exclaims Number Four. "I work as hard as any. Example as well as precept is my motto, and I carry it out, and never tell any one to do that to which I am unwilling to put my own hand. Gladly should I take a little relaxation, or breathe a breath of fresh air now and then, if I felt it consistent with my duties to others to enjoy for myself what I deny to them."

Then go to the Bees, thou slave to the desk and the wheel. Where would be the sweetness of their honey, the strength of their wax, the buoyancy of wing, and that cheerful hum—where the elasticity and health of their busy insect life, were not fresh air and sunshine a part of their daily life, a material element in their existence? Nay, more, the first hours of their daily round of activity are spent outside the hive; and with the nectar of the flowers they imbibe the breath of early morn and the first freshness of reviving nature. The day-spring is to them truly the spring of the day's activities. Their first flight is upwards, their earliest effort heavenward. Those pupils of nature's instincts alone return to their indoor life about the hour when worn and exhausted men and women creep forth from their imprisoned life, to breathe the noxious dews of evening, the staleness of an atmosphere from which the vital energy of sun and oxygen have been withdrawn, and return to add the fatigue and oppression of night to that of the day before resuming the diurnal toil in the narrow hexagonal cell of their daily labour. Yes, fling that stone from your airy height, ye happy living things, that mount on the morning breeze and scent the early odours of the dewy flowers. Ye may find a vulnerable point in many a son and daughter of leisure as well as in the person of the mechanic, the student, or the man of business.

"Not at me, not at me, for I hate artificial life and all that belongs to it," gladly exclaims the child of the country. "Give me the simple Humming Bee in its home in the wild woods, where it makes honey for its own wants alone, enough for its family requirements, and never invites the covetous propensities of rapacious man to stifle their young lives and rob their winter store."

Ah, there you remind us of the best that can yet be said of these faultless creatures. No useless hermit colony, with self and selves for the end and object of their labours; they never stop short when they have built their comb and gathered honey for themselves alone. As the summer is prolonged, and their sphere is enlarged, so are their efforts expanded, that other beings, often their most cruel enemies, may share the sweet results of their toil. Unspoilt by artificial life, and, so far as we can see, utterly unchanged in their simple habits and lives of active usefulness by all the refinements of cultivation, the Bees inhabiting the most delicate of glass hives in the loveliest of gardens are as busy as the Bees before the Flood, as united in their action, and as ready to quit the homes of luxury at the call of their leader, as if they were the first Bees on whom the necessities of exertion, of union, and of forethought had devolved.

Humbly and contentedly they betake themselves to the rudest shelter, and seek in the wildest retreats of nature as in the richest garden the nectar concealed alike under the simplest petals and the fragrant cluster.

Too active, too happy, and too kindly to fling a stone metaphorically or to use literally their sting while unprovoked, their pleasant humming falls on our ear as one of those soothing sounds in nature, like the plashing of the waterfall, the sough of the wind among the trees, or the music of the

> "hidden brook
> In the leafy month of June,
> That to the sleeping woods all night
> Singeth a quiet tune;"

which harmonize with good and sacred thoughts, and suggest alike to the contemplative and the active mind the combination of their respective gifts in a useful yet not thoughtless existence.

But the Bee was not born to hum only; it only hums to beguile its work. Let us watch the little tribes as they pass to and fro from their hive this morning. Fear not their stings if we stand aside and do not put ourselves in the way of the busy citizens. If some human monster will obstruct their passage and come between them and their storehouse gate, and does not move on at the sound of an angry buzz, the way *must* be cleared, even at the expense of an occasional sting. So let us keep to the side, and they will be far too busy to turn from their labours to examine, still less to molest us.

First of all we see some half-dozen loitering idly about the door. No, not idly, for see how they scrutinize each fresh arrival, as if to say, "Have you brought home a proper load of honey, or have you only been at play?" But this is not their chief duty. They may be only making a passing salute, inquiring of their returning friends about the state of the weather or the flower crop, whether the white clover is plentiful or the lime-trees are exuding honey well. They are there to warn off intruders. If we approach too near in front, one of these sentries will dash forward with an angry buzz, and if we do not wisely take the hint, the brave little soldier will soon return with a reinforcement from the guard-room to enforce the command. Horses, dogs, and other animals understand this threatening buzz very well, and soon retire. But their smaller foes are not so easily repelled. The sentinels touch with their antennæ every creature that tries to creep in, exactly like a soldier on guard demanding the pass-word. Now hornets, wasps, and moths, who, like human beings, do not make honey, have a very sweet tooth, and know where the nectar is stored. They often try to pass the barrier, and, being individually stronger than a Bee, would succeed were not the sentinel speedily reinforced. We may often see dead wasps laid in front of the hive, and sometimes can witness a pitched battle, though the intruder is generally driven off and seeks safety in flight, like a robber with a bad conscience. When the guard is relieved at night the door is often barricaded with a wall of *propolis* and wax to keep out the night-flying moths. Slugs are not so easily warned off, and will sometimes creep in more from stupidity than from mischief. When this happens the Bees kill it, and, unable to drag it outside, plaster it over with a coating of *propolis*, the resinous wax with which they line their dwelling, and leave the mummy at the bottom of the hive. If it is a snail which has intruded, they send him within his shell by a single sting, and then wall him in, and cement his living tomb to the floor with its mouth downwards.

But the Bees who are passing and repassing the sentries are not all laden alike. Some of them have little yellow or red tufts on their legs, others have none. But all who return are laden. There are three substances required in the hive—pollen, or bee-bread, the food of the youngest larvæ; wax to make the combs; and honey for the support of the community. Those with tufts on their legs have been collecting the pollen from the stamina of flowers, which they carry worked into lumps, and retained by the hairs on their hind legs. The purveyors of honey and wax carry their stores, drawn from the nectar of flowers and the sweet juices of trees, in their throats.

To understand how the pollen is carried, we should examine the hind leg with a microscope. We shall then see that the upper joint is flattened, and its edges surrounded with stiff hairs, which form a sort of basket, into which the powder is put by the action of a sort of brush of short hairs which covers the lower joint. When the Bee enters a flower, it takes a plunge among the pollen, covering its whole body, and then brushes itself down into the basket on its thigh, till a good-sized ball is formed. If it cannot complete its load in one flower, it will always seek out another of the same kind; but never on any account will it mix the pollen of two different kinds of flowers, unless by accident. Thus we always see the ball of a uniform colour, red, yellow, and white pollen being never mixed. When the pollen-bearer has entered the hive, it pushes its burden into a cell, and another Bee follows, and kneads up the mass with its jaws, packing it tightly down.

The honey-gatherers and the wax-gatherers—for these are really the same—draw in the juices from the flowers by their trunk, which serves as a mouth and a pump, through which the liquid passes into the first stomach, and thus is carried to the hive. But often the labourer does not wait to disgorge itself into the cell, but, on arriving at the door, opens its throat, when another Bee, perhaps one more aged and feeble, and less capable of field work, though perfectly fit for domestic toil at home, receives the sweet load and discharges it into the storehouse. Of course the workers feed themselves while they are out, and often give a supply to their friends by the way. They also feed those employed on the combs by going to the place when they are working and stretching out their trunks. The other Bee inserts the end of its trunk, and sucks up

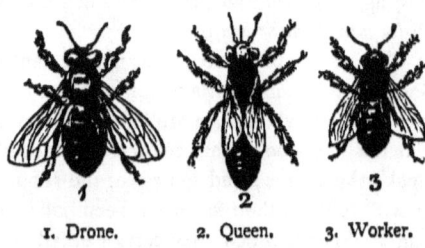

1. Drone.  2. Queen.  3. Worker.

the offered honey without having to leave its work. But how is the wax supplied? This was long a problem, till it was discovered that wax was a secretion, or rather an exudation formed in very thin layers between the plates of the abdomen of Bees. That it is in some way made from honey Huber ascertained, because Bees fed only on pollen did not secrete it, and those fed on honey or syrup did so.

Let us now follow the workers inside the hive. And here, if we have not got a glass hive through which to watch, we must be content with a peep by the eyes of others. Just beyond the sentries are stationed those who relieve the purveyors from the field. Others are busy in cleaning and sweeping out the bottom of the hive, others in storing honey or bee-bread, more still in forming new combs, and many others in tending and feeding the young larvæ in the breeding-cells, or waiting on the queen. For all these working Bees, industrious though they be as labourers, assiduous as nurses, are toiling not for their own—for they never are either fathers or mothers—but for their brothers and sisters. As with the wasps, so among Bees, there are three sexes—the drones, or males, who are only hatched in summer, and neither work nor sting; the queen,

of whom there is only one at a time in each commonwealth; and the
mass of the community, or workers, who are in reality females stunted
in their growth, and differently fed and housed in their infancy. So far
they resemble other Hymenopterous insects, as the ants and wasps;
but in the origin and government of the little commonwealth, which
each hive in reality is, they differ much from their nearest cousins, the
wasps.

The female, or Queen Bee, is far less active than the queen wasp.
Very few people not bee students have ever seen a Queen Bee. Unlike
the lady of the yellow bands, she takes no share in the founding of a
new colony. She never works from the day of her birth to her death.
She is worshipped like an Eastern potentate, in the strictest seclusion,
indulged and petted, instead of going forth with the first warm rays of a
spring morning, like some hardy Norseman of old, to found new colonies, and lay the foundations of a busy city. Nor is she to be blamed
for this. Her form and nature forbid the effort. Though her body is
twice the size of that of a working Bee, her wings, unlike those of the
wasp, are very short, and can only bear her up for a little time with
great effort, while her abdomen is far too heavy to enable her to move
about with ease.

But how, then, is a new colony to be founded? Here comes in a
wise provision of Nature's God to meet the case of the Bees. They
build no houses for themselves. The time which the wasp must devote
to the preparing, fortifying, and enlarging of the walls of its house, the
Bee, relieved of this labour, expends on the collection and preparation
of food for the winter. The one perishes with the early frost; the others,
huddled together, and securing warmth by their crowded numbers, are
ready to recommence their labours with the opening of the first crocus
of spring. In a state of nature the Bees find hives in clefts of the rocks,
in hollow trees, and sometimes in holes in a dry sandy bank. There they
find dwellings ready made to hand, and quite as convenient as the most
comfortable straw hive or the most neatly finished wooden box which
their owners provide for them in servitude, though not in captivity. We
see how Bee nature remains the same through all generations, how the
new swarm will get into a chimney, a hole in the wall, the eaves of a
house, or under the thatch, the hereditary instinct having never been
lost, through thousands of generations of hive-homes. Still the Bee has
no strong prejudices, and when the bee-keeper has provided a comfortable hive, and smeared it well with sweet syrup, the queen, if once she

has dropped into her quarters, and found them warm and sweet, is not disposed to assert her right to choose her own residence, but settles down at once, the monarch of all she surveys within.

In many respects the hives which man provides suit the Bee taste better than most of the homes they could find for themselves. In the first place the Bee likes neatness and symmetry, and the combs can be formed more evenly and regularly than in a shapeless hole. Then the holes in the rocks have often large openings, which it is very difficult to build up sufficiently to prevent the intrusion of many unwelcome visitors. Experience, too, has taught men in different countries to provide hives suited to the climate. The English straw hive is made for warmth, and is well thatched with an extra covering in winter, to prevent the frost benumbing the little prisoners. I do not think the Bees like the wooden hives so well as the old-fashioned straw ones, unless they are double-cased and very well sheltered, for they are much colder. But they have no objection to a wooden box to work in in summer, with a thatched house above it, to which the whole family retire when the summer season is over. In North Africa, where warmth is not required, and where the Bees are in fact wild, and allowed to roam and choose for themselves, the Arabs hang up in the trees rolls of cork bark with a cork lid, and quite open at the bottom. The Bees have the instinct readily to choose these, because they are safe from prowling intruders, who, unless they are winged, cannot get into the hive, which is suspended from a bough; while if any winged thief attempt to fly in at the bottom, there is an army of defenders ready to dash down upon him, and give him a lesson in honesty. Thus I have seen an incautious bird, which has earned its name of bee-eater from its partiality for devouring Bees and wasps (*Merops apiaster*), skimming like a swallow and snapping up the workers as they returned heavy-laden with sweets, till at length, hovering too closely under the hive, a myriad of indignant soldiers dashed out together, and whether they stung him or not I cannot say, but he soon sheered off thoroughly humiliated, and came back no more. These cork hives are also cool, so that the combs do not melt under the shade of the tree in summer, though they keep out the winter rains.

In Palestine, where the climate is still hotter, the bee-keepers have devised a yet cooler fashion of hive. They make a large pipe or cylinder of clay, about two feet in diameter and more than a yard long, open at the ends. They smear it inside with honey, and when they have shaken the swarm into it, they lay it flat on the ground, and plaster up each end with clay, leaving only small front and back doors, into which no mouse can creep. They generally heap about twenty-one of these hives in the shape of a pyramid (for they keep vast numbers of Bees) in a tier of six at the bottom, diminishing by one each row. The whole are then plastered over with earth and clay, and as they stand in the yard look very like a hen-house. At both ends are stuck up a number of boughs, the more prickly the better, for the double purpose of assisting the laden bees to alight, and of protecting the entrance and the neighbourhood from "bee-eaters," winged or creeping; for lizards as well as birds are among their enemies there. When summer is nearly over, the Syrian bee-master begins to help himself. This he does on the principle of "live and let live." With face and hands well muffled, he removes the clay from one end of the tube, and with an iron hook pulls out the combs one by one, handling them carefully, as the hook detaches them from the top of the hive. If there are any young or bee-bread, he carefully cuts off that portion of the comb and replaces it in an upright position. He takes care to leave enough honey for the winter store, only removing the combs at one end. The next year he opens the other end, so that the Bees are compelled to renew them every two years, and they never become clogged, as in our old hives, with the cast-off skins of larvæ till they are too small for use. These Bees must have the bump of locality largely developed, for though I have seen a pyramid of seventy-eight hives, I never noticed the busy Bees at a loss to find their own. Lighting on the bushes every minute in swarms, each, after a minute's pause, went direct to its own home, though there was as little to distinguish one from another as in the rows of houses in some new suburban street.

In tropical countries again, as in India, there are Bees which dispense with hives altogether, and which hang their combs openly under projecting ledges of rock, generally in the deep ravines of rivers, where they are secure from all enemies except winged ones, and these we must suppose they manage to keep off by a large standing army, for the soldiers must be increased in proportion as the position is exposed.

But how is the colony formed? How is a new kingdom established? Here the natural increase of the population acts along with the instinct of the queen. No queen can endure a rival near her throne; but the working bees, careful to provide against accidents, and maintaining that "the king never dies," take care each spring to rear a few female eggs in cells on the edges of the comb, very solid, and much larger than the others, and to feed the larvæ in these cells with food different from that supplied to the workers, being heavier and sweeter. This alone, along with the greater space of the cell, is enough to form a queen instead of a worker. When the queen larvæ are nearly ready to leave their cells, they make a peculiar noise, which very much disturbs the peace of mind of the reigning queen. She rushes over the combs in a fury, endeavouring to tear the young queens out of their cells; but each is guarded by a body of workers, who, at other times so respectful to their monarch, now venture firmly to resist her. This is more than any lady, accustomed always to have her own way, can endure. She rushes about distracted, and even frenzied; drops eggs anywhere, regardless of the use of waxen cells; and runs over the bodies of the workers, as they cluster

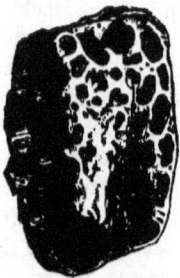

Queen Bee Cell.

on the combs. None so mean to do her reverence—none draw aside and stand respectfully in file on either side as she passes, after their ordinary habit—she has lost her guard of honour, and is indeed a deposed sovereign in her own palace. But though her subjects seem for the moment to be rebellious, it is rather a panic than a conspiracy which has seized the community. They so far forget themselves as to strike their royal lady—she so far descends from her dignity as to run a Malay muck, striking every one she meets. This only aggravates the tumult. Every worker as it returns laden from the field is seized with the excitement, and runs about with pollen on its legs or honey in its stomach, never thinking of depositing its burden, but, smitten by the epidemic of confusion, joins the general scramble. At length the queen finds her way to the door, and rushes forth to cool her fury in the open air. It is only the second time in her life that she has ever left her palace. But now it is for ever. But she is not alone. She finds thousands of her subjects still devoted to her, chiefly the elder and more experienced, who prefer to follow the fortunes of their self-expatriated sovereign rather than run the risks of republican anarchy in their native hive. One breath of fresh air seems enough to calm their ruffled spirits. The queen cannot fly far, and, following the guidance and example of some of her devoted attendants, she settles on a branch or in a cavity of a tree, rock, or building. The swarm collects around her. That extraordinary cluster is formed of one row of insects hanging on, with another and another suspended by their fore legs hooked in the extremities of the hinder legs

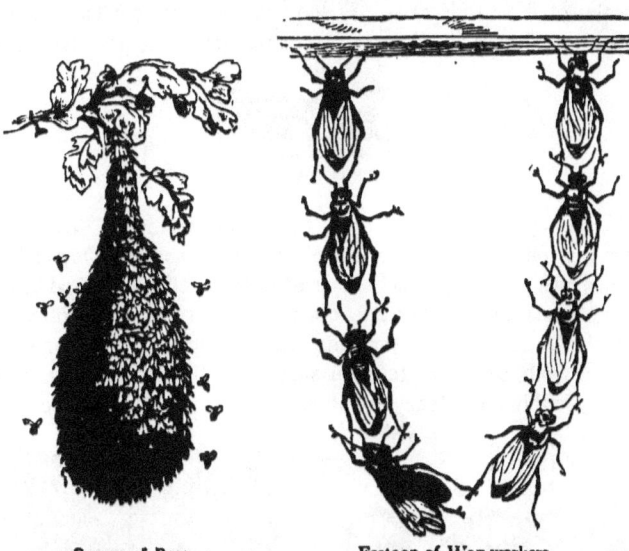

Swarm of Bees.    Festoon of Wax-workers.

of those above them, till the first row seems to have to bear a weight a hundred times that of their own body. Then when the cluster has been shaken into the hive provided by their owner, if only the queen have been enclosed with the others, without the delay of an hour they betake themselves to form combs and to arrange the furniture of their colony, evidently feeling that, so long as they have their queen with them, they are no exiles. They can quit their country, but not their allegiance. If, however, they find that too many have left their old home, and that the swarm is needlessly large, a portion will return in two or three hours.

But now how are they to go to work, supposing we have them safely encased in their empty hive? The old queen has many eggs to lay, and these must not be wasted. There is no time to lose, and combs must at once be found. Here comes into play the most marvellous part of the Bees' constructive instinct. The workers who have fed themselves on honey attach themselves in a row by their fore legs to the hive, and hang motionless, while others hook themselves on to their hind legs by their own fore legs, and others again hang on to them, till a long chain is formed, and looped up by the different chains joining at the bottom. Thus they all patiently wait, till they have secreted plates of wax between the scales of their abdomen. As soon as a Bee finds it has a plate of wax ready, it leaves its place in the ranks, takes the wax between its

Commencement of Comb.

mandibles, kneads it, and fixes it to the top of the hive. This done, it at once goes off to the fields for a fresh supply of honey. Others follow with their wax; and when there is a sufficient quantity deposited to work with, the masons come forward, hollow out the shapeless lumps, and mould the bottom of the first cells. Fresh supplies are brought, and soon the comb begins to take its shape; the hexagonal cells, of the thinnest possible consistency, being all laid horizontally, back to back.

 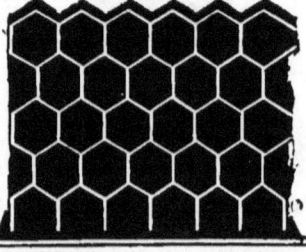

Sections of Honeycomb Cells.

While some are extending the foundation of the comb right across the hive—for Bees always lay their foundations at the top and work downwards—more wax-producers are adding little lumps of wax, like masons' labourers bringing materials for the comb-builders, who without rule or compass fit all with the most minute exactness, and the work proceeds with wondrous rapidity, several thousand cells being sometimes formed in a day. One set of Bees shape out the bottom of the cells, roughly moulding the six sides. Others, whom we may call finishers, succeed them, and beat out the wax with their mandibles till it is as thin as tissue-paper, and plane down all the roughnesses. In a day or two there is accommodation provided for the eggs which are the hope of the colony, and pollen and honey are being collected for their support so soon as they hatch, which is in about seven or eight days after. But the cells are not all of the same size. About a tenth of them are larger than the others, though of the same shape. These are for the drones or males, and it is strange that the instinct of the queen, who creeps over the comb attended by her servants, and lays an egg in each cell, knows at once the difference between a male and a female egg, and drops each into its intended place. But the eggs of the Queen Bee no way differ from those of the workers. It is not till the swarm has been housed for a week, that the Bees build a few thick circular cells projecting from the edges of the comb, to receive eggs which are to be royally reared, so as to provide against any calamity befalling the reigning sovereign. It is a large hive which contains twenty of these cells, and the queen only drops an egg into them at intervals, lest too many claimants should emerge at once.

But let us now return to the old hive, and see what has taken place there since the departure of the old queen on her voyage of discovery. The young queens whose threatened advent so alarmed her, have not yet left their nursery. Nor will the nurses who courageously defended their charge allow them to do so yet. As they break through their waxen lid, they build it up again, but pass a little honey through an opening, to keep the young lady quiet till they see fit to let her come forth. No sooner has she emerged, than, like some Eastern potentate, she inaugurates the new reign by searching out and slaying her nearest relations, rushing from one queenly cell to another, and trying to kill the imprisoned inmate. The nurses vigorously resist, and if the hive has increased sufficiently to cast off a second swarm, she is allowed to follow the example of her mother and lead forth another party of emigrants. If not, she is permitted to glut her jealousy on those furthest advanced towards maturity. She tears open the cell, and at once stings the helpless prisoner. As soon as she has gone in search of another victim, the workers also enter the cell, and drag out the carcase of her slain rival. When all are slain, the queen remains quiet, and devotes the rest of the season, till the approach of winter, to the laying of eggs, at the rate of two hundred a day. But in a month or two a new massacre begins. At the approach of autumn, economy becomes the cry. There must be no eaters who are not workers, and the helpless, stingless males, whose work is done, have to be got rid of. The drones are pursued from one part of the hive to another, set upon by the workers, and stung to death. Their carcases may then be seen strewn by hundreds on the ground near the hives, where they have been dragged out and dropped. Not even the larvæ or pupæ of the drones escape. The cells are torn open, the young ones pierced, their juices sucked, and the bodies thrown out. It is only if any accident has happened to the

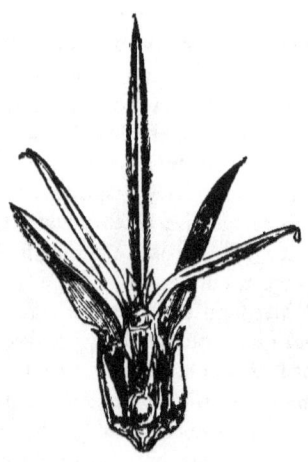

Tongue of Worker Bee.   Sting and Poison-bag of Worker Bee.

queen, and the nurses are carefully rearing some half a dozen in royal cells, from whom to choose a successor, that the drone pupæ are allowed to live. If a queen perish or is lost, and there are no royal cells, or none with eggs in them, the workers at once build several proper cradles and remove some of the newest laid female eggs into them to be reared for the throne. The moment the first hatched comes forth, she is at once surrounded by crowds of courtiers, who brush her, lick her, serve her with honey, and attend her every movement, while all stand aside respectfully in file as she passes to and fro.

Such is Bee life. And who can throw a stone at either its policy, its industry, or its architecture? Yet these busy insects are not toiling for themselves alone. Their labour is consciously for the community, unconsciously for us also. The supply they store in summer is far more than is needed for their own wants. Why, then, should we not be content with a share, and leave the hive to work for us another year? We all laugh at the man who killed the goose that laid the golden eggs. Yet this is exactly what too many bee-keepers do in England. Instead of levying a tax on all their hives, they leave some untouched, and destroy the heaviest and the best worked by suffocation. This is one reason why we have so few Bees compared with most foreign nations. To smoke their Bees is a barbarism which would horrify the Syrian or the Greek. It is very easy to get the honey without killing a Bee, or running the risk of a sting. Seventy years ago it was discovered that the smoke of powdered puff-balls lighted under the hive will stupefy the whole swarm for some minutes, so that the hive can be examined, and the combs with virgin honey cut out, the pieces with young Bees and pollen being left or replaced. The same effect is just as easily and safely produced by chloroform; and the bees recover without any injury so long as they have not been too long exposed to it. Let us hope ere long bee-keepers will learn to combine humanity with profit, and will as soon think of smoking their Bees as of burning their hayricks.

## BUTTERCUPS AND DAISIES.

### I.

MOST people know a Buttercup when they see it. Grown-up people, I mean—for, of course, boys and girls do; but there are not many of the G.-U.P. who care for flowers, except the lovely ones that grow in gardens and greenhouses; those we are going to talk about are generally left for the benefit of the cows and the children. Do not be alarmed and think you have a great deal more to learn about these when you know something of a Buttercup. Not so, for the general principle is the same in all the plants, from the great oak to the tiniest weed. The differences of classes and species will not come under our notice, as we are only going to find out how they grow. How a Buttercup has a power of discernment is tolerably well known to little folks; but in case any of you should not be aware of it, follow my directions carefully, and you can prove for yourselves. Gather the largest and brightest yellow Buttercup you can find, then hold it closely under the chin of your companion (supposing it is not papa or your big brother, who may have long beards), and if the colour is reflected there, you will find out that he or she is very fond, probably too fond, of butter.

It has always puzzled me to know why things grow, and why dirty-looking bits of stick should come up out of the ground changed into beautiful flowers and fruit. When I was a wee child I would ask, "Why does that little plant grow into a big bush, mamma?" and I always received the same answer, "G d makes it grow, dear." This was, of course, true enough; but what I really wanted to know was this, "God tells that plant it is to grow, and it grows; how does the plant do it?" When I was old enough I tried to find out for myself, and, with the aid of a large microscope and some difficult books, succeeded in learning a few of the wonderful ways in which the dark little root begins the work of forming pretty flowers for us to look at. But before I go on any further I shall stop to say, that what I am going to teach you is both wonderful and interesting, and because it is both of these things, I shall not try to tell it you in the form of a fairy story, or in any of the clever pretty ways in which you learn so much; for I know I am talking to really sensible children who do not want to have knowledge given them as though it were medicine covered up in marmalade, at least this is what I always think when I find pretty stories written for the sake of teaching things that we should all be glad to know about without having them given us dressed up in a tale. I don't find fault with the tales, you know, only it seems making believe the knowledge is nasty, like the medicine, if it is put into one. Now, as it is nothing of the sort, you shall have a scientific lesson all to yourselves, just like grown-up people, and with no pretence that you will not take it unless covered up with the sugar or jam of a fairy book.

Before I could learn anything about the flowers, I had to get some, and I have three Buttercups with roots here on the table. Nothing very wonderful in that, you say: the one I hold in my hand is just like others you could find,—it has a root, a stalk, and leaves, that is all. Nothing very unusual, certainly; but somebody says that it has a flower, and that the flower has yellow leaves, and little yellow things in the middle. This looks true, so I must not contradict you until I give you a good reason for understanding that every part of a flower is a *leaf*,—in fact, everything that the stalk and branches bear are leaves. You all admit that a plant is pretty, and only the root is at all repulsive; so our first business must be to find out what the delicate green leaves and stalk want with such an ugly appendage.

If I give you a pot of mould and this Daisy, and say, "Plant this," I believe every one of you would do it; probably not one of you would put the flower in the ground, and the root upwards. "How silly!" those boys think; "any muff knows plants won't grow unless the roots are in the mould."

"Exactly so," I reply; "but, then, why not?" The three Buttercups before us must help us to find out. The first that is lying on the table I pulled up out of the croquet lawn this morning. It is withered and dried; the leaves look unhappy and shrivelled up; the yellow petals have fallen off, and, to all appearance, poor thing, it is quite dead.

No. 2 was taken up yesterday and put in water. This looks quite bright and bonny: the buds have opened, and the flowers are capable of finding out the important fact I mentioned, if any one likes to come here and have their chin examined. "Of course, the water keeps it alive," say the boys, who always know so much more than girls—or think they do. But I quite agree with them this time, that it is owing to the water the one plant is in better condition than the other; and it would be possible to revive No. 1 by supplying it freely with water. We have thus learnt one fact: plants drink. How they drink is our next business. Do I proceed with the Buttercup as I should with a thirsty child? If you were dying of thirst it would not be much use taking you near a glass of water, unless some one was there to put it into your mouth; and that is just the case with the fainting Buttercup; but I do not put the water into its mouth, "contrariwise," as Tweedle-dee would remark, I put the mouth or mouths into the water. Thus we have found another fact: plants have mouths which are all over the roots, but these mouths are not like yours; for you bring water or tea or coffee close to you, and open your mouth to receive it—that is all; the action is on your part, and with all animals it is the same; but you might throw Pussy's saucer of milk over her for a long time, and it would not do her so much good as if she drank it. Observe, I say "not so much good," because she would have some benefit, although it would probably be very little. People who are very ill, and who from some reason cannot eat or drink, may be kept alive for a time by having milk or soup baths, but of course do not live long upon it. Now, if our skins were *quite like* the roots of plants, the nourishment supplied to them from the outside would get into our bodies, and we could exist, perhaps, without solid food.

The root of a plant, before it is put in the ground, is like a dry sponge, which soon almost doubles its size when put in water. A root also absorbs the moisture which surrounds it, by means of the tiny holes with which it is covered; these are the outside cells, and as soon as they are filled they burst the thin wall that separates them from the next, which, when filled, does the same to the next cell, and this goes on until the whole root is softened. But, you will say, merely absorbing moisture would not make new roots and leaves; it would only make larger parts that already exist, there must be something beyond passages for water.

If I cut this sponge in any direction, it is full of tiny cupboards, something like a honeycomb; and, indeed, the root of a plant cut open is more like honeycomb than sponge, for every cell is full of a substance, or rather a liquid, which is food for the plant. This is composed of different elements that are for the most part too difficult to come into my work at present; but one of them is simple enough, and this we will try to understand. I cut open a large potato, and a very thin slice of it will show you, in the cells not burst with the knife, numbers of tiny things that scientific people call granules, and these are starch, which is the food of all plants. I hear you making another objection. "Plants can't live on starch," you say. Perhaps they cannot entirely, just as you could not live on arrowroot altogether; but then you like it sometimes, and it is very nice and very good, too, for sick people. "Oh! but then arrowroot *is so* different." It may be, but I *know* that arrowroot and potato starch are the same thing, and contain the power of helping the plants to live throughout the cold winter; that it is found in *every root*, and is kept in tiny cupboards which vary in size and shape with the growth and habit of the plant. We have now found another fact: plants require food. The starch stored up during the winter dissolves in the warmth of the spring, and pushes the fibres out into the moist earth to find more nourishment than is contained in the simple root. Now for the answer to the question I asked when I was a little girl, for it is that I am telling you.

Here is a root-fibre, but it has to be magnified before one can at all see what the inside of it is like. The real fibre, this brown thread, looks too delicate to push its way into the hard earth, and one expects it to break very soon in working its way round the stones in the ground. So it is, and but for a method of restoration that renews the outside sheath, there would be little hope of our having many wild flowers.

You see round the outside edge the layer of cells that bear the hardest work; and as this gets worn and withered, by forcing through the moulds, it is replaced by a fresh set of cells from the interior of the fibre on the "growing point," represented by a dark spot. This dark spot is a collection of cells packed closely together, and of slightly different forms from the outside ones.

We have already noticed that a root has neither leaves nor buds, and avoids the light, and that where the fibres start from there is a thickened part of the root, called by gardeners "the stock," and the stem and lower leaves spring from the upper part of this. One could almost fancy the little cells that are swelling and growing here hold solemn councils as to which of them shall make root-fibre and dive into the deep dark earth, and which of them shall come upwards to seek the light and be dressed with pretty green leaves and yellow flowers. This is only fancy, however, for the fact is that the earth's moisture passed into the root of the Buttercup, filling up the outside cells, and then bursting the thin partitions that separated them from the next, and causing new divisions to grow up. This process is perpetually repeated as the plant pushes forward. We have now partially learnt what goes on in the root and root-fibre, so we will now examine the stem that soon makes its appearance aboveground.

A piece of wood shows, on a large scale, what I want you to see: that is, bundles of fibre with soft matter between them. The soft matter is called cellular tissue, and is the part of plants I have been endeavouring to make clear to you. We will now take a piece of boiled rhubarb to examine how veins similar to those in the trunks of trees are formed; for this will show us a part of the process of development that may be traced from the poorest weed to the most lofty tree. Now, if I ask you to tell me any particular characteristic of a piece of cooked rhubarb, you will most likely say it is "stringy;" and this is what we want to find out, namely, the difference between the cells which are distributed over the whole plant, and the "stringiness" of this new condition which we find in the stem.

I now divide one of the thick strings into separate parts, just as I should a piece of grass or straw, and I find that this, like the rhubarb pulp or the potato root, is also full of little cells; but, besides these, there are long tubes having their sides curiously marked with delicate threads, that run in a spiral direction, or in the shape cotton will form that slips off a new reel. But rhubarb tubes are a thousand times more delicate than cotton, so how tiny must be the veins and cells of the Buttercup! Just at first, in the baby plants, there are none of these veins or fibres to be seen; they are in this respect very like human babies, who only have gristle for bones. These fibres are formed in plants by a set of cells placed one above the other, and, as these break, the passage is formed through which the moisture passes, and collections of these tubes close together form the woody fibre of trees. This you may easily understand by watching a man chop wood: he is careful always to notice which way the fibres go, and then puts his hatchet so that it shall split between them and not across them, for he knows it would cause him double labour to try to cut through the outside coating of the fibre, when his hatchet will go easily through the soft part or *cellular tissue*.

We have now found out one of the things with which we started, namely, that a flower lives longer in water than on the dry table, because it requires drink, and it absorbs this in its cells and vessels; also, we have learnt that the root being the cupboard where the food is kept, and which cannot be used by the plant until it is dissolved, it would be no use to leave that outside and put the flower in the ground, moisture being necessary to soften the starch contained in the cells. There are many plants that can live and grow entirely in water; but, as we are only talking about a general principle, they do not need any special description here, as the cellular and vascular system is the same throughout the vegetable world.

To complete this little sketch of the growth of the Buttercups we must ask why they die. For, if continually absorbing water would support life, there would be hardly a limit to their length of days. Even flowers in pots do not live long in a room, but soon turn yellow, and shrivel up, although they are continually kept moist. Let us look about for the cause, and almost without an effort the reason of the short life of plants in close rooms will reveal itself.

Why is the sunny side of the greenhouse steamy? Why is the shade over the ferns wet inside? Again, there is a similarity between you and the plants, for these perspire profusely, and this process is called by botanists "transpiration," but it means sweating or perspiring. The moisture taken up by the root is given off all over the plant, just like perspiration from the skin of human beings, and it is when this happens to too great an extent that flowers die. It is also this which causes the vapour on the inside of the glass, and accounts for the sunny side showing it first. The hot dry air tempts the plants to put forth their own moisture as rapidly as possible; and this transpiration goes on so quickly in a dry room, or in the open air when it is very hot, that the power of absorption in the root cannot keep pace with it.

In this chapter we have only spoken of the food and drink of the plant, and why from these being supplied it lives and grows; but there is yet another resemblance between animals and plants that has not been mentioned. *Plants breathe.* They have on the surface of the leaves tiny holes, through which the air enters into the cells; and this and the action of the light upon them makes the pretty colours we are all so fond of. The green of the leaves is contained in the cells, and can be separated with care from the other things I have spoken of. Plenty of fresh air is as necessary for a flower as for a child; and, indeed, not only are the former particular about its being fresh air, but even then they only just take in that part of it which they like best. This is called *carbon* by the chemists. But, dear me! it is time to leave off; for if chemistry is added to the botany, I am sure everybody will be quite puzzled. So I shall leave the breathing and colouring for another time.

## II.

IN our last chapter I promised to find a way of explaining to you where the plants keep their pretty colours, and also how they use the cells on their leaves as mouths to take in air.

To begin with the first of these things, it is tolerably clear to us that green is the favourite dress worn by the plants during the spring and summer.

But these green clothes are not put on for prettiness, nor even for tidiness, but serve a much more wonderful purpose, that we shall find out when we talk of the breathing of the leaves.

Just now, however, we will try to learn what this green really is, and will look at ordinary leaf-cells under the microscope. When these are killed in such a manner that the contents of the cell separate from the cell-wall, you find it is not an entirely green thing at which you have been looking. In the midst of a fluid comparatively colourless you see minute green specks, which is the true colouring matter. These little specks or granules are called by the botanists *chlorophyll*, and are contained in the leaf-cells of almost all the ordinary plants. Chlorophyll is rather a long word and not English, so we must ask the boys, who always understand

Greek, to tell us what it means. You see it is a very good thing sometimes to have boys and girls together, because then they can help each other. Two short Greek words make up our long one, the first, χλωρος, meaning green, and the second, φυλλον, a leaf. These tiny green granules that make the world such a pretty place to live in are entirely dependent upon the sunlight for their existence.

If you try to grow a plant in the cellar you can prove this, for it will be a most sickly-looking affair. Not only does it refuse to grow into a healthy flowering plant, but it obstinately objects to be even green.

The roots manage very well without light, and wear dingy brown colours, or we should be obliged to plant them in that topsy-turvy fashion I once suggested. Chlorophyll in the form of granules is not the only way in which it is developed. There is another kind that you may as well know of, although it is unlikely you will find out the plants who prefer this material for their clothes; they are principally water-plants that have the green part of their cell contents as a jelly-like substance. This has no very positive shape, although it sometimes lies like a ring across each cell, and now and then will arrange itself as a spiral ribbon. But in the granules I first mentioned is found the common form of chlorophyll, and it is to the action of the light we owe the green beauty of the fields and woodlands.

I talk of these green specks very much after the fashion in which I might describe things as large as an elephant, and not at all like the atoms they are, so I must try to give you an idea of their insignificance. Did you ever think how little leaves weigh in comparison with the trunks and branches? Yet the difference between these is infinitely less than that between the separated grains of chlorophyll and the rest of the cells. The whole of the green granules that make the foliage of a great tree so beautiful could be put quite easily into a tea-spoon, and are so light you could almost blow them away.

Flowers are not green: where then do they keep the scarlet, the blue, and the white that we all love to see and to gather?

The cell formation we have talked of so often is not confined to one part of the plant, so you know that in the petals or flower-leaves these little cases will be found just as in the root and the stalk. It will not surprise you then to hear that the brilliant colours of geraniums and roses result from nearly all the cells being filled with one kind of colouring matter. Paler-coloured and striped flowers have colourless cells intermixed, which are consequently less noticed than the others.

Some of you may ask how the leaves of a begonia growing in a greenhouse could show two colours at the same time on one leaf, not in patches of colour but apparently mixed up together. Cells being arranged in layers, this is easy to find out, because one layer of cells is filled with red and another with blue, so that the two give a pretty effect.

As I want you to notice the things you are near to, I shall not stop to describe any of the wonders that live in the conservatories. These will come to us in the books we shall be able to read when we are grown-up people too stiff and rheumatic to pick up Buttercups or to make Daisy chains. It will be very nice then to remember that we learnt all about these when we were very little, and it will help us to know always that there is sometimes the most to learn when there is nothing very remarkable to be seen.

There are some wonders in every garden, however, so before we give up the flowers I will tell you of one of them.

Everybody has seen what is called a *white* blossom. I could easily mention half a dozen that are known as white. What do you think I have to do if I want to know whether a flower is or is not pure white? Put on your spectacles, some of my pert little people think. Well, perhaps I shall when I am old enough, but at present my eyes are a pair of very useful blue ones, so I do not put anything of the sort upon them. I place the white pink or the rose on white paper instead, and very quickly discover that to obtain a pure white blossom I must go somewhere else for it. Delicate, pure white flowers are so scarce, however, that I only know of one that has the merit of perfect whiteness. It is of this one that I am going to tell you.

Unlike the half green or pale cream-coloured ones that pass for white, and that have colouring matter contained in their cells, the lily has nothing but air in the midst of its cellular tissue. I wonder if the people who lived years ago, and who were always trying to find something upon the earth to realize their idea of perfect purity, knew of this wonderful fact? You might easily believe they did, because they always dressed the altars dedicated to the Virgin Mary with lilies, and even now the children prefer to carry these flowers in the Roman Catholic festivals to any others. Whether the ancients were scientific enough to know that the lily had only cells filled with air to make it so beautifully white I cannot tell, but it is quite evident they were artistic enough to discover the difference between the pale-coloured flowers and the white ones.

How plants breathe is almost a prettier thing to tell you about than the little green specks in the cells, and is, so far as I am going to explain it, quite as easy to understand.

Have you ever pulled a leaf to pieces by splitting it from edge to edge? I had before I was very old, so I shall suppose that there is hardly any one reading this who has not often pulled the skin off the geraniums. Very often you have, no doubt, taken out the little caterpillars who live between the upper and under surface, so that you know very well what I mean by the skin of the leaves. This skin I shall call for the future the *epiderm*, and you must try to recollect that by this word I only mean the very outer covering of the cells within. This epiderm keeps the rest of the leaves protected, and is generally quite smooth. There are many varieties of leaves, and your eyes, I am sure, are quite bright enough to distinguish the rough from the smooth ones without any of my help. Scattered over this epiderm are pairs of small cells not touching each other closely throughout their entire length, but having just their ends in contact. By this means there must naturally be a part of the surface opening freely to the cells containing the chlorophyll and starch underneath.

Let us cut the leaf through in another direction and see to what this opening leads. Only to a larger cavity underneath, as we expected. This little chamber, that reminds one of a mouth behind two lips, is a space between the growing cells, where the air first enters on its way through the rest of the plant. These openings are named stomata, and, like the mouths of animals, are able to open and shut. I think if these stomata were a little less particular it would be more convenient to the people who want to take care of their flowers. Just imagine, for instance, the stomata shutting themselves up as tightly as possible when the season is too damp or too dry to please them. I am so sure, however, that they know what sort of air the plant requires upon which they grow, that I shall say good bye to the Buttercups and their breathing fancies.

But wait, little plant, for if I let you off like this, where are the next year's flowers to come from? You must promise me that next year the meadows shall grow Buttercups, or what will become of the cows? The Buttercup gives no answer, so I must tell you how the flowers take care that the fields shall never be without them.

When we were looking at the root of the plant, we knew very well that the brown little thing was not the very beginning of a new plant. We guessed that this root had in some way grown out of something, but it was not any part of our work then to find out what it had really started from.

Now, however, the case is altered, and the lovely way the little plants are first formed and carried to their resting-place is the prettiest of all the pretty things that I have taught you. When the Buttercup was freshly gathered, you told me that it had a yellow flower and tiny little things in the middle of it that could not possibly be leaves. You are wiser now, though, and understand that whatever the root and branches bear are leaves. Of course I do not mean that yellow and scarlet petals are green leaves cut into a different shape or dyed a different colour, but that all parts of the flower, whether petals, stamens, or pistil, have cells, vessels, stomata, and epiderm, just like the foliage of the plant. I can see that you still have a little doubt about the yellow things in the white lily with which you made baby's face so ugly; and even if you admit the things upon which the yellow dust grew to be leaves, you are not at all inclined to do the same for the yellow dust. Well, you shall keep doubting if you like until I have finished, and then you will have to give it up altogether, for the little powder has the means within it of making the next year's plants.

The Buttercup *stamens* with their *anthers* are too small for us to see, so we will take the lily to show us the pretty good way in which the flowers are helped to take care that the world shall never be without them. The *pollen* or yellow dust is a collection of cells differing from those in every other part of the plant. Instead of being enclosed all together in a skin or epiderm, they rest on the outside quite separate from each other and from the plant. They are apparently useless, and it is difficult at first sight to find for what purpose these atoms came. You see, all we have learnt at present has seemed to prove that it is the sole business of cells to fill up with air and moisture, break into each other, and make a growing plant. This the cells that make the *pollen-grain* cannot effect in their present state, because there is nothing to keep them together. But, just think of it! all next year's flowers depend upon what becomes of this yellow dust. If it were not for these the flowers would never have fruits or seeds. At present each little grain is covered while the flower is opening with a delicate skin, and then as the flower expands another covering grows. This one has in some plants little rough things upon it, that help to hold it in the resting-place it finds. But in spite of having a slight external protection, they have no power of growth by themselves. They might grumble with some justice, one would think, at their hard fate in being exposed to the windy weather, when their brothers and sisters are all carefully covered up. But the pretty soft dust knows very well that if it could find its way to the centre of the flower, there would be a home for it, and that moisture to feed it and protection from the scorching sun are waiting to help its growth. The difficulty is to find the road to this cosy place. Who should you think helps these little atoms to plant themselves where they may grow into real seeds?

The insects who steal stores of honey from the fully-opened flowers are the true nurses to the *pollen-grains*. Bees are busy collecting their winter food, so they fly upon, and in, and round about the blossoms, never considering for a minute what an untidy mess their furry bodies will be in. They are very soon all over dust, and while they are enjoying their breakfast and looking after their own business, I will tell you what kind of home it is to which the baby cells are carried.

Growing from the centre of the flower is a green tube, and at the lower part of this there are little egg-shaped bodies, which in time become seeds. And at the top this green stalk or tube spreads into pointed leaves called the *stigma*.

But you must understand, the cradle at the base of the tube, and the egg-shaped bodies or *ovules* in it, are of no more use by themselves for next year than the dry-looking little grain upon the anthers.

The bees, we may suppose, have by this time finished their work, and by their fussing buzzing way of doing it have carried some of the yellow atoms on to the spreading *stigma*. This removal does not at first seem to have improved their position, but after a little the moisture on the *stigma* induces the yellow visitor to come into the tube, and presently, safely away from the wind and the rain, it finds the little *ovules*, which at once begin the business of growing into real seeds,—a thing which could never have happened if the bees had objected to spoiling their clothes.

As the *ovule* and the *pollen-grain* grow up, the flower-leaves die away. This is clearly the case with the peas and beans, for when they are fit to gather, the blossoms are all gone. The cradle, or ovary, and the true seeds are in this case easily separated, and I dare say you know pretty well what happens to the latter.

Sometimes, however, seeds and ovary are both eaten, as with French beans and gooseberries.

In many of the common flowering plants the ovary does not grow fast enough in proportion to the impatient seeds within it. So the poor thing, which is now very little better than a dried skin, bursts open, and the grown-up ripe seeds are scattered abroad by the wind. Once upon the ground, there is no guessing where they may spring up; and although I get cross with mere weeds for starting in my garden beds, I am forced to admit that this method of planting is a very good thing. Little folks who love wild flowers must be very thankful that there is this way of managing, for I do not believe any one would take the trouble to gather and plant the pretty things in the hedgerows. Fruits which have this propensity to let their children break away from home and start in business for themselves are called *dehiscent*.

Having told you how plants drink, dress, breathe, and take care of the years to come, I ought to tell you how they talk. I do not really know that they do talk anywhere but in fairy books, and as I could give you no idea how they do it, except that reason which the Tiger Lily gave to Alice in Wonderland, I think it will be as well not to try to explain; for I am determined to tell you only those things that I know to be true.

## ANTS AND ANT-HILLS.

"MOTHER! I have been to the ants, and I sat down on a little mound to consider their ways, and—and—they 've—stung me!" exclaimed a youth of small dimensions, rushing into the house, rubbing his arms and ankles, and trying to brush off the angry insects from his neck and ears.

"As might have been expected, with less wisdom on your part than on theirs—when you have been sitting, like a huge giant, on the top of their newly-built house, crushing it in, with all its long galleries and beautiful gateways, its upper and lower stories, its nurseries, cellars, and grand central hall. Poor little insects! how could they think that a child who was able to destroy their work of weeks had really come to learn a lesson, and to take an example from their tiny selves?"

"Yes, indeed, mother; but I did want to know about them. Was it not enough to provoke any one? I was so vexed at being wakened up when I was sleepy; and then for nurse to be always saying the same thing, 'Go to the ant, thou sluggard!' I thought at last I would go and watch them, though I am not a sluggard, and see if they did not get a good sleep when they wanted it, without being forced to get up like me.'

"And so you went, when you had been out of bed a few hours, and had had a good breakfast, to see whether the ants were as early risers as yourself! If you had been up with the sun, and could have seen the inside of that busy little ant-hill, instead of crushing it with your weight, you might have seen the careful and diligent nurses who live in the upper stories of that wonderful house, very early astir, and going to call their young masters down below. You might have seen them tapping them with their little antennæ, which look like horns, as much as to say, 'Time to get up, the sun has risen!' and then waking up the little baby ants, and carrying them up through the long galleries that lead to the top of the ant-hill; and then so carefully laying them outside that the bright rays of the sun might warm them and help them to grow, as to be an example to many human nurses, who stand talking and leave their babies in the cold."

"Now, mother, I do believe you are only inventing a fable to tease me. How can anybody tell what those little creatures do underground? And, besides, I am very sure you would be angry with nurse if she took me and baby out in the sun, and without a parasol, too, to put over baby's head."

"I assure you the nurses of the little baby ants are very careful not to leave them in the heat of the sun after the very early morning. As soon as the air gets warm and the sun is hot, they carry them into the rooms near the top, where the rays have penetrated, and where the warmth can still reach them. But the older ants can bear the sun, and like to feel its rays; and though they are very industrious, yet, as they begin work with the dawn, they take a little rest sometimes in the heat of the day, and lie heaped together in the sunshine. But do not be surprised if I still tell, by-and-bye, about some ants who do walk about with green parasols, and——"

"Stop, please, mother, do! I am sure you are only laughing at me. But I really should like to hear something true about ants. Are they like bees that make honey?"

"They make no honey, nor do they build such curious combs as we see in the bee-hives, but in their own way they are just as wonderful; though, as we are told, they have no guide, overseer, or ruler. I always think they are intended rather as an example to older people, who have to provide for themselves and their families with diligence while they have health and strength, than for children, who are guided, ruled, and overseen by their parents, and have everything provided for them without any trouble of their own."

"Oh, but the little baby-ants must be just like us, because you say they have nurses too."

"Yes, in one way they are. Just so long as they cannot run alone, they are dressed like the young children in the East, or like the babes which the Indian squaw hangs behind her back or on a peg in her tent. They have natural swaddling-clothes. They are wrapped up so tightly in their larva-covering that no legs can be seen; only a head and wings can be traced through the transparent skin in which they are folded. Of course you know they begin life by being an egg; but they are hatched in a fortnight, and then the nurses take such care of them—to keep them clean, to brush and comb and shampoo them—that very soon they begin to be ready for the next change. If you could only look at the tiny insects, the nurses, through a microscope, you would see on all their legs some very fine soft hairs, which they use as brushes, and a spur close by, which, if needful, we may imagine can do the work of a comb. The shampooing is done by working about, kneading, and distending the thin skin which covers their limbs, till it is ready to open and let them go free. Then they wind a curtain of silk round their own little bodies, and go to sleep, to wake up full-grown ants, without guide, overseer, or ruler."

"Then does every little ant do everything for itself?"

"In that respect they are the most wonderful animals you ever heard of. They do everything so exactly in order, and all together—all knowing their own business and doing it—that one would imagine they had a commanding officer or a king to order them about every movement."

"Indeed, mother, if they can work as well without a king as they wound without a sword, I think they are very well off without one. They look just like a regiment of soldiers."

"A regiment of officers they are, for each one understands as well as another the order of march and a soldier's duties. They never fear danger, but advance in their order of battle with the greatest firmness, the advanced guard wheeling round to the wings every five minutes to make room for others to come forward in their place. Myriads may be sometimes seen pouring forth from two rival cities, and meeting half-way between their respective habitations, equalling in numbers the armies of two mighty empires. Though they occupy only two or three square feet, yet the picture they present is that of a field of battle between contending nations of men."

"But what have they to fight with? Do they sting each other, as they stung me?"

"The ants, like men, have different kinds of weapons. But though there are many kinds of ants in foreign countries that have stings in their tails, and are called *Myrmica*, our common ants have no sting, but they have large mandibles or nippers, with which they bite you. They are called *Formica*. There are many hundreds of species both of Myrmica and Formica spread over all the countries in the world. We have a great many kinds in England,—the Black Ant, Brown Ant, Red Ant, and others. The way in which our ants fight when they come to close quarters is by seizing each other with their nippers, and when they have hooked themselves on to each other, struggling till the weaker is dragged away. If another soldier comes up, he will seize his comrade, and so help to pull away the other. They are so bitter against their enemies that they will sooner suffer themselves to be torn in pieces than let go their hold. Some kinds will attack others that are twice as big as themselves, trusting to their superior numbers, and going two against one. In these battles, when the strength of the two soldiers is equal, they will tug away at each other, and, each squeezing his enemy, will roll in the dust, and lie till reinforcements come up. Sometimes six or eight may be seen tugging in a chain on each side, pulling with all their might, till some more come up on one side than the other, and the weaker are dragged into captivity.

"But they have other weapons besides such force. When two ants grapple, they raise themselves on their legs, and turn their bodies up in front, squirting a venom from the extremity of their abdomen against the face of their foe. This poison is well known to chemists, and is called formic acid. Thousands of ants may be seen in battle shooting this poison at one another, which has a strong odour, and is as destructive among them as gunpowder is to us.

"This sort of battle is like crossing bayonets, but very often the army throws out skirmishers before coming to close quarters. When they see their enemies but cannot reach them, they stand up on their hind feet, press their abdomen between their legs, and shoot simultaneously and with force some jets of their formic acid at the foe. This is exactly like the archers of old, or the musketry of modern battle. After the engagement, thousands of dead and mangled strew the ground, but far more are led away as prisoners; for the ants are very fond of making prisoners, as you will hear soon, and all the time of the fight crowds are to be seen hurrying up with reinforcements on each side.

"They chiefly attack—after the fashion of the wicked slave-traders among men—a kind of ants called, from their colour, the Negro Ant; and when they succeed in making them prisoners, they bring home their slaves, and employ them in all menial offices; only with this exception, that the ants are always their own dairymaids. But that I will tell you about by-and-bye. At present I will give you the account of those who have seen the attack and defence, and the droves of slaves being conducted to the ant-hill of the successful combatants; only telling you, to begin with, that they are also like the old Highlanders and the Border marauders, or cattle-lifters; and that these attacks are frequently made with a view of possessing a herd of cows, on the milk of which they feed with so much delight."

"Oh, tell me first about the dairy, and then about the fighting for the cows! What do you mean by that? The cows must be very tiny ones."

"You have often seen the little green insects that crawl up the stems of the rose-trees. They are called *aphides*, and these little creatures are the cows, which yield a sweet juice much delighted in by the ants, which keep their cows in all sorts of ways.

"There is a species of Yellow Ant, which does not roam much about, but lives chiefly on the milk of its herds, which it keeps underground—like the unhappy cows of some of the London dairymen—at the bottom of its citadel; and an ant-hill is more or less rich in proportion to the number of its flocks. There are many other kinds of cowherd ants. Some take less trouble than others with their cows, and, being active and good climbers, run themselves up the branches on which are the aphides; and milk them there. Others take so much pains as to make a little tunnel of earth from the foot of the tree to their nest, in which they carry home the cows underground, without being seen or disturbed by other ants. Other kinds again make sheepfolds or stalls for their cattle, apart from their own nests. They build with earth round the stems of plants little houses, round within, and as smooth and hard as these ingenious little plasterers can make them. These folds are of the shape of a funnel, sometimes of a ball, with a very small hole at the bottom for the ants to go in and out. Other ants will make a little hollow ring of earth and decayed wood mixed into hard plaster round the branch of a tree on which are their *aphides*, which they carry down to this prison, and then visit them from the inside of the tree, by passages through the bark without coming outside.

"Their way of milking is very curious. The body of the aphides or plant-lice is very soft and tender, and they have a proboscis by which they adhere to the leaf or plant. For fear of bursting them, the ant strokes them and caresses them with its antennæ very gently, until the creature loosens its hold, when the ant gently carries it away. There are two horns near the tail of the aphis, which exude the sweet juices of the plant on which it has been feeding. The ant begins by stroking down its captive and flattering it with its antennæ or feelers, and then strikes these horns gently, when a little honey-like drop is voluntarily exuded. This the ant takes up with the end of its feelers, and conveys to its mouth.

"But they not only capture, they actually breed their cows as well. They take the greatest care of their eggs, gather them up carefully, keep licking them and moistening them, and glue them together with a sort of gum from their own saliva, as the parent would have done if she had been free, and so they hatch generations of captives within their ant-hills. They also collect food and bring it to them, lest their cows should go dry for want of grass."

"But you said the ants kept slaves too. I should like to hear about their negroes."

"I do not wonder, for I think of all the marvellous things that have been discovered in the ant-world, this is the most marvellous of all. But I must tell you how it was first discovered. It had puzzled many people who had gone to the ants and considered their ways, why it often happened that there were two kinds of ants—Black ones and Red ones—together in the same ant-hill. Huber, a great French naturalist, who made many wonderful discoveries about ants, at length discovered that one particular kind, which he called the Amazon Ant, did nothing but fight, and he found also that in the nests where there were two kinds, there were never any male or female ants, but only workers of the Black sort. For I should tell you that just as there are three kinds of bees,—the queen, who always stays at home, the drones or lazy gentlemen, and the workers, who are are females that never lay eggs,—so it is with the ants also, only that their ladies are not queens, and there are a great many of them in one nest.

"Now, the Amazon Ants have mandibles or pincers, which have no teeth of curved shape like those of most other ants, but are straight, and are consequently more like spears or swords than hoes and rakes, such as the other kinds have. Their business is fighting, and they want servants both to build their houses and take care of their children. Accordingly, every evening, a little before sunset, they set out like the kidnappers in Africa against a negro village, to surprise some industrious ant-hill in the neighbourhood, which their scouts have reconnoitred and reported on. They surround the fortress, and then all rush upon it together. The few Black Ants that stand sentry at the entrances are soon overpowered, and the robbers rush at once to the rooms where the eggs and young ones are, seize them and carry them off, never taking any males or females. They bring them home and hand them over to the slaves they already have of the same kind. These slaves have evidently got quite accustomed to their life and fond of their masters, for they show great joy when prisoners are brought, and are very sad when the Amazons have failed. They run to meet them, relieve them of their precious load, take care of the young; they shampoo them, undress them, take off their swaddling-clothes at the proper time, and feed them. The Red children of the Amazons and the Black captives are brought up together and live like brothers. Not only do the slave ants do all the work of the nursery; they build and repair the castle, and they are sent out after the cows, and bring back the captives for their masters to milk. However, they always get their share. In fact, they keep the key of the pantry, for they open and shut all the doors of the castle and of the rooms in it every night and morning; and they will sometimes help the Amazons, with whom they have lived, in fights against their own kind."

"But how can ants know one another? They look all exactly alike, at least all Red Ants do, and all Black Ants the same."

"That is another curious thing about the ants. Not only do ants of different kinds fight, but often ant-hills of the same species will go to war, yet the soldiers never mistake between a comrade and an enemy. Huber once kept a number of ants from one nest for four months in his house. At last some of them escaped and met some of their old companions. They were seen to make all sorts of signs of joy, to kiss each other, stroke one another down with their feelers, take hold of each other with their mandibles, and then, after visiting their old nest together, they came out in great crowds, the escaped prisoners showing the way, till they found the place where their old comrades were confined, got at it, and took the whole of them home with them."

"But surely if all this be true, ants must be able to talk to each other, and tell their friends all they know? How can they make one another understand?"

"Well may you ask, and in fact they have such strange power of communicating information to every friend they meet, that the word 'antennate' has been invented to express the dumb language of the ants. They are seen to touch each other in all sorts of ways by their antennæ, and it is no doubt by signals thus made, and instinctively understood by all, that they communicate. When one ant touches the antennæ of another, it instantly stops, and goes in the direction its informant wishes. But we have not yet been able to read this dumb alphabet."

"Well, I really think after this I could believe that ants carried parasols; though they seem to be always toiling, fighting, or working."

"There are a great many kinds of ants, and they all have different habits and ways of building. The Parasol Ants only live in South America, and make nests of a different kind of architecture from any we see here. They use leaves to thatch the domes which cover the entrances to their underground homes. If they did not thatch these portions, they would be washed away by the heavy tropical rains, which would enter and drown the young ones. In order to provide leaves for thatch, they go out in immense hosts, select a tree to their fancy, and ascend it in long files. Each one places itself on the surface of a leaf, and with its sharp scissor-like jaws makes a nearly semicircular incision on the upper side; it then takes the edge between its jaws, and by a sharp jerk detaches the leaf. Sometimes they let the leaf drop to the ground, when they are gathered and taken away by another relay of workers, but generally each marches off with the piece it has cut out, holding it over its body by its jaws; and as they follow file close to each other, not an ant can be seen, but the procession looks like a long line of animated leaves on the march. Sometimes a great heap may be found of circular pieces of leaf about the size of a sixpence, left on the ground away from any ant-hill. But if we wait long enough we shall see a whole relay of workers come back to the place, and not a leaf will be left behind. When they reach their homes, they cast the leaf down on the hillock, when another set of workers place it in position, fastening each leaf down with a little pellet of fine earth kneaded by themselves, and which act like pegs to keep the leaf fixed. These ants make such enormous nests underground, that they have been known to undermine and destroy the embankment of a large reservoir.

"Other ants make hillocks in the woods several feet high. Most ants in this country make the lining of their rooms and passages of blades of stubble, small fragments of wood, minute pebbles, and whatever substance they meet with which they can carry, and so they often pick up grains of corn. The ants which make hillocks still have the greater part of their nests underground. Some sorts make only one entrance, with

long winding passages to their halls and nurseries. Others, like our Red Ants, have many entrances open to the air, in the turf."

"But does not the rain get in?"

"No; for they take care to have doors, and every evening, or whenever it is wet weather, they barricade themselves. They bring little beams and lay them across the gallery, then they place others on the top of them crosswise, and finally they employ pieces of dry leaves, very broad, to cover the hole. When the last gates are shut, sentries are placed behind them, to guard and watch over the safety of the rest. At sunrise the barricades are removed, and the passages opened for the workers; but if the weather be rainy, the gates remained closed. Their earth roofs are laid upon little beams; and as they carefully knead the earth in pellets and moisten it with rain-water, and then dry it in the sun, it is exactly like brick-making, and the sun-dried bricks turn the rain very well. Even if it should penetrate in very wet weather, the ants are pretty safe in their larger rooms, which are lower down, and where they generally live.

"Other ants are called Mason-ants, because they use only earth pellets or sun-dried bricks, without mixing stubble and wood, and hollow out large vaults in the ground, often many stories deep, with a labyrinth of galleries and passages. Their chisel is their teeth, their compasses are their antennæ, and their trowel their fore feet. The larger rooms are supported by solid pillars of earth. Those different stories have their different uses. The upper ones are chiefly reserved for the *larvæ* or young, that they may be near the warmth of the sun; but when its heat becomes, as they think, too great for the little ones, they are carried downstairs to the halls. If the rain gets in, the upper stories are still dry, and all the colony mounts to the higher chambers. There are sometimes forty stories in one nest.

"But Masons, Bricklayers, and Thatchers are not the only handicraftsmen among these marvellous insects. There are Carpenter-ants, which hollow out the inside of trees. They will seize upon an oak or a willow, and completely scoop out many square feet without the life of the tree being at all affected. The stories and galleries are innumerable and very small, separated only by partitions left in the wood, not nearly so thick as fine cardboard, and here and there a little column standing. These supports are thickest at the top and bottom, just as in human architecture. Every pillar has a base and a capital.

"And now we may forgive the ants for all their bites, after the wonderful facts we have learnt from going to them and considering their ways."

I think my readers may not object to stay a little longer, and hear one or two stories about the ants, some of which have come under my own knowledge. In some countries ants are very numerous. The Fire-ant of Brazil in South America, not larger than our Red Ants, sometimes multiplies so as to drive the people out of the villages; and when the rivers rise, or the wind blows the swarms into the water, their dead bodies may be seen washed on shore in heaps, looking like a deposit of black earth, for many miles. They undermine whole villages, and fill the houses like an Egyptian plague, disputing every fragment of food with the people, and even destroying clothes to get the starch out of them. The only way to keep anything safe is to hang it in baskets by cords from the ceiling, and to steep these cords in a very strong solution of disagreeable oil, which the ants cannot abide. To sit at peace, the legs of the chair must be rubbed with this oil, and the legs of your footstool must also have been steeped in it.

In some seasons, in the island of Bermuda, there is a little Red Ant which is as great a plague, only that it does not bite very severely. At these times you cannot take a step anywhere without crushing hundreds of these little insects. The cedar-trees of Bermuda, a kind of juniper which grows all over the island, are covered with a sort of gum, of which the ants are very fond, and every tree is covered with long lines of them, one line marching regularly up and the next as regularly down, like files of soldiers. Sometimes I have counted more than a hundred files on the stem of a single tree. Those ants are scattered all about the rooms of the houses; but when there is nothing particular going on, you can only see an ant wandering about here and there on the floors, the wall, or ceiling of the room. But put a plate of butter on the table; if it is not very far from the side of the room, you will see a lonely ant on the wall stop. It will turn itself in all directions, as if to calculate how to reach the plate. When it has reconnoitred for a few minutes, it runs down, touching every ant it can meet on its way. Every ant that is touched passes the message to others. Some of them will run across the floor and out of the room. Very soon you will see an ant climbing the leg of the table, followed by two or three more, and in less than half an hour there is a long line of hungry insects streaming incessantly from the door, up the table, and on to the butter-dish. I have often cleared the ants away, and then put the plate on the stand which hung over the table from the ceiling. The ants would run about, look up at the dainties out of reach, make circles about as if very much confused, all touching each other with their feelers, and then on a sudden they would make a line of march down the table-leg. But they were not so easily got rid of. In a few minutes a black thread was seen rising up the wall, along the ceiling, and down the cord on to the swinging stand. Happily for our dinner, this ant dislikes train-oil very much. So in the autumn each leg of the dinner-table stood in a little leaden cup of train-oil, and our chairs were likewise planted in saucers of train-oil. When we did this, our dinner was safe.

But even this troublesome little ant did us many a good turn. The most annoying pest in Bermuda was a great red cockroach, four times as large as the English one, and with a very disagreeable smell, with which it scents everything it touches. It eats far more than the ant, and devours leather, cloth, and every sort of animal substance. But the ants are very fond of eating cockroaches; and though it would take two hundred ants to weigh as much as one cockroach, they kill and devour thousands of them. Of course one or two ants could do nothing with it. But whenever they find a cockroach standing still, or eating, or in its hole, they collect in myriads, and without disturbing it completely surround it. On a sudden the little army rushes on its prey. The victim is instantly covered. For a few minutes a struggling mass of ants is seen being moved along, but each is hard at work with its teeth: the cockroach is soon eaten alive, and nothing left but a horny skin, and the hunters are off to search for another.

In other countries, among the forests of Sweden and Switzerland, where there are many lofty ant-hills, they serve for a compass to the traveller who has lost his way by night or in the fog. Their nests are always made from east to west, with their peak at the east end, which is very steep, while the ridge slopes gently down to the nest. Thus when there is no sun to guide him, the wayfarer knows in what direction to travel by considering the ants. The Swiss also make lemonade from the Yellow Ants, by putting a piece of sugar into their nests, on which the insects at once squirt their acid to melt it, and it is taken out thus steeped in formic acid, and tastes like lemon.

There is a story told of an ant, which reminds us of the story of Robert Bruce and the spider, and which teaches us the same lesson set forth by an ant which lived 600 years ago. It is in the Life of Tamarlane, the Tartar prince, written by an Arabian historian. That terrible conqueror was once forced to take refuge from his enemies in a ruined building. As he sat alone there many hours, and was almost in despair, his attention was attracted by an ant carrying something larger than itself up a high wall. He counted the efforts it made to gain its end, and found, that sixty-nine times its burden fell to the ground, but the seventieth time it reached the top.

"This sight," he said, "gave me courage at the moment, and I have never forgotten the lesson it taught me."

So when we have anything to do which is difficult or troublesome, but which we ought to do, let us go to the ant, go on trying, and we shall generally succeed at last.

However, the ants can play as well as work. A famous traveller, who considered the ways of the ants in South America, says: "Their life is not all work, for I frequently saw them leisurely employed in a way that looked like recreation. When this happened the place was always a sunny nook in the forest." He had been watching an army of ants on the march, and had noticed that while the main body carried burdens, the pioneers went before to make the road, while others with larger heads than the rest were the officers, and trotted alongside without even carrying anything. But they kept a sharp look-out, and often went out on either side to see that no enemies were lurking near. "The main column of the army and the branch columns at these times were in their ordinary relative positions; but instead of pressing forward eagerly and plundering right and left, they seemed to have been all smitten with a sudden fit of laziness. Some were walking slowly about, others were brushing their antennæ with their fore feet; but the drollest sight was their cleaning one another. Here and there an ant was seen stretching forth first one leg and then another, to be brushed and washed by one or more of his comrades, who performed the task by passing the limb between the jaws and the tongue, and finishing by giving the antennæ a friendly wipe."

Some ants are like drones, and never work at all. These are the male and female ants, and have wings, which none of the workers have; but they are very kindly treated, and are not turned adrift like the drones. They may often be seen in September flying about in great swarms, and tumbling to the ground together; but the female ants, which are the largest of all, as soon as they are going to lay eggs, lose their wings, which they loosen off their corslet, and then their kind nurses carry them home again to their nests. Sometimes, when they think the ladies are too fond of gadding about, they take hold of them in the nests and cut off their wings, so that they cannot escape. But they are very affectionate to them: they carry them from room to room according to the weather, bring them food, and each lady ant has about a dozen servants, who are always stroking and kissing her, and when she dies they will remain for several days brushing and licking her body before they will take her out and bury her, or rather bear the body to some distance from the ant-hill.

Ants, wonderful as they are, have many enemies. There are ant-thrushes, ant-eaters, and ant-lions. Quadrupeds, birds, spiders, insects, all join in waging war upon their armies. The ant-lion, however, is not a lion, but an insect, which makes pitfalls for the ants in the sand. It chooses a place where the sand is very dry and loose, near the ants' track. Then it scoops out a funnel with steep sides, and buries itself at the bottom. It has a large pair of jaws, which it sticks up just at the point of the funnel, but on a level with the sand, so that they cannot be seen. The ants in their travels pass the pitfalls. One of them slips on the soft sand, and comes scrambling to the bottom, as we should do in running down a gravelly hill. The moment it is there, the jaws are lifted and it is seized and devoured, and the trap repaired for the next incautious wanderer.

I have said nothing here about the White Ants, of which we often read, but of which there are happily none in England, because they are really not ants at all, but belong to another class of insects altogether, the *Neuroptera*, of which the dragon-flies are a family found in this country. The true ants belong to what are called *Hymenoptera*, and are in the same class as bees and wasps.

## ROOKS AND THEIR RELATIONS.

"WHY *will* you always draw your woods and trees with such flat tops? Look out of the window, and take a lesson from nature, and the trees in Dr. Yesterday's young copse. *They* have all *round* tops, if they don't stick right up into the sky, in a lively aspiring sort of way. But you have drawn on your imagination for yours, which look as if they had had their crowns stolen or their heads bitten off," said a pert brother, peering over his sister's shoulder, while she sat busily sketching an old manor-house, half hidden by forest trees, enlivened by a flight of dark birds, and occasionally raised her head to look dreamily, not at the landscape before her, but apparently at something in the clouds, or far beyond the horizon.

Very provoking, doubtless, she felt it, that her work was so little appreciated by the fidgety youth by her side, who generally seemed to feel equal to the task of criticising, when he could not rival, the work of others.

He now wound up his remark by the cruel suggestion, "I suppose after all it's because they are easiest to draw so—and you're quite right, Polly,—take it easy."

He probably knew by experience how to elicit an answer when other methods failed, by rousing his sister's feeling on a tender point.

"Who takes it easy now, by jumping at a conclusion instead of coming at it by a little roundabout common sense?" replied she. "And while you *are* about it, you might give me credit for a good active reason, instead of a lazy one. Do you suppose *I* would waste my time over those modern half-grown twigs that never saw the last century? *Mine* are good old ancestral trees, of a sublime aristocratic contour, that *have* had their heads duly bitten off year after year."

"Then I was right, after all; only, as you have not made a bargain with me to ask no more questions, as the American gentleman whose leg had been *bitten* off did with his companion of the irrepressible curiosity: like his questioner, I would give my head, if it was worth your acceptance, to know *who* bit them off. Our aristocratic old grandfathers never kept slaves with shears, to mount the trees year after year and trim them all so straight, and at the same time so *easy for you to draw.*"

"The nibbling process was doubtless far easier than my task under the battery of your nonsense. It went on from generation to generation of those black gardeners—trained to the business, who have been there, father and son, for six hundred years. They bite off the twigs every autumn and spring, and use them for repairs."

"Sound teeth the old fellows must have had! But I'd thank them not to meddle with my trees. I would give them their dismissal very soon, tooth and nail, if they had been my gardeners for twice six hundred years."

"Sooner said than done would that ceremony have been, I fancy," said the sister. "They seem, at least, *my* Rooks—I mean the *family* I was thinking of, that had the care of the trees I was trying to draw—seem to have had a kind of freehold residence, that none could dispute with them, though their title-deeds might be rather unintelligible to you and me. But I like old times and old places, old families and old gardeners too, so long as they act as these have done, doing credit to their old traditions, and keeping up the family honours by bringing in with every new generation a fresh supply of young energy and new industry to add freshness and vigour to the old ancestral *prestige*. They have never gone down in the world, and I don't think they ever will for ever so short a time, except to rise again the better for it; and they never think it beneath them to provide for themselves and their young ones by honest labour. There is not a busier, happier, or more lively family in the world, who keep together better, or who, having sense enough to do nothing without *caws*, are more likely to make a noise in the world."

"The ancient gentlemen might have more taste, I think, and better use for their teeth, too, I should think, if they are in so good a way of earning their bread, than to nibble off the tree-tops. I suppose they do it all the more, now that the fashion is exploded of trimming the yew-trees and hollies into peacocks; but if they can show cause for that, I'll forgive them, and——"

"Listen to them—listen and look there," said his sister, suddenly, as a flock of noisy Rooks, which had just been disturbed, came sailing by, and the eyes of the youth, mental and bodily, opened a little wider than they had done, to take in his sister's meaning.

"All very well, the romance of Rookdom and flight of fancy. I'll forgive all that, including the very decrepit pun, and give the old fellows credit for being very aristocratic, and holding their own, and keeping themselves up in the world, and even looking down upon poor wingless bipeds like ourselves. But why call them gardeners, or talk of honest work in the same breath with such a set of mischievous, rapacious thieves? I should call them freebooters, outlaws, banditti, Bashi-bazouks—anything in the irregular cavalry line. Don't you remember, even the old song makes it a settled matter between the human and the winged Robin Hoods to part the day and night between them?—

> 'The chough and crow to roost are gone,
> The owl sits on the tree;
> Arouse ye then, my merry, merry men,' &c."

"A very unequal division of the spoil they made, nevertheless. What would Robin Hood or Friar Tuck have said to the feast on a few grains of corn or slices of potato—or even a savoury caterpillar with red spider sauce? And yet I fancy the British public would have gladly compromised with the archers of Sherwood Forest, for the loss of a fat buck whenever he wanted it, and a few of the gold pieces of which travellers used to be disendowed by him for the benefit of the poor, if he had turned his talents as successfully as our black friends up there to ridding the land of noxious creatures, whose depredations are far more to be dreaded than their own."

"Now, if there is any knight-errantry of that sort among them, I shall try to look up to them a little more, and perhaps take a leaf—a green one, I suppose it must be—out of their book by-and-bye. But you've fought well, you've earned a truce, and I'll leave you to your flat tree-tops and tall chimneys, and I'll find out if there's anything worth reading to you about these champions of the give-and-take principle which Robin Hood and the Rooks seem to have hit upon, though in rather a lawless way."

"I think you'll find they are not the only impromptu takers and givers in nature. It is a principle that prevails largely in creation—and yet not lawlessly. Laws that man can neither make nor decipher keep up that wonderful balance in nature, and enlist an obedient compliance from the creatures, which neither human laws, prisons, nor revolutions have ever accomplished among reasonable beings."

"On the contrary, *reasonable* beings like to amuse themselves, and leave society and their account-books to balance themselves the best way they can. But here are some stories about Rooks, Jackdaws, and other cousins of theirs, which will be a very good accompaniment to your romantic and aristocratic pencil.

"The wise men of Greece used to think the *owl*, Minerva's bird, the emblem of wisdom and learning. They had a far higher opinion of the owl, whose effigy adorned their coins, than had the parish clerk, who, with wizened face buried in his stout rector's cast-off wig, gave out—

> 'Like to an owl in ivy-bush,
> That frightful thing am I.'

But if the Greeks had been better naturalists, and known a little more about the Rooks and their relations, they would certainly have honoured them before the moping solitary bird of the night. The owl keeps himself and all his wisdom, if he has any, to himself. He is like a solitary savage in the forest, who knows nothing of society, its laws and regulations, but lives only for himself. The Greeks ought to have taken the Rook, for if ever there was an aristocratic republic in the world like their own, it is to be found in the tops of those tall trees. The Rooks understood the laws of property, and acknowledged hereditary settlements long before man had discovered feudal tenure or forty-shilling freeholds. The old folk at home maintain undisputed possession of the same forked branch which has been the flooring whence many a family has hopped into the world. The young folk have to seek a settlement for themselves, and must build their new home by their own labour. But young Rooks, like young men, very often make a bad start in life, and invest their labour on bad security. A gale of wind dissipates their fortune, and the sticks they have toiled to gather are scattered in a moment. They try to start again with borrowed capital. 'The old folk have plenty of sticks; we may as well take a few whilst they are away at work.' But father Rook keeps a good account of his building materials. Listen to that solemn 'caw, caw,' from the topmost bough when he returns about an hour before sunset. See now how the old parliament Rooks gather round him, and listen how they groan forth their caws in chorus as mother Rook tells how she has been robbed; then there is a pause—the jury are considering their verdict. On a sudden there is a universal jabber, the assembly darts off to the neighbouring tree, and, in a few seconds, the nest of the dishonest young pair is scattered to the winds. Depend upon it, Rookdom knows nothing of the law's delays, but their republic is administered with prompt justice between Rook and Rook. There is no toleration among them for the doctrine that 'he should take who has the power, and he should keep who can.' It is even said that an incorrigible offender has been strangled by his fellows; but, as I have not seen this, I will not assert that the republic admits of capital punishment, though I have often seen audacious offenders pertinaciously driven into banishment, and compelled to settle apart in a penal colony.

"We said the Rooks were true aristocrats, and they have shown their dislike of human revolutions, for they were so dissatisfied with the overthrow of the old *régime* in France, that at the Revolution they nearly all quitted the country, and comparatively but few have returned to it. This is really true, only prosaic people have explained it by the fact that the trees which surrounded the old *châteux* were nearly all cut down, and so their inhabitants had to seek for new quarters. In the same way the Turks tell us that the storks are true Mohammedans, because they nearly all left Greece after the War of Independence, the reason being that the stork, like the Rook, knows his friends, and that, while the Mussulman cherishes him on religious grounds, the Greek, with no such scruples, dislikes his litter, and robs his nest, when he claims to share the roof with its proprietor.

"The Rooks appear to have some strange law as to continued occupancy. As soon as their young are fledged, unlike their cousins the Jackdaws, they desert their homes, and take for a time to a vagrant life, like the civilized Red Indian, who cannot forego his three months' hunting in the year, or like the Londoner, who rushes down to Margate or Ramsgate. In the summer the Rook loves his country ramble, and, Arab-like, roosts at night wherever he has happened to find food and sport. But as the days begin to shorten he revisits the ancestral trees, and by the end of September the whole republic has gathered at head-quarters, and with deafening cawings continued till past the sunset hour, we may fancy the rival story-tellers are recounting their summer adventures, each striving to outdo his fellow in tales of prowess and of wonder. And now, in assertion either of freehold or tenant right, each begins to repair his nest. It can surely be for nothing else, for when the spring bids them prepare for domestic cares, not a shred of the old nest is left, but the new home is carefully formed from its very foundation with fresh tough twigs, judiciously selected, and twisted off the growing trees long before the owners of the soil have thought of opening *their* eyes, and beginning *their* morning work.

"But the Rook knows the proverb, 'The early bird catches the worm.' Early to bed and early to rise, he is content with four hours' sleep in summer, though he has no objection to a siesta at noon. But he seems drowsy when he first gets up; and as he leaves his perch before sunrise for the potato-field or the meadow, he sails sluggishly along, too sleepy to utter a single croak. But he has a long day's work before him; he has many miles to travel before his household and himself are supplied The farmer need not be jealous of him, for though he may swallow a few ears of corn, or munch a new potato, he never yet destroyed a crop, and his vegetable food is merely sauce for the thousands of grubs which he destroys. Worms and caterpillars are his staple, and he walks quietly about the field, always facing the wind, lest his feathers be ruffled, piercing the soil for a worm, or digging up the root of a plant at which a caterpillar is gnawing. He has done no harm to the agriculturist by uprooting his grass, for wherever he has plucked it there was a worm at the root, and his thrusts have checked further mischief.

"Where the land is tilled, and consequently looser, so that the grubs are more deeply buried, the Rook, from his habit of thrusting his bill into the earth, wears away the feathers of his face, and thus, while in England the young Rooks are feathered to the nostrils, the old ones are bare up to the eyes. But in countries where subsoil ploughing is not in vogue, and where the ground is consequently so hard that the insects do not bury themselves deep, the Rook has no opportunity of shaving his face by rubbing it in the earth, but continues, as in Asia Minor and Syria, to grow his natural beard and moustache.

"The Rook seems to consider that he is a friend of man, and ought to be treated as such. The finest trees will not induce him to nestle far from human habitations; and, where trees are scarce and men are many, he will put up with rather indifferent and even unlikely quarters. There are, at the present moment, four or five rookeries in London itself, and, until a few years ago, there was a little rookery between St. Paul's and the Thames, in the very heart of the City. But there the Rooks' last retreat has yielded to the advance of improvement, for a new street has been cut through his quiet refuge in the garden of Doctors' Commons. There, in the centre of the busy City, I have counted thirteen kinds of birds secure from guns and gamekeepers. Attached, however, to the society of law, the Doctors' Commons Rooks have accompanied their unfledged neighbours, and have settled down in the Temple Gardens. When attracted by human society, the Rook will sometimes adopt a more artificial foundation for his nest than his native tree. For years a pair established themselves on the vane of the tower of the Exchange, in Newcastle, and supplied Bewick with one of the favourite subjects of his pencil. A similar attempt was lately made in the City of London, on the vane of St. Olave's Church, and Rooks have built between the wings of the dragon of Bow Church. They had no fear of the City churches being demolished then. But, like the herons, the Rooks understand a notice to quit in the shape of cutting down trees. Baron Ravensworth lost his heronry through the cutting down of a single tree amongst the many on which these noble birds had established themselves. But the Rook does not wait for the tree to be felled. He has noted human manners and customs, has pondered on the laws of cause and effect, and either the instinct or the inherited wisdom of Rookdom has ascertained that, when a piece of bark has been pared from a tree, the axe will shortly do its work; and consequently, when an elm has been thus scored, the rooks will at once cease from building on it.

"The Rook is not without his enemies, and the chief of these is the agricultural economist. He sees the Rook in a corn-field, and shakes his fist at the 'thief' who is robbing his granary. But have patience: the Rook, like other tax-gatherers, is not popular when he calls for the rates; but if the tax-gatherer has soldiers, and policemen, and judges, and ironclads, to show for his money, so our friend reminds us that if for one month he tries a vegetable diet, he has saved you acres upon acres of corn by his unwearied consumption of grubs and wireworm for eleven months of the year.

"Like most old families, the Rooks have various relatives and hangers-on not quite so respectable as themselves. Foremost among these is the Carrion Crow, who imitates his cousin so well that a careless observer might easily mistake him for the gentleman himself; but his coat wants the beautiful purple velvet gloss, and, though as shining as the other, is of a more sombre hue.

"When he opens his mouth, his note betrays him, for he has abandoned the deep croak of the Raven, without attaining the cheery caw of the Rook. Again, he is a skulking, sneaking fellow, a solitary, lonesome ghoul, who seeks his unclean diet without a fellow, and gorges himself whenever he has a chance.

"The Rook never acknowledges the relationship, and the Carrion Crow builds a lonely and untidy hovel, unlike the Rook, heaping up any rubbish that may come to hand, generally on a fir-tree in a secluded corner of a plantation. On a foundation of rotten sticks, it plasters a layer of fresh earth, and then—for her young seem less hardy than the Rook's—adds a thick lining of wool and hair, which it plucks from the backs of sheep and cattle.

"But, though devoted to his young, he is a cruel fellow, this Carrion Crow; he will watch the new-born lambs on the hill-side, and, for once, calling in the aid of two or three brother ruffians, will tear out its entrails, and pick out its eyes, before the mother has time to defend it. But he carries a guilty conscience with him, and seems aware that man is every where his enemy. Look at those Rooks marching with a dignified gait at the ploughman's heels, and picking the grubs out of the fresh-turned furrow,—they are conscious of their merits then, and know that the political economist has no jealousy of their presence in *that* field. But where is the Corby, as the Carrion Crow is called in the North? *He* is skulking on the other side of the hedge, or more likely three or four fields off, more wary than even the Rooks in harvest-time, for he knows that *he*

will never have a friend's welcome. And yet he is a wag in his way, for when he is once tamed, and become familiar with man, there is no bird in nature, not even the parrot or the magpie, more fearless, or fonder of a joke.

"One in Edinburgh was evidently in the interest of the shoemakers, for his delight was to peck at the heels of every barefooted urchin he came across, and the more frightened they were, the more delighted was he. They have also a wonderful memory: like the bear, they are in the habit of burying the portion of the carcase they cannot eat at the time. A tame Crow was once seen cunningly burying a dead mole in a garden: he smoothed the earth so cleverly over the spot, that the sharpest eye could not have detected the grave. He was then shut out of the garden for a week, when, on the door being opened, he instantly hopped to the spot, and exhumed the savoury morsel.

"Far better known is another poor relation of the Rook—the impudent and familiar Jackdaw. He seems to be aware that there is a sort of immunity afforded to the Rook, of which he has not the smallest scruple in availing himself. You scarcely ever saw a flock of Rooks unaccompanied by a number of their chattering cousins, whose sharp 'chak, chak' may be at once distinguished from the more dignified 'caw' of their leaders.

"Where towers and rocks are scarce, the Jackdaw often builds his nest under the protection of the Rooks, and I have seen in my garden a Jackdaw's nest thrust in the fork immediately under the platform of the Rooks, while a few inches below the starlings had secured a snug hole. In fact, the Jackdaw is a pert and loquacious little fellow, ever cheerful, always on the alert, and ready either for business or frolic. He is not so respectable as his big relation, but is at least the most pleasant of the family, and very fond of society. But he prefers towers to trees, and is particularly addicted to our English cathedrals. He has established himself in St. Paul's Cathedral; he once succeeded in setting fire to York Minster, and he inhabits many buildings in London, Edinburgh, and all our great cities. He can, however, dispense with human society, and is equally fond of ruined castles or desolate sea-side cliffs. If need be, he can even descend to a rabbit-hole, and, among other unlikely spots, he has established a colony on the giant stones of Stonehenge. He nibbles no twigs for the perennial repair of his nest, which, when he gets inside a tower, is an enormous cumbrous structure. I fear he has robbed many a poor washerwoman of her character, for he is particularly attached to caps and lace for the lining of his home. In one nest was found a piece of lace, a worsted stocking, a silk handkerchief, a frill, a child's cap, and various other articles. In fact, he has an ungovernable propensity for carrying off articles which are of no use except to the owner, certainly not to *him*. For his foundation he prefers as much ready-prepared timber as he can find lying about in the haunts of men.

"The late professor of botany at Oxford prided himself upon a magnificent collection of grasses. Now, as grasses are, of all plants, the most difficult to distinguish when out of blossom, each was carefully marked by a neatly painted label; but from time to time the labels disappeared. So useless a theft perplexed the worthy professor, and, for a time, the deed was attributed to undergraduate mischief. A watch was kept; but, when the gardener appeared in the morning, the labels that were safe overnight had disappeared, evidently by ghostly hands, for there was not a foot-mark on the beds. At length, in the interior of Magdalen Tower, it was discovered that a pair of Jackdaws, wishing to raise their nest two or three feet above a disused staircase, had established it on a pile of several hundred labels, which they had collected long before the gardener had turned out of his morning watch.

"The Rook has one more relation, the prince of his clan, very different in his habits, and whom we in England rarely see. How many in a thousand ever heard a *Raven's* croak? To see him at home now, we must travel to the Orkneys or the Hebrides. The pairs that remain in England might almost be counted on the fingers, but so long as he was allowed to remain, the Rook was not more faithful to his breeding-place, and many a 'Ravenscliff' and old 'Ravenstree' remind us of the spot where for ages, year after year, their brood was reared.

"In this country at least they are not fond of the society of their fellows, for the young are invariably sent out into the world to seek their fortune far away from the parental home, and, if it were not for the enmity of gamekeepers, the Raven would soon again become a familiar sight in most districts of England. It must be necessity, and not moroseness, which makes him so unsociable, for his carrion food is here but scarce; not so in warmer climates, where he becomes as sociable as the Rook. Thus, about the Mosque of Omar, in Jerusalem, hundreds of Ravens, of our species, nightly congregate. They seem to have learned that the Moslem veneration for sacred places makes them there quite secure. Of all the birds of Jerusalem, the Ravens are the most characteristic and conspicuous. They are present everywhere to eye and ear, and the odours that float around remind us of their use. The discordant jabber of their evening sittings round the mosque is deafening. All the cousins are collected. The caw of the Rook and the chatter of the Jackdaw unite in attempting to drown the hoarse croak of the old Raven, but, clear above the tumult, rings out the more musical call-note of the Brown-necked Raven. We used to watch this great colony, as, long before the city gates were opened, they passed, in the grey dawn, in long lines, over our tents to the northward, the Rooks, in solid phalanx, leading the way, and the Ravens, in loose order, bringing up the rear. Before retiring for the night, popular assemblies of the most uproarious character were held in the trees of Mount Olivet and the Kedron, and not till after sunset did they withdraw in silence, mingled indiscriminately, to their safe roosting-places in the sanctuary. On a wet day—and there was some wretched weather at Jerusalem—the Rooks would determinately set out on their travels, but the Ravens stayed at home, sitting about by twos and threes among the olive-trees, generally in silence, but now and then croaking a doleful remark on the weather, or warning from their neighbourhood the draggled Jays, whose soft plumage was no better protection, in such a downpour, than a lady's evening muslin. Posted as sentries round the down-trodden city, they seemed like the ghosts of old patriot heroes groaning over its decay."

## THE LUNAR HALO.

THERE are some phenomena of nature which suggest false ideas. For instance, when we look at the broad expanse of ocean on a moonlit night, and see a path of glory on its surface, directed towards the moon's place, we seem to be assured by the sense of sight that that broad track is illuminated while the waters all around are dark. A little consideration, however, assures us that the impression is a false one, that in this case seeing is not believing. The moon's rays really illumine the whole surface which lies before us, and we fail to receive light from other parts than the track below the moon, *not* because they receive no light, but because the light which they receive is not reflected towards us. An observer, stationed a mile or two towards the right or towards the left of our station, sees a different track of light, while the part which seems bright to us seems dark to him.

The rainbow is another phenomenon of this deceptive kind. We seem to see an arch of many colours suspended in the air,—and when we learn that it is due to the presence of drops of water in the air, we are apt to infer that where we see the red arch there are drops lit up with red light; where the yellow, green, or violet arch, that the drops are aglow with yellow, green, or violet light. But in reality this is not so: the same drops which seem green to us will seem red to another observer, violet to another, and to yet other observers will show none of the prismatic colours, but only the dull grey colour of the cloud on which the rainbow seen. We have here a pretty emblem of the varied aspects which events of the same real nature present to different persons, or according to the different circumstances under which the same person may see them. One shall see events in rosy tints, or with the freshness of spring hues, or with the melancholy symbolled by the

"deeper indigo (as when
The heavy-skirted evening droops with frost);

while to others the same events shall show only the ordinary tints of commonplace life.

The lunar halo is one of the phenomena thus deceptive to the view. We see all around the moon a circle or arc of light, nearly white, though sometimes faint tints of colour can be perceived in it, while the space within the circle seems manifestly darker than the space outside. The appearance of the halo as seen under favourable conditions is shown in the picture on the next page. In this country the dark space round the moon is not generally so well seen as in countries where the air is clearer. But this is in reality the characteristic feature of the halo, as its name shows; for the name is derived from a Greek word signifying threshing-floor (the old threshing-floors being round), and thus naturally describes a round space relatively clear, surrounded on all sides by a ring of aggregated matter.

We seem in looking at the lunar halo, then, to see the moon at the centre of a dark space, surrounded by a ring of bright particles, outside which again are particles not quite so brightly illuminated as those forming the ring, but more brightly than those within the ring.

But in reality this impression, which, so far as the sense of sight is concerned, seems *forced* upon the mind, is entirely erroneous. There is no real distinction between the space which looks dark all round the moon, the space beyond which does not look dark, and the ring between the two spaces which looks bright. These are all equally illuminated by the moon, in the same sense, at least, that we say the surface of a moon-lit sea is all equally illuminated, neglecting slight differences which do not concern the point we are specially dealing with. Precisely as the path of light on the ocean is not a real path of illumination, bounded on either side by dark spaces, so the ring of light round the moon is not a real ring of light, bounded on one side by a less bright region, and within by a dark space.

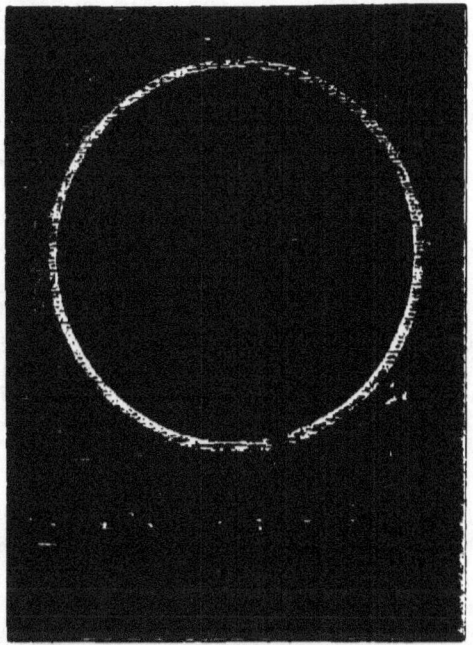

The Lunar Halo.

Here it may be worth while to notice how the particular illusion here considered has deceived even scientific men.

It had been noticed by Tyndall, in certain experiments, that a very sensitive measurer of heat, when placed under the moon's rays, gathered together by a powerful condenser, seemed to indicate cooling rather than heating, as we should expect. On this a French student of science pointed to the darkening under the moon where the lunor halo is seen as evidence that our satellite possesses a certain power of clearing away vaporous matter from the air. "*On peut dire*," he said, speaking of the dark space within the halo, "*que la lune ouvre alors une porte par laquelle s'échappe le calorique que l'action solaire a emmagasiné dans les couches inférieures.*" "One may say," that is, "that the moon then opens a door through which the heat escapes, which the sun's action has stored up in the lower layers" (of the air). It will be manifest, if we remember that a lunar halo can often be seen at the same time from stations hundreds of miles apart, that there can be no such opening of clear air. For the cloud layer in which the halo is formed is but a few miles above the observer; and, therefore, if one observer saw a circular opening in this layer, with the moon at its centre, another, a hundred miles from him, would see the space in a very different direction. The moon would not only not be at the centre of the space for this second observer, but would not be visible through the space at all. Moreover, the space could not possibly seem round to both observers; if it seemed round to one, it would look like a very flat oval of darkness (almost a mere line) to the other.

The real explanation of the lunar halo is very different. When you see such a halo, you may be certain that there is, high up in the air, a layer of light feathery cloud—the cirrus cloud, as it is called—composed of tiny crystals of ice. These crystals, as we know from those which in winter sometimes fall (not as snow, but as little ice-stars), have all a definite shape. They are in fact little prisms of ice, with angles like those of an equilateral triangle. These little prisms deflect the light which falls upon them, just as one of the drops of a chandelier deflects any light which falls upon it. If you hold a prism-drop of a chandelier between the eye and a light, you will see that the prism looks dark; it is really lit up, but it sends the light away in such a direction that the eye receives none. Now move it gradually away from the line of sight to the light, and at a certain distance it appears full of light; or, to speak more correctly, it sends the light it receives directly towards your eye. Beyond that position it again looks dark, but not so dark as when it was nearly between the eye and the light.

The little crystals of ice perform the same part with respect to the moon, when we see a lunar halo. Those between us and the moon, or within a certain distance from the line of sight to the moon, are, in reality, lit up by the moon's rays; but they send off those rays in such directions that we do not receive the light. Thus, all the space lying towards the moon, and for a certain distance all round, looks dark. But, at a certain distance, these little crystals send us light. If we could see them separately, they would seem to be full of light. That is the distance where ice-crystals of their known shape act most favourably in deflecting light, —that is, send off most for all the varying positions (not places) they can be in. At greater distances, a small proportion send us light. Thus, at that distance we have a ring of light, and outside the ring we have a gradual falling off in the quantity of light.

But the reader will be apt, perhaps, to say, How can all this be proved? No one has ever been among the ice-crystals of the feathery clouds when they are performing this work. When Coxwell and Glaisher made their highest ascent, the feathery clouds seemed almost as high above them as ever. Nor, if any one could reach those clouds, could he see the ice-crystals at their work. Yet there are few points about which science is more certainly assured than about this explanation of the halo. For we know the shape constantly assumed by ice-crystals: we know according to what precise law ice bends rays of light falling upon it; hence we can calculate quite certainly where, if ice-crystals make the halo, its rings should be seen. And the halo has the precise position thus calculated from the known law of optics, and the known facts about ice and ice-crystals. The diameter of the halo should be, and is, about eighty times the apparent diameter of the moon, or somewhat less than half the arc which separates the point overhead from the horizon.

There is, however, yet stronger evidence. Haloes form around the sun as well as round the moon,—in fact, more frequently. Solar haloes have so much more light in them that we can recognize varieties of tint. Now, it follows from the laws of optics that, for the red part of the sun's light, the halo ring should have a smaller diameter than the halo ring for the violet part, intermediate colours having their corresponding intermediate halo rings. Thus, the halo ring, as a whole, should be rainbow-tinted, red on the inside, then orange, yellow, green, blue, indigo, and violet; and these colours are shown (under favourable conditions) in this order.

The student looking out for haloes, solar or lunar, must be careful not to confound them with solar and lunar coronas, that is, not the corona of astronomy, but rings of light around the sun and moon, much smaller than the true halo rings. What I have said above about the size of the true halo will suffice to prevent such a mistake. Coronas are not nearly so *easily*, though they have been quite as thoroughly, explained by science, as haloes.

It is singular to observe how utterly unlike the interpretation of the halo by the science is from the natural interpretation. The observer would say, There surely is a dark space all round the moon, and round that a ring of light,—I see these things, and seeing is believing. Science says there is no dark space, and there is no ring of light; while the eye of science perceives something where the lunar halo shines which ordinary vision cannot recognize. Up yonder, many miles above the earth, science sees millions of crystals of ice, carried hither and thither—so light are they—by every movement of the air. Science sees these ice-crystals deflecting the rays of moonlight, sifting the red rays from the orange, and these from the yellow, yellow from green, green from blue, blue from indigo, and indigo from violet. Science, in fine, perceives processes taking place in those higher regions of air compared with which the most delicate analyses of the laboratory are utterly coarse and imperfect.

There is a purer and nobler poetry in the lunar halo as thus understood than in its mere visible phenomena, attractive and beautiful though these are. Idle indeed is the fear that the interpretation of this special mystery of nature will leave the number of nature's mysteries diminished by one. On the contrary, for the one mystery explained many deeper mysteries are suggested. The phenomena discernible by the sense of sight are explained, but only by bringing into the range of a purer and more piercing vision phenomena infinitely more wonderful. If one could see through some amazing extension of visual power, or if even the imagination could adequately picture the rush of light-waves of all orders of length upon the line of crystal breakers, their deflection in all directions, their separation into their various orders of wave-length; if one could perceive the actual illumination of the ice-crystals, even where they seem dark to us, and the continual fluctuations of the troubled sea of ether between the crystal breakers and the earth below,—the scene would infinitely transcend in interest and mystery, the picture would be infinitely more suggestive of solemn thoughts, than the scene—beautiful though it doubtless is—presented by the halo-girt moon to ordinary vision. Truly they know little of the real meaning of science who regard it as depriving natural phenomena of their effect on the imagination, as robbing Nature of her poetic influence.

## COALS AND COLLIERS.

THE exact date when coal began to be used as a fuel is very uncertain. The appearance, it not the use, of the mineral must have long been known in districts where it was already exposed, and presented itself on the surface. According to good authorities, it was used in England in 852. In 1259, King Henry III. granted the privilege of digging coals to certain parties in Newcastle. In seven years more, coal had become an article of export, and was termed sea-coal; and in 1306, so extensive was the use of coals in London, that Parliament complained to the King of the various vapours by which they polluted the atmosphere; and, in consequence, proclamation was made against their further use during the sitting of the House, lest the health of the knights of the shire should suffer during their residence in the metropolis.

In Scotland, coal was known and probably used at a very early date. Chalmers, the antiquary, informs us that coal was worked at Bo'ness by William de Verepont before the end of the twelfth century, and that a tenth part of the coal was paid to the monks at the Abbey of Holyrood.

The principal coal countries in the world are six—Great Britain, Belgium, the United States, France, Russia, and Austria. In Great Britain the coal formation amounts to 11,856 square miles, yielding 35,000,000 tons a year; in Belgium the area of coal is 518 square miles, and the number of tons annually produced is 4,960,077; in the United States the coal-fields are not less than 133,132 square miles, but they are worked only to the extent of 4,400,000 tons a year; in France the space is 1,719 square miles, and the yearly number of tons 4,141,617; the extent of the coal formation in Russia and Austria is not known, but the yearly out-put in the former is 3,500,000 tons, and in the latter 659,340. Thus there are produced from the great coal-fields of the world no les than fifty-two million six hundred and sixty-one thousand and thirty-four tons. It will be seen at once how large is our own share of this gross total. Of the whole thirty-five million tons yielded by our collieries, we export 2,728,000, leaving the remainder for domestic and manufacturing purposes. The number of persons employed in our collieries is about 170,000; and the capital invested is not less than £12,000,000.

It will also be at once seen how much Great Britain owes to coal as the means of its material prosperity. The astonishing increase of Glasgow,

Manchester, Birmingham, Leeds, Sheffield, and Newcastle-on-Tyne, must be attributed to the presence and working of coal in their vicinity. Nearness to coal-fields has exercised a far more important influence on cities and towns than many would at first suppose. Take only one example—that of the small central coal-field of England. Its area scarcely equals that of one of the larger Scottish lakes, and yet that limited coal-field has made Birmingham a great and flourishing town, the first iron depôt in Europe, and filled the surrounding country with crowded towns and villages. One is amazed as he thinks only of what the coal from the Staffordshire fields has done—how many thousand steam-engines it has set in motion, how many thousand railway trains it has propelled, how many million tons of iron it has furnished, raised to the surface, smelted, and hammered.

What is coal? It is now generally admitted to be the product of decomposed vegetable matter; and there are two ways of attempting to account for the fact that such immense quantities of it have been brought together. Some suppose that the plants from which it was formed grew and died on the spot where the coal exists, and that a bed of coal must at first have resembled a peat-bog. Others are of opinion that the vegetable matter was swept from the land into estuaries or lakes by floods and streams. This latter is called the drift theory. It is difficult to account for such immense accumulations of vegetable matter spread over surfaces so extensive in either way. But the vegetable origin of coal, at all events, has been almost universally admitted.

Hugh Miller, in his own wonderful manner, thus supposes the formation of coal. He says—"Imagine a low shore, thickly covered with vegetation. Huge trees of remarkable form stand out into the water. There seems no intervening beach. A thick hedge of reeds, tall as the masts of pinnaces, runs along the deeper bays, like water-flags at the edge of a lake. A river of vast volume comes rolling from the interior, darkening the water for leagues with its slime and mud, and bearing with it to the open sea reeds and fern, and cones of the pine, and immense floats of leaves, and now and then some bulky tree, undermined and uprooted by the current. We near the coast, and now enter the opening of the stream. A scarce penetrable phalanx of reeds, that attain to the height and well-nigh the bulk of forest trees, is ranged on either hand. The bright and glossy stems rodded like Gothic columns; the pointed leaves stand out green at every joint, tier above tier, each tier resembling a coronal wreath or a ancient crown, with the rays turned outwards, and we

see atop what may be either spikes or catkins." And so proceeds the eloquent geologist. The decay of all this great growth was the origin of coal.

This amazing vegetable growth lived before man was made. It flourishes and decays on the face of the earth age after age, and furnishes thus the means of comfort and useful art for man when he shall in countless years be formed. The plants and trees of those periods were of enormous dimensions, as may still be seen from fossil remains of some of them.

When it is attempted by geologists to explain how coal was formed, they are obliged to hazard certain speculations or opinions. The sandstone, shale, and limestone connected with it, must all have been formed under water, and the coal on land—perhaps in a marsh—the bottom being raised by the depositing of four fathoms of sandstone and shale, which were formed of the material swept down from the neighbouring land till a bay or gulf appeared, which shoaled into a tract of marshy ground. Upon this forests grew for a thousand years, and the decay of these made a thick stratum of vegetable material. The land then sunk suddenly or gradually under water to a great depth, and remained there for ten thousand years, perhaps, till a fresh deposit of sandstone and shale took place, the pressure of the weight of which, aided by water, turned the vegetable stratum into coal. The bay may be supposed to have been raised again—for the process of making the earth as we now find it was going on, and many underground fires were burning, and occasioning numerous convulsions—there was again a marsh, resulting from the decayed vegetation which had once more flourished. By a process similar to the first, another bed of coal was formed—and so the work proceeded for thousands of ages. It is believed from the testimony of the rocks that the land must have sunk in this way at least thirty times, and descended more than 3,000 feet.

The total thickness of coal in the English coal-fields is generally about fifty or sixty feet. This total is in most districts divided into twenty or more beds, each varying from six feet to a few inches thick; and the coal alternates with from twenty to a hundred times as great a quantity of sandstones and shales—these deposits being very variable in all coal districts, while the most regular are those of coal and ironstone. The entire thickness of this mass of coal, iron, sandstone, and shale, in alternating strata, may be estimated at about 1,000 feet, which is about its average in South Staffordshire; but the difference is very considerable in the various coal-fields. The great coal-field of Northumberland and

Durham, for instance, consists of eighty-two beds, of alternate layers of coal, sandstone, and slate-clay; and these, taken together, make an average of 1,620 feet of thickness, which, however, varies in different parts. In this mass the distinct beds of coal number between thirty and forty —the thickness of the whole being about 45 feet—there being eleven seams which are not workable on account of their thinness. The two most important beds are the High Main and the Low Main, the one being 360 feet below the other, and the thickness of each averaging about 6 feet.

The great coal-field which stretches from South Yorkshire, Nottingham, and Derbyshire very much resembles that of Newcastle. It extends from the north-east of Leeds nearly to Derby, a distance of more than sixty miles. The number of seams is about thirty, and they vary from 6 inches to 11 feet. The total thickness of coal is about 80 feet. The South Lancashire coal-field commences in the north-west of Derbyshire, and goes on to the south-west of Lancashire, having Manchester nearly in the centre. The South Staffordshire or Dudley coal-field is specially important on account of the extensive ironworks which it maintains, and is in length about twenty miles from north to south, and, in its greatest breadth, about seven miles. One portion of it is distinguished by the presence of one continuous bed of coal 30 feet thick, which is styled the main, or ten-yard coal, which is made up of thirteen different seams, some being close to each other, and others separated by thin bands of shale. Which is called the carboniferous limestone, and the old red sandstone, are entirely wanting in this coal-field; and this is believed to indicate that the coal strata of Dudley must have been formed in a fresh-water lake.

The various coal-fields of the country are to be found in widely different parts of England and Wales, Scotland, and Ireland—the working of them being largely dependent upon the industrial character of the population, and the spirit and enterprise of the proprietors. If one were to take a map of Great Britain—a geological one, of course—he would find four great patches in England, Wales, and Scotland. In Ireland the coal-fields are considerably scattered. But, in the parts of the country which have been named, he would see one in Scotland, lying between Fife and Ayr, and Edinburgh and Stirling—the Fife and Ayr portion naturally including the district around Glasgow; another in England, lying between Berwick and Durham, and far reaching into Northumberland; the third would occupy a great part of the centre of England;

and the fourth would extend from Monmouth towards Pembroke and St. David's.

It must be understood by the youngest of our readers that coal is not always found lying in flat layers, as the leaves of a book or the "courses" of stone or brick in a building. No. By means of many underground commotions, the coal and the other stony material with which it is always associated have been so broken up that this valuable mineral is to be frequently found descending aslant from the surface almost, and may then be met with in the form of a basin; but coal may always be counted upon in certain company. Just as in a book of twenty volumes you may always know where to look for the fifteenth, so among the various rocks which accompany coal one may be always sure, if it be there, to find it in its own place among them. It may not be there, or there may be no great quantity of it; but if it can be discovered in that locality, it will, without fail, be found in its own place. In many coal-fields there are "slips" and "dykes," which stand nearly, if not altogether, upright, and separate the portions of the seam from one another, thus occasioning much labour in carrying on the working of the mine. Yet these walls of separation are often useful. In case of fire in a pit, for example, they not seldom stop the conflagration, and so prevent a vast amount of mischief, and great loss of life and property.

A coal district is not generally the most beautiful—not even by nature. When Providence gives wealth below ground, the beauty above is usually but scanty. But the working of the mineral necessarily makes the face of nature even less comely. In such a district one sees, as the most conspicuous objects, tall engine-houses, and taller chimneys, pouring into the sky dense clouds of smoke, which hide the whole landscape, such as it is. There are groans and whistlings and unearthly sounds everywhere, which noises proceed from pulleys, and gins, and railways employed in the raising and carrying away of the coals. As you approach nearer to Newcastle or to Glasgow—in which latter place coal is largely employed in the smelting of iron—these smokes and fires increase upon you, and you feel as if you had passed into another world.

"Sinking" for coal is an interesting process; and, from certain sure indications which cannot be mistaken, the men who are employed in it can understand, with great accuracy, what progress they make. They can, indeed, tell what strata they have gone through in their boring, and what likelihood there is of trouble and expense from water in the working of the future pit. The pits are usually worked in chambers or rooms, with pillars of coal left to support the roof, or on the panel system, which requires less of the valuable mineral to be allowed to stand. The floors of these chambers are, it may be well believed, by no means level, neither are such apartments always well ventilated. Towards the bottom of the shaft ponies are employed, in almost all pits, to drag the "hutches" to the point from which they ascend, and the invaluable "Davy lamp" has been the means of saving very many lives from the fatal effects of foul air. Water, breaking in from old workings, has been one of the most disastrous causes of death, harrowing accounts of which, notwithstanding all precautions, now and again distress the community in connection with this dangerous employment.

The extent of the galleries of some of the older coal-pits of the North is very great. The Killingworth mines must have nearly 160 miles of galleries; and at the Howgill pits, not far from Whitehaven, the mining has been carried more than 1,000 yards under the sea, and at least 600 feet below its bottom. The deepest pit in the kingdom is that at Monkwearmouth, near Sunderland. To the Bensham seam it is 265 fathoms, and to a reservoir of water fifteen fathoms more—in all 1,680 feet; that is to say, it is seven or eight times as deep as the London Monument is high, or in depth as much as four times the height of the dome of St. Paul's.

The principal work in the coal-pit is that of the hewer, and his toil is very great, his confined position making it worse. He is generally bathed in perspiration and enveloped in coal-dust. The best hewers have learnt to do their work quickly, and it is curious to watch them shifting their postures, and strangely adapting themselves to the exact form or figure suitable to the bringing down of the coal with advantage and speed. The pits are generally hot, to the extent of a degree of temperature for every fifteen yards down, as compared with that at the surface, and this, too, makes the toil of the workmen more exhausting. Pits must be worked, in some instances, both night and day, and this is managed by so changing the sets of men that the night-shift, which is generally disliked, is fairly distributed amongst them all. In the discipline of pits much care is usually taken of the condition of the air, the amount of water, and other particulars in regard to the workings themselves, as well as in respect to the lamps and the conduct of the men; and yet, in spite of all, frightful accidents now and again occur.

The houses of the pitmen are generally erected either by the proprietors of the colliery, or by companies who speculate in the building and let them to the coal-owners at from £3 to £4 per annum. Of these cottages there are three kinds, as a rule—those which contain two rooms on the ground floor only; those which contain one room on the ground floor, and a loft above; and those which have two rooms below and a loft above. It being necessary that the dwellings should be near the pits, when a pit is worked out the houses are abandoned, and then present a most dismal appearance. Generally speaking, these dwellings are in rows, and the rows are again to be found in pairs. Such villages are not usually remarkable for their cleanliness. The furniture in the cottages themselves is very dissimilar, some of them having showy and apparently costly articles, and others only what is extremely dingy. The personal dress of both men and women is often loud and gaudy. The pitmen having an unlimited command of coals by merely paying a nominal price for carriage, the fires in their houses are tremendous.

Education is still far behind among the colliers, although, in this respect, a vast improvement has set in, since the prohibition of very young children from working in the pits. This has frequently been aided by the generosity of the masters; and the Education Act may be expected to carry the progress still further. There are Sunday schools and chapels at most villages, the same building being frequently used for all such purposes. The population of a village varies from 500 to 3,000. As to religion, colliery villages are cared for, in some instances, by clergymen of the Church, and in others by Independents, but most of all by the various sections of the Methodists. This is the case in England more than in Scotland. In the latter, the usual denominational distinctions are more prevalent. The attendance, both North and South, is moderately commendable. Temperance societies have done much for these people.

The character of colliers, as a rule, is reckless. Men who are exposed to great risks, like sailors and colliers, seem apt to be so. Colliers everywhere are given to intemperance and gambling, but especially in the South. In the North it is by no means so much so; and there a man, not rarely either, may be found willing to forego his self-indulgence that he may be able to buy a book. Some colliers and colliers' children have creditably risen in the social scale by their own efforts. George Stephenson was a collier, and went into the pit at six or seven years of age; the celebrated mathematician, Dr. Hutton, was originally a hewer of coal; and not to mention others who might be named, Thomas Bewick, the celebrated wood-engraver, was the son of a collier. He used to say that his earliest recollection was that "of lying for hours on his side, between dismal strata of coal, plying the pick with his little hand, by the glimmering light of a dirty candle."

The strikes of colliers have been well-nigh their ruin. The accidents they suffer from are manifold and disastrous. There are consequently many poor families in almost all colliery villages—poor headless families, and poor disabled men—and these are generally well cared for by their neighbours. We ourselves have known such men, and they have been kept in comfort: among them one man who, by the falling in of the roof, was confined, along with others, a prisoner for eleven days, his only subsistence being the oil of his lamp. With care he was brought round, but never could work; yet he never knew want.

---

## WILD FLOWERS.

MAGNIFICENT as florists' flowers are, I am of Perdita's opinion, that the

"—— art which in their piedness shares
With great creating Nature"

makes them far less interesting than Nature's wildings. I love to wander wild-flower hunting. Not to carry off the pretty creatures in captivity in a candle-box, partly because I am too lazy to carry one, partly because if I had one crammed with specimens, I should not be learned enough to give them dog-Latin names in a *hortus siccus*, which seems to me a *very* dry garden; but to see them where Nature has sprinkled them—smiling up from the green grass, instead of crucified on whitey-brown paper.*

Very few townsfolk, I fancy, except professed botanists, are aware of autumn's wealth in wild flowers. Spring and summer are supposed to be almost exclusively their seasons. On this clear, calm, balmy morning, when the little wind that wanders is warm and soft as the down beneath a swan's wing, and the island can be seen resting like a lavender cloud

---

\* " And 't is and ever was my wish and way
To let all flowers live freely, and all die
Whene'er their genius bids their souls depart
Among their kindred in their native place."
—LANDOR.

on the verge of the gently heaving, spangled sea, let us, then, stroll from the Downs to the shore, and note what wild flowers we can find upon our way, now that the year is "growing ancient," although not yet "on the birth of trembling winter."

If we were in Ireland, in Kate Kearney's country, we might claim those arbutus-blossoms dangling over the shingle wall of the parsonage garden as wild flowers. How exquisitely the white blooms—like fairies' waggon-bells—and the round, rough, reddening "strawberries" contrast with their own evergreen foliage, and the glossy leaves of the fig-trees making "a green shade" behind. If the strawberry-tree is a doubtful wilding in the parsonage garden, an undoubted native has invaded it— the white bell-bind. To an outsider its enamelled vases seem as noble a wild flower as our isles produce; but John, the parson's gardener, in spite of his quasi-clerical character, swears fiercely at the strangling "weed" which twines and chokes like a legion of Laocoon-snakes, however he may try to grub it out. This evening primrose, with its last night's yellow blooms, not yet aware that ere another night comes they must die, no doubt is a truant from the garden; it has planted itself, certainly not for concealment's sake, in a roadside patch of lilac-blossomed mallow. Up one little flight of worn wooden steps, and down another into the churchyard, which in some places is almost choked with yarrow, rubbing its white and pink heads against slanting gravestones, black-grey as if they had been blasted, or nodding its heads at one another over long, green, rotten grave-boards, buttoned with little brownish-yellow funguses. In one corner, with a party of the parson's bees buzzing round it—their hum almost the only sound in that still place—there is a clump of hairy-leaved, blue-blossomed borage; and in another a pigmy fir-plantation of nettles in bloom, with here and there a miller-like—fleshy, juicy, floury —goosefoot lifting its stiff flower-plume above them. Tiny buttercups are scattered over the graves, and the rabbits on the adjoining Downs must find pasture there to their hearts' content, or surely they would have nibbled down, long before this, the sow-thistles in blossom at the base of the chancel's yellow-plastered wall. On the older church walls, of stone and shingle, patched with unplastered brick, tiny blue speedwell peeps, fairy-eye-like, between its downy leaves, the straggling wild lettuce displays at all angles its little yellow flowers, and creeping mother-of-thousands drapes the hoary masonry with its purple leaves and blossoms. The ivy which now helps to hold together the old walls it has loosened, and which has climbed up to the weather-board turret, in which we can

see the bell waiting silent until next Sunday, has just began to flower. The robin that lit upon the inky yew, and began to sing upon that gloomy perch, flies off as we pass, even before its brief, plaintively-cheerful song is finished. We sidle through a V-like opening in the green, grey, silver-and-orange lichened churchyard wall, and find ourselves upon the springy Downs, where the only roads are tracks whose turf has lost a little of its elasticity. It would be almost impossible to take a step here without crushing a harebell, did not the slender wires of stem and pale blue drooping blooms often refuse to be crushed, and rise up again from beneath the heavy tread like the frail human creatures who so often bear affliction better than their stronger fellows. Daisies, and dandelions, and groundsel are out here as they are everywhere else almost.

Lady's-tresses (why so called I do not understand) lift their spires of tiny white blossom. The scabious, whose rootlets the Devil is said to have bitten off, seems to get on very well without them, if we may judge from its show of flourishing blue heads we see about us; and sheep's-bit has similar flowers upon its hairy stems; and if we were to seek for it, no doubt we could find the hairy shepherd's-purse in flower, and, *perhaps*, we might find the hairy greenweed's and the hairy thrincia's yellow blossoms. The yellow-wort's pale blooms are drinking in the sunshine, as if they wanted to improve their complexion. Mint and thyme are pouring out their warm fragrance. The gentian's purple blossoms are basking in the genial heat like the yellow-wort's pale ones. Out of pink, purple, and white-blossomed heather, and golden-bloomed furze and broom, rabbits scurry, hares gallop, with laid-down ears, carrion crows rise sullenly, and moorfowl fluster in startled and startling haste. Pick that sprig of furze which seems to be covered with red cobweb—mind you don't prick your fingers. Look into the red-tangle skeins, and you will see the buff blossoms of the dodder. Here is another parasitical curiosity, this leafless brown stick of broom-rape, with its brown flowers that could scarcely have appeared less flower-like when in full bloom than now when, wilted, they make their stem look like a blighted Aaron's rod. Where the Downs swell up into abruptly scarped cliffs, we find the shrub-like tree-mallow in imperial magnificence of purple bloom, and the bloody cranesbill also arrayed in rich attire. We turn inland again, after looking out on the laughing sail-flecked waters, and craning over to have a peep at the still moist sands they have left, turned by the alchymic sun into dazzling gold, and crunching plentiful fat mushrooms beneath our feet, wander again, but downwards, over the deliciously lonely Downs. Whom have we met, or rather have we seen in human form? One white-smocked shepherd, carrying a poetical crook across his shoulder, and smoking an unpoetical black pipe—with his shaggy dog sauntering beside his sauntering master, but glancing up at him for orders, like an officious usher eager to prove his *raison d'être* by pouncing upon his "principal's" pupils. But no pouncing was needed. Tranquilly spread, strayed, stopped, and again gathered themselves together the grazing sheep, mildly tinkling their musical bells. We watch the shadows of the few passing clouds flit over the Downs as we watched them change the colour of the less undulating sea.

In our downward course we suddenly come upon a thatched, shingle-built cottage, and discover that we have reached the limit of unenclosed ground. Hawksbeard waves its yellow blossoms on the cottage's dark thatch; and the hedge-bank that divides the garden from the Downs is bright with the big golden blooms of hawkweed, foiled by toothed purple-spotted leaves. "There's *no* road this way," says the deaf old man working in the garden, at last turning round from his spade—of whom, in *crescendo* tones, we have been asking our way. "There ain't no road; but you can come in by the house, if you like, and then go along the field, and through that bit o' wood, and that 'll take yer into the lane." In the garden hedge, on the other side, honeysuckle is opening its second blossoms, and in the ditch a bristly teazel lifts its terrified-looking hair-on-end white head. In the fields we pass feathery-leaved, brilliantly red-bloomed pheasant's eye, hemp-nettles, red, white, yellow, and pale purple, strong-scented lilac-blossomed mint, and purple-blossomed wound-wort.

On the bank up which we clamber, to creep through a hole in the hedge into the wood, we brush against the downy-leaved mullein's spikes of purple-stamened golden flowers, and the yellow blooms of the "washed-out"-looking ploughman's spikenard.

Inside the wood a little clump of golden-rod makes sunshine in a shady place; bees have found out their autumn favourite, and hum around it a cheerful hymn of praise. Put your arm before your face, and charge up the bank through these serried ranks of purple-headed plume-thistle, and now for a leap down into the lane. Here the ditches are choked with the rough leaves of the forget-me-not, sprinkled with blue-and-yellow stars. Farther on we see the yellow buttons of the bur-marigold, the bright blooms of the untidy-looking rag-wort, the persicaria's rose-red spikes, the green spikes of water-pepper, the stars of star-wort floating just beneath the surface of the standing water, the pinkish-white panicles of the arrow-head, and, though these are not flowers, but only flower-like in their look, the clustered coral berries of the cuckoo-pint, spotted arum, wake-robin, or lords-and-ladies, to give it its dear old childhood's name. The flossy seeds of the traveller's-joy that drapes the hedges look still more flower-like. The white "lilies," as the country folks call the flower in this part of the world, of the convolvulus, also hang from almost every other spray over the yellow toad-flax.

Up and down stepping-stones projecting from the ivy-mantled grey wall into a pasture tufted with the gracefully serrated leaves and white blossoms of dropwort, and patched with purple-mottled medick out in yellow bloom. In the damper pasture beyond the burnet waves its mulberry-like flowers, and the sodden banks of the big pond in the bottom, which is almost choked with pond-weed and plated with duck-weed, are red with streaks of water-wort.

Over a superannuated axletree doing duty for stile into a turnip-field, at the farther side of which two sportsmen and their pointer are solemnly at work, looking, if one may judge from their attitudes at this distance, as if they thought the marking and bagging of game the whole duty of dogs and men. We, however, are on the look-out for flowers, not birds. We find the blue-bottle blossoming again as brightly as when the vanished corn rustled over its faded fellows in the next field, and the shepherd's-needle's white flowers, set off by its vividly verdant leaves. The reapers and the gleaners are gone, but we can still find a harvest in the next great stubble-field, in the middle of which an old black mill with canvas sails stands idle, restfully watching the great stacks around which it may have to grind; tiny white blossoms of slim corn parsley, gleaming corn marigolds, purple and yellow toad-flax, and purple snapdragon.

We cross, by a plank bridge, a little stream. Moneywort—not the kind with golden blossoms, set off by silk-purse-like green, which gave, I suppose, the plant its name, but one with almost pallid little flowers—grows here and there upon the banks in ravelled skeins, lilac-blossomed malodorous hairy mint grows in thick clumps, and blue-blossomed skullcap vies with it in height.

The plank takes us into a close-cropped croft, on which red cows are tranquilly chewing the cud. About the trees which shade the green churchyard, from which the meadow is divided by a low grey wall, towers a twisted spire with a rusty lightning-conductor, its gilt weathercock flashing, its grey shingles gleaming in the autumn sunshine. The grey vicarage, a grey farm-house and grey farm-buildings, that have a look in their antiquity of not having been always dedicated to secular purposes, hem in the croft on another side. Round the old well bloom a few purple heads of meadow-saffron, and just outside the wall of the farm-garden some yellow-bloomed St. John's-wort has planted itself. Those dirty-white umbels by the hedge yonder are the blossoms of the unfragrant meadow-pepper, which is said to give a bad taste to butter.

Through the quiet churchyard, with a peep into the quiet church, standing, hushed from Sunday to Sunday, in the midst of the graves of the many generations who have worshipped in it; into the thatched, gabled, pargeted village-street, in which the only living being that we can see is a dog sound asleep in the sunny middle of the road. The place makes one think somehow of an Arctic summer midnight. Up such another lane as has already been described, and through a white gate at a level railway-crossing. The metals glitter in the sun, so trains must pass to keep them polished; but the place is so still, the white-haired, brown-skinned toddlers who stand staring in silence at the door of the pointsman's cottage, with eyes and mouth vieing in wide-openness at the passing strangers, are such genuine little rustics that it is hard to realize that panting, screaming locomotives and rumbling, rocking, jerking long lines of carriages rush past here from busy, noisy London.

Over a run-dry canal, its swelling green banks dotted with daisies, and on to a somewhat dreary spread of waste ground. Wormwood grows here in bushes of smooth-surfaced leaves, plumed with dim yellow flowers and fluffy-leaved, purple-blossomed, unfragrant horehound. Here, too, we find the white blossom of the black nightshade.

Where two roads cross upon this bit of waste there stands what was once a four-armed finger-post. One of its arms has been broken, or has fallen off, almost close to the post. The fragment that remains gives merely the undirecting direction, "To ——." The one inscribed "To the Sea" might have been better spared, since, before we follow its guidance, we can begin to smell the brine.

On the right there is a bit of bog. The green and white leaves and pink blossoms of butterwort, and the milk-white-veined blossoms of the grass of Parnassus, beautify the normally dismal quagmire. Thence our road runs between marshes on each side, dotted with ragwort and purple, crimson-blossomed red-rattle. Where the marsh becomes salt-marsh we find flower-spiked arrow-grass, green-blossomed glass-wort, purple and yellow sea-starwort, sea-wormwood in blossom, and the delicate pink-passing-into-red blooms of the marsh-mallow.

Where the sand that has blown over the sea-wall has been flooded we find white-bloomed wild celery and shiny-blossomed cudweed, and round about on the sandy mound that for the present protects the land grow pink-flowered hare's-foot-trefoil and fleshy-leaved flowering sea-plantain.

If we live to visit the place thirty years to come, perhaps we may find seaweed on the flat where now we have found land flowers. The sea, breaking itself so gently against the almost sand-buried "groynes," looks harmless enough now, but in time of storm it rushes over the sea-wall on which we stand, and for many and many a year has been eating away this coast, as a silkworm scallops a mulberry-leaf, or a hungry boy a slice of bread and butter.

---

## BEES AND THEIR BUSINESS.

I WAS very idle; there is at this present moment no manner of doubt about it; but why Dr. Watts should have written that tantalizing poem, to be the delight of governesses and the torture of children, I know not. I am sure his little story has not at all the effect the good old man intended, either from the manner of its application or the general dislike people have to insects. There is one comfort, however, in this order *Hymenoptera:* we have got to a real sting at last, for if bees and wasps are despised, they are also very much feared, and can certainly make themselves felt if they are injured. Before describing the family of Bees, I shall tell you an amusing incident, which made a lasting impression on my mind when a child. One warm summer afternoon, when even the birds could only get up enough courage to make a plaintive "tweet," "tweet," and children could not find any to work sums or write dictation, our governess was unusually energetic. It was a way she had to be blind to physical incapability, and to put every ill-learnt lesson to the score of idle wilfulness. Finding us all incorrigible, she went to the desk, and lectured the whole school on the advantages of gathering mental honey when we were young, and quoted solemnly the hymn I have mentioned. We were all listening, or looking out of the window, when a burring, whirring sound, like the last vibrations of an organ-pipe, came on our ears; the instrument producing this sound flew in at the window, and went straight to the desk, and made an effort to find honey in the flowers of the lecturer's cap. No sooner was the innocent creature perceived than the oration ended, and our governess uttered a scream which might have been heard all through the street, covered her head up in the first antimacassar she could find, and, seizing the hearth-broom, buffeted the "nasty thing" out of the window. She resumed her exhortation when this was accomplished, but had too much respect for our common sense to again revert to the Bees as an illustration. Philosophers say that example is better than precept, and the result of the lecture was that we despised the industry as much as our teacher the insect.

The advantage of "improving each shining hour" was not very clear at that moment; but when we went home to find unlimited supplies of honey for tea, there seemed a bare chance that Bees were not such very bad things after all; and it is the honey that brings me to the chief business of this history. Amongst all the useful things that insects make for us, there is certainly not one which has been longer known and better appreciated than honey. But the curious part of the story is, that the children everywhere seem to like it. Indians and Africans, French babies and English children, can one and all eat the food which the Bees provide, *not* for them, but for their own use in the winter. Now, there is a general opinion that only one kind of honey supplies all this feasting. But this is not so, for the Bees who make honey for us visit different species of flowers, and honey from Yorkshire moors is different in flavour from that which the southern Bees make from the cultivated garden blossoms. Honey from the northern moors has a *flavour* of the *smell* of the heath-plant that supplies it. That expression sounds like nonsense, but nothing else describes the relation between the flowers and the honey. Green honey has a curious appearance, but when first collected is prettier than English honey. The Isle of Bourbon gives this curious product, but I do not think any one has found out why it should be this colour. It is sweet and fragrant, and is as useful as our own.

I have said that honey depends for its flavour upon the fact that it is collected from different flowers, and it is grievous to relate that Bees are utterly unscrupulous about the choice of flowers: with their strong legs and pushing heads, they thrust themselves into the very centres of a rare petunia or a delicate iris, and the most anomalous results follow in the variation of the plants. To force their way into open blossoms is fair and reasonable enough, but if our friend the Bee cannot get in one way, he finds another for himself. Over fifty buds of a magnificent fuchsia were destroyed in one day by a party of Bees, who found their way into the greenhouse through an open window. Without the smallest hesitation they bit the bud half through with their mandibles, and abstracted the honey. It is perhaps a familiar fact to every one, that the buds of these flowers are perfect nectaries until completely opened. The juice or nectar in the cups of many other flowers can be easily obtained; the so-called "tears" of the crown imperial, and the drops which hang from the petals of the cactus, are useful reservoirs for the Bees. But intelligent as they are, they make sad mistakes sometimes, and collect material from poisonous blossoms. In 1790 the Bees of Philadelphia gathered honey from a plant in the neighbourhood, which had been more than usually uxurious, and such serious illness followed when the honey was eaten that it was generally reported the city had been attacked by the plague.

The drink called mead, which is made from honey, is getting so very countrified and old-fashioned, that by the time the children who read this work have changed into men and women it will be almost forgotten. Nevertheless, many labouring men of my acquaintance in a country village know no other beverage throughout the winter than this, and are very much stronger, and much more sober, than their neighbours. I would not wish to imply that mead has not the power to intoxicate, for it is quite different, and a wine-glassful of strong mead will have an effect —upon people who are not used to it, that is—as powerful as brandy. When one remembers how much fighting and quarrelling went on in olden days amongst our Saxon forefathers, there seems a strong probability that the Bees were indirectly to blame. Bees made honey, the people made mead, and the drinking of this had not always a soothing effect. But it is not the mead alone which does this, for there is somewhere in England a specimen of honey from the Black Sea that retains the powers of ntoxication.

The only well-known product of a bee-hive, besides honey, is the golden wax which forms the comb; this the Bees make by some peculiar process within them, and hold in plates between the rings of their abdomen. A Bee laden with wax shows golden-coloured stripes upon the under surface of its body. Cells of many plants contain waxy secretions from which Bees may obtain the materials for this, but it is supposed that the buds of the hollyhocks supply wax, as well as the bark of the poplars, from which they collect a sort of resin that they convert into paper to line their nests with. This is a substance made of wax and resin, and is perfectly waterproof. Wild Bees cover their nests with it, but the domesticated Bee knows better than to take so much trouble. These have found out by experience that the hives provided for them are satisfactory, and therefore but a small proportion of *propolis*, or bee-paper, is found in them.

The fourth material which these indefatigable little insects gather is the pollen from the flowers. This is not to supply them with honey, which is the usual fib told by ignorant nurses to little folks. This yellow dust is only gathered to be mixed with the honey to feed the tiny larvæ. Bee nurses know that too much honey is not good for children, and set us human folks an example, by giving their own babies a mixture of pollen flour and honey, called bee-bread. These little pellets of paste are carried directly into the stomach of the Bee, and are then disgorged to put in the cells where the young larvæ are hatched. This is a curious fact when we know that the honey becomes honey without ever entering the stomach of the Bees.

If we attempt to watch a hive as soon as a new swarm is swept into it, we shall see how they commence operations. There seems at first only confusion; in a few minutes they arrange themselves. One party flies out of the hive and begins the business of collecting propolis, another party brings back to the hive wax, and a third keeps watch over the queen. The occupants of a hive are as various as those in a white ant-hill, but they do not look so different. A queen Bee, a worker, and a male Bee, cannot be distinguished at a glance, but a very little examination shows the difference between them. Then there are the larvæ and the pupæ imprisoned in the tiny cells. All these individuals exist by hundreds in the hive, except the queen, and she is alone in the glory of reigning over such a busy community. But in our new hive there are at present only the workers and the queen; of drones or male Bees there are but very few, if any, and they very soon die. After a few hours, a shapeless but thick-looking packet of wax is formed from the side of the hive. This is the commencement of the new comb. Two sculptor Bees begin upon one side of it and form the first cell, and two others work at the back of it, and the two cells are thus placed with their bases to each other.

But this comb, which is first formed, is not to contain honey, but the eggs which the queen Bee is generally prepared to put in them within thirty-six hours after the first swarming. She goes over about 12,000 cells, and after putting her head into each one to see that there are no intruders, she deposits an egg separately in each. Several worker Bees attend her and close up the cells and the tiny egg within them. In the meantime other cells are completed, which are to form storehouses for the food of the young larvæ. So careful are the Bees to make sure of their pollen-paste or bee-bread, that they do not attempt to collect honey until a sufficient amount of pollen has been gathered. As soon as the little eggs are hatched, and the worker Bees perceive it, they take some pollen-paste and a little honey, and transform it into a little pellet. One is dropped into every cell that contains a living larva. After them come other Bees, each bearing the pellet of bee-bread, and they may be seen to put their heads into each cell as they follow the others, so that no one shall be overlooked or forgotten. In those cells on the edge of the comb, the pollen-paste which is given to the larvæ has an acid taste, and the Bees which are brought out from these cells are the new queens. It is one of the most wonderful things in any history that the food which is supplied to the insect causes it to become either a female or a worker Bee. Would it not be a comfort if all the idle people in the world could be made workers by having the right sort of food given them? I know many people to whom I should supply it, if I could first find it for myself.

Finally, a third class of cells are built, which have thickened edges, and in these the winter store of honey is deposited; each cell is then sealed over with wax in a manner familiar to every one who has seen new comb. Some cells are left open, and these contain the honey for public use, and are refilled when empty.

The young larvæ are fed until the time arrives for them to change into the pupa state. Then they perform their first work, and spin a delicate silken case for themselves. No sooner does the worker Bee discover what is going on than it closes the little insect, who is now a *pupa*, in a waxen cell, by putting a lid upon the top of its home. It is enclosed until its perfect metamorphosis is accomplished, and the mandibles of the perfect insect are first used to bite through the waxen lid of the cell. The new Bee then comes out and joins with the other workers in the labours of the hive. But in the royal cells the proceedings of the workers are only the same up to a certain point: when the larva in one of these cells attempts to bite its way out, the workers forbid it to appear, and seal it with a delicate transparent membrane. The little prisoner makes a distinct sound with its wings as a protest against such cruel treatment, but of course it has to submit. After a time it puts out its tongue against the sides of the cell, and the workers, perceiving this, cut a little hole in the lid, insert some honey, and again closely wax up the cell. This proceeding is sometimes continued for several days, and then the prisoner is set free, but only partially, for if she attempts to go near the cells of other workers, she is bitten and pulled back by the attendant nurses. This is a necessary precaution, for it is rather a bad feature in Bee history that the females of the hive are so quarrelsome. The reigning queen would kill her successor if she found the slightest chance of doing so, and it is only the lookers on (the neuter Bees) who can prevent one queen killing the other.

The new queen is kept a prisoner until a sufficient number of workers are set free from the cells; and then, the hive being too hot and too crowded, a party sallies forth, led by the new queen, or the old; and a swarm on the nearest tree is the result—the queen, remaining in the hive, having peaceful possession until another female is hatched, when the old proceedings are repeated. But, in saying that the hive is too hot, I am reminded that Bees have as great an idea of ventilation as any other animals, vertebrate or invertebrate. A party of Bees stand always on the floor with their heads towards the interior of the hive; these move their wings backwards and forwards in one direction; and another party stand with their heads towards the door, also with their wings in a constant state of vibration. Thus the air is kept constantly circulating in the hive, and the whole of it is thoroughly ventilated. Perhaps a prettier thing to watch is the care which the Bees have for each other. When a Bee returns from its wanderings after a circuit of two or three miles, and when its tiny wings ache with flying, and its legs are stiffened by the load they are bearing, it finds plenty of help in its home. Loving friends come clustering round it, and, with the tiny brushes on their front legs, take the pollen from the weary Bee into their own mouths and quickly lighten her of her load. When the hive is very busy, if the new home-comer is not noticed, she will plant her two fore feet firmly on some object, and summon the assistance which is at all times willingly rendered.

The chief character by which the Bees may be known, and, indeed, the whole order *Hymenoptera*, is their wasp-like waist: this many people call the stalk. It is a character so marked in the wasps that I need not dwell upon it, except to point out that it may be distinguished in swarms of tiny insects who are very commonly called flies.

*Hymenoptera* have four wings—two large front ones, and two smaller at the back; when in flight, the hooks on the front margin of the back wings catch the edges of the front ones, and they have, when flying, apparently but one wing on each side. As soon as they alight on the flowers, or are at rest, the wings are unlocked and fold on each other. One very important addition in the structure of the Bees is that they have a sucking and a biting mouth. All the other insects have two mandibles and a moderate under lip. Bees have, on the contrary, sharp but not very large jaws, and a long projecting proboscis formed by the under lip. This is rolled up, when they fly, in a little cavity between the thorax and head, and it is wonderful to any one who first discovers the length of it, how so perfect an instrument can be packed in so small a compass

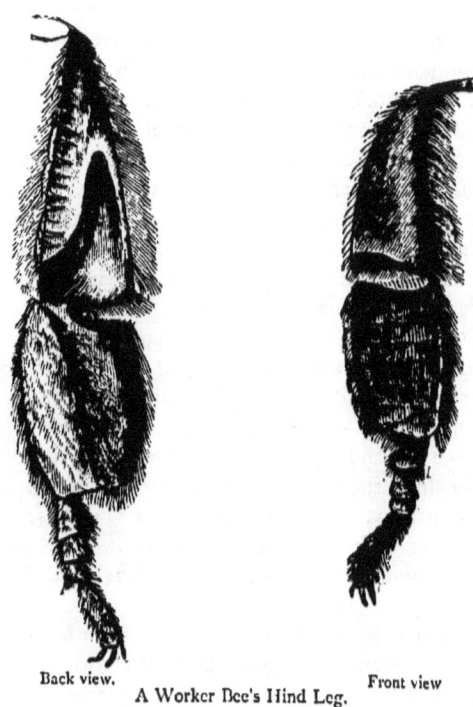

Back view.   Front view
A Worker Bee's Hind Leg.

The last and most important appendage is the triangular plate formed upon the third joint of the worker Bee's posterior leg. This plate is polished and smooth at its base, is fringed upon each edge with long hairs, and forms a pocket to retain the gathered pollen. Who would have expected an insect to have had a pocket to use, and one that is of as much importance to them as a pouch to a kangaroo?

There is a difference in the use, however: the kangaroo carries the baby and the food for it in the same bag, while the Bee, for good reasons, leaves the babies at home, and carries only the food for them in this elegant little pannier.

In spring days, when the insects first come out to visit the crocus-blossoms, they may be often picked up in a half-torpid condition from the cool winds. This little plate is seen loaded with yellow pollen, which the hair fringe supports, and the whole insect may be closely examined by curious people before the warmth of the hand has restored the Bee and enabled it to fly away.

## STEAM AND THE STEAM-ENGINE.

STEAM is the name given in our language to the visible moist vapour which arises from all bodies which contain juices easily expelled from them by heats not sufficient to burn them. Thus we say, the steam of boiling water, of malt, of a tan-bed. It is distinguished from smoke by its not having been produced by burning, by its not containing any soot, and by its being condensible into water, oil, or inflammable spirits, or liquids composed of these. We see it rise in great abundance from certain bodies when they are heated, forming a white cloud, which diffuses itself and disappears at no great distance from the body from which it has been produced. In this case the surrounding air is found to be loaded with the water or moisture which have produced it; and the steam seems to be completely soluble or lost sight of in the air, composing, while thus united with the air, a clear atmospheric fluid. Yet, in order to its appearance in the form of a cloud, its mixture with the air seems to be necessary. If a tea-kettle boils violently, so that the steam is formed at the spout in great abundance, it may be observed that the visible cloud is not formed at the very mouth of the spout, but at a small distance from it, and that the vapour is perfectly invisible when it first escapes from the kettle. This is rendered still more evident by fitting to the spout of the tea-kettle a glass pipe of any length, and of as large a diameter as we please. The steam is produced as abundantly with the pipe as without it, but the vapour is transparent and colourless throughout the whole length of the pipe; nay, if this pipe communicate with a glass vessel terminating in another pipe, and if the vessel be kept sufficiently hot, the visible steam will be produced as abundantly at the mouth of this second pipe as before, and the vessel will remain quite transparent. The visibility, therefore, of the matter which constitutes the steam seems to require that it should be mixed with the external air.

When steam is produced, the water is gradually reduced in bulk in the tea-kettle, and will soon be totally expended if we continue to keep it on the fire. It is reasonable, therefore, to suppose that this steam is nothing but water, changed by heat into an aërial, or, as it is called, an elastic form. If so, we should expect that the abstraction of the heat from it would leave it in the form of water again. Accordingly, this is fully verified by experiment; for if the pipe fitted to the tea-kettle be surrounded with ice, or any cold substance, no steam will issue, but water will continually trickle from it in drops; and if the process be conducted with the proper precautions, the water which we thus obtain from the pipe will be found precisely equal in quantity to that which disappears from the tea-kettle. Steam is, therefore, the matter of water converted by heat into an elastic vapour.

We are most familiar with steam when it rises violently rom heated water in the process of ebullition, or boiling. The observance of steam at this stage is highly instructive, and its production and appearance may be studied with advantage by examining it in a glass vessel placed over a strong lamp. When heat is first applied, a rapid circulation of the fluid is the result. The water at the bottom, being first heated and expanded, becoming lighter than the rest, rises to the top, and is replaced by the current of colder water descending, to receive in its turn a further increase of heat. By-and-bye, small globules of steam, formed on the bottom and surrounded by a film of water, are observed adhering to the glass. As the heat increases, they enlarge. In a short time several of them unite, form a bubble larger than the others, and, detaching themselves from the glass, rise upwards in the fluid. But they never reach the surface; they encounter currents of water still comparatively cold and descending to receive from the bottom their supply of heat, and, encountering them, the bubbles are deprived of their heat, shrivel up into their original bulk, and are lost among the other particles of water. In a short time the mass of the water becomes more uniformly heated; the bubbles, becoming larger and more frequent, are condensed with a loud crackling noise, and at last, when the heat of the whole has reached the point which is known as 212° Fahrenheit, or the boiling-point, the bubbles from the bottom rise without condensation through the water, swell and unite with others as they rise, and burst out upon the air as steam, of the same heat as the water from which they are formed, and, pushing aside the air, make room for themselves.

The singular sounds produced from a vessel of water, exposed to heat previously to boiling, have attracted attention. The water is then commonly said to be simmering or singing, and when this takes place it is because the vessel is boiling at one place and comparatively cold at another. This noise is most distinctly heard when the fire or flame applied is small and its heat intense, when the vessel is large and the water deep; for in that case the entrance of the heat will be more rapid than the circulation can convey to the remote particles of fluid, and so bubbles of steam will form more quickly at one place, and be rapidly condensed, than can be the case at another. The degree of speed with which such bubbles follow one another will regulate the pitch of the singing tone. This peculiar phenomenon is to be observed in the greatest perfection when we have attached a slender pipe to a close boiler producing steam, and carried its open mouth, of the diameter of one-eighth or three-sixteenths of an inch, down below the surface of cold water in a glass jar. When the mouth of the steam-pipe is held just below the surface of the water, the steam issues in great rapidity in small bubbles, producing an acute tone; and, on the other hand, when the pipe is held at a considerable depth, the concussions become more violent and louder, their intervals of succession greater, the tone is lowered, and finally, the shocks become detached, and so violent as to shake the glass and surrounding objects with much force. On this subject Professor Robinson observes, that a violent and remarkable phenomenon appears if we suddenly plunge a lump of red-hot iron into a vessel of cold water, taking care that no red part be near the surface. If the hand be now applied to the side of the vessel, a forcible tremor is felt, and sometimes strong thumps. These arise from the collapsing of very large bubbles. If the upper part of the iron be too hot, it warms the surrounding water so much, that the bubbles from below come up through it uncondensed, and produce ebullition without concussion. The great resemblance of this tremor to the sensation which is experienced during the shock of an earthquake has led many to suppose that earthquakes are produced in a similar manner—an opinion which is by no means unlikely. Any obstruction on the bottom of a boiler on the inside—for example, a piece of metal or a stone introduced among the water—may produce a succession of small concussions by the sudden condensation of gas collected around it.

The permanence of the boiling-point is one of the most remarkable of the phenomena of ebullition. When water has once been brought to boil in an open vessel, it is not possible to make it sensibly hotter, however strongly the fire may be urged or its intensity increased. This fact is very striking, because we know that heat continues to be thrown in exactly as fast as before the boiling-point; and that in that case the heat of the liquid also rose rapidly, whereas now it has altogether ceased to increase. If a thermometer of mercury, air, oil, or metal be placed in this water when it is cold, the temperature will constantly increase, and expand the matter of the thermometer, as heat is supplied, until the water boils; and then, whether it boil slowly or rapidly, with a strong fire or a gentle one, the thermometer will continue to stand at the same point. This point is so well known that it furnishes our standard for the comparison of temperatures, and is the same on all thermometers, being called the *boiling-point*, although it is differently numbered on each, being called 212° on our common thermometer, or Fahrenheit's, 80° on Reaumur's, and 100° on the Centigrade thermometer.

It is also to be observed that the temperature of the steam arising from boiling water is the same as that of the water from which it is produced, and remains equally invariable, so that all the steam issuing from water boiling at 212° is itself at 212°. The knowledge of this fact assists in accounting for the disposal of the heat which the fire gives out during the time of ebullition or boiling, for it is manifest that the heat is all the while carried off by the large volumes of steam, at a temperature of 212°, that are diffused through the air; and therefore an increase of heat in the fire, instead of increasing the heat of the water, only increases the volume of the steam thrown off and the quantity of heat carried away. This is proved by a very simple experiment. Take a strong glass flask, place water in it, and a thermometer among the water, and let it be held over a lamp until the water boil, and the thermometer will be observed rising till it reaches 212°, when the steam will begin to escape rapidly from the neck of the flask. These remarks apply only to steam in open vessels. Let the flask now be corked tightly and the heat continually applied, and it will be observed that the thermometer does not now stand at 212°, but rises rapidly from that point up to 220° and 230°, showing that the free escape of the steam into the open air is what keeps down the heat at 212°, or the boiling-point. If the heat be still applied, the experiment may be made still more instructive by suddenly pulling out the cork of the flask, when the vapour will instantly rush out in a large volume and the thermometer sink down to 212°, showing that all the excess of heat was formerly, and is now again, carried off by the steam into the air.

Thus, water placed on the fire soon rises to 212°, and a thermometer plunged into it remains at this point, however long it boils, if the vessel be open; but if the vessel be provided with a steam-tight cover, the temperature of the liquid may be much increased, according to the strength of the vessel. The force of the confined steam is in such cases enormous, and experiments of steam under high pressure or confinement are hazardous, unless the vessel be of great strength. The Marquis of Worcester burst a cannon by this means, and the frequent explosions of steam-engine boilers is a familiar instance of the same fact. Dr. Black and Mr. Watt heated water in a strong copper vessel to 400°, and in some of Perkins's experiments lead was melted in water subjected to strong pressure and closely confined, yet in an open vessel we cannot heat water to more than 212°.

In one of Dr. Black's beautifully simple and conclusive experiments a vessel containing some water at a temperature of 50° was placed on a red-hot iron plate; in four minutes it began to boil, but it required twenty minutes to convert the whole into vapour. In the first four minutes it had acquired an increase of 162° of temperature, and as the heat was uniform during the whole time of the experiment, it must have received an equal quantity of heat during the whole interval, or, during the other sixteen minutes, 810° must have flowed into it, yet during the whole time a thermometer in it rose no higher than 212°. Black naturally inferred that this large quantity of heat which disappeared had entered into the vapour in a *latent form*, or in a form which is hid from the thermometer. Dr. Black was the discoverer of the doctrine of latent heat.

It appears that the reason why water boiling and exposed to the air does not reach a higher temperature than 212° is that the steam which is raised by any additional heat carries that additional quantity of heat along with it into the air. The intelligent inquirer may naturally here ask why water requires to be heated up to 212° before it will throw off its heat and vapour into the air—why steam does not rise equally strongly from water at 200° or 180°. The answer is that the force of the heat is not sufficient to enable the steam to make its way against the pressure of the air until it reaches this point. When the pressure of air on the surface of the water is artificially diminished the steam does actually rise, and the water bubbles and boils with great violence at temperatures far below 212°. It is only when the surface of the water is exposed to the full pressure of the air in an open vessel that it is prevented from rising in vapour at temperatures lower than the usual boiling-point. If the surface of the hot water be protected from the pressure of the air by being placed under a glass shade, and the air be removed from the inside of the shade by an air-pump, the water may be made to boil at all temperatures below 212°.

Dr. Black, by means of a series of experiments undertaken by him and by his friend Mr. Watt, inferred that when water is converted into steam, it unites with 940° of heat, which the thermometer does not indicate; or, in Black's mode of expressing it, that quantity becomes *latent* in the steam. This determination nearly agrees with the experiments of Lavoisier, who estimated the quantity which thus disappears at 1,000° Fahrenheit.

The invention of steam as a moving power is claimed by various nations; but the first extensive employment of it, and most of the improvements made upon the steam-engine, the world indisputably owes to England and America. Steam-engines in their infancy were known as "fire"-engines. As early as 1543, a Spanish captain, named Blasco de Garay, showed in the harbour of Barcelona a steamboat of his own invention. An Italian engineer, G. Branca, invented, in 1629, a sort of steam windmill. The steam, being generated in a boiler, was directed by a spout against the flat vanes of a wheel, which was thus set in motion.

In England, the first notices we have of the idea of employing steam as a propelling force are contained in a small volume, published in 1647, entitled "The Art of Gunnery," by Nat. Nye, mathematician, in which he proposes to "charge a piece of ordnance without gunpowder," by putting water instead of powder into the gun, ramming down an air-tight plug of wood, and then the shot, and applying a fire to the breach "till it burst out suddenly." But the first successful effort was that of the Marquis of Worcester. In his "Century of Inventions," the manuscript of which dates from 1655, he describes a steam apparatus by which he raised a volume of water to the height of forty feet. This, under the name of "fire-waterwork," appears to have been actually at work at Vauxhall, in 1656. Sir Samuel Moreland, in 1683, submitted to Louis XIV. a machine for raising water by means of steam. The first patent for the application of steam power to various kinds of machines was taken out in 1698, by Captain Savery. In 1699 he exhibited before the Royal

Society a working model of his invention. His engines were the first used to any extent in industrial operations. They seem to have been employed for some years in the drainage of the mines in Cornwall and Devonshire. The essential improvement in these engines over those which had formerly been in use, was the introduction of a boiler, separate from the vessel in which the steam did its work. One vessel, in all former engines, had served both purposes. He made use of the condensation of steam in a close vessel to produce a vacuum, and thus raise the water to a certain height, after which the force of steam pressing upon its surface was made to raise it still further in a second vessel. In all attempts at pumping-engines hitherto made, including Savery's, the steam acted directly upon the water to be moved, without anything intervening or coming between. This was now to be changed. To Dr. Papin, a celebrated Frenchman, is due the idea of the *piston*. It was first used by him in a model constructed in 1690, when the cylinder was still made to do duty also as a boiler; but, in an improved steam-pump invented about 1700, he used it as a dividing medium, which floated on the top of the water, in a separate vessel or cylinder, and the steam, by pressing on the top of it, not on the water itself, forced the water out of the cylinder at the other end. The next great step in advance was made about 1705, in what was called the "atmospheric" engine, and which was invented conjointly by Newcomen, Cawley, and Savery. This machine held its own for nearly seventy years. The naming of a few particulars respecting this engine will enable the least mechanical of our readers to understand the application of steam as a propelling force. In this engine the piston, which is a steam-tight valve, moved up and down in a cylinder, into which steam was admitted from a boiler directly below it. This cylinder was connected with the boiler by a pipe provided with a stopcock, for cutting off or admitting the steam. The steam, rushing into this receptacle, drove up the piston, or steam-tight valve, to the top of it, and the piston, being connected by a rod with a beam, moved it upwards, and thus worked as a pump, or, by accommodation and other arrangement and application, as a propeller of wheels, or pinions, or machinery. A dash of cold water was then thrown into the cistern, into which the steam had thus been permitted to rush. This condensed the steam, and created a vacuum, and this brought down the piston to the bottom, under the pressure of the external atmosphere, and, of course, with it the rod and the beam with which it was connected. The cock was then turned to admit fresh steam below the piston, and thus the motion began anew, and

the working of the engine proceeded. The opening and shutting of the cocks was at first performed by an attendant, but subsequently a boy, named Humphrey Potter (to save, it is said, the trouble of personal superintendence), devised a system of strings and levers, by which the engine was made to work its own valves. In 1717, Henry Beighton, a Fellow of the Royal Society, invented a simpler and more scientific system of "hand-gear," which rendered the engine completely *self-acting*. During the latter part of the time which elapsed before Watt's discoveries changed everything, Smeaton brought Newcomen, Cawley, and Savery's engine to a very high degree of perfection.

The next essential improvements in the steam-engine were those of Watt, which began a new era in the history of steam power.

James Watt produced from the atmospheric engine of Smeaton the pure steam-engine, which he left to us in its present state of high improvement. He was the man who lowered the scale of expense in favour of the fire-engine, when it was a more costly power than that of horses, except when fuel was extremely cheap. In his hands it ceased to be an atmospheric engine, and became wholly a steam-engine, capable of being applied to any purpose, on a much larger scale, and at much less expense than the power of horses. He reflected that, "in order to make the best use of the steam, it was necessary, first, that the cylinder should always be maintained as hot as the steam which entered it; and secondly, that when the steam was condensed," or again reduced to water, "the water of which it was composed, and the injection itself, should be cooled down to 100°, or lower where that was possible." The means of accomplishing those objects occurred to Mr. Watt in 1765. The separate condenser was to be distinct from but in connection with the steam-cylinder, and into this the steam from the cylinder was to flow—all the operations of condensation being performed by surrounding it with cold water, or by injection, or both. The water which would necessarily accumulate in the condenser, Mr. Watt proposed to remove by means of a pump. "It next occurred to me," he says, "that the mouth of the cylinder being open, the air which entered to act on the piston would cool the cylinder, and condense some steam on again filling it. I therefore proposed to put an air-tight cover on the cylinder, with a hole and stuffing-box for the piston-rod to slide through, and to admit steam *above* the piston, to act upon it *instead of the atmosphere.* There still remained another source of the destruction of steam, the cooling of the cylinder by the external air, which would produce an internal condensation whenever steam entered it, and which would be repeated every stroke. This I proposed to remedy by an external cylinder, containing steam, surrounded by another of wood, or of some other substance which would conduct heat slowly. When once the idea of the separate condensation was started, all these improvements followed in quick succession."

It would be inconsistent with our limits to enter into any description of the constructive details of steam-engines. The principle of their power is here indicated—all besides is an affair of rods, and cranks, and beams, and pinions, and adaptation to the purpose which is intended to be served. Since Watt's time great improvements have been made in the construction of engines, but the application of the steam power has been essentially the same down to the present day.

The uses to which steam power has been directed are very various and manifold. At first this force was employed chiefly in the pumping or raising of water. Then it was used for purposes of manufacture and locomotion. Very early in the history of steam locomotion projects were formed for running steam carriages on common roads, each carriage to have passenger accommodation as well as steam power. In this department of invention there has been much ingenuity displayed by many skilful men in England, France, and America, but commercially all the different modes of promoting steam passenger traffic on common roads have been failures, and therefore within the last quarter of a century such inventions have been intended for the traction of heavy loads rather than passengers, and even in this direction such engines have not come into extensive use. The passenger carriages cannot, however, be said to have themselves failed. Sir Charles Dance ran such a carriage between Gloucester and Cheltenham in 1831, doing the nine miles in fifty-five minutes, having made 400 trips without an accident. In the same year, Mr. Hancock began running his steam carriage, called "The Infant," regularly between London and Stratford; and some time afterwards, Mr. Scott Russell ran his invention between Glasgow and Paisley. The probability is that, especially in connection with tramways, we shall ere long see a renewal of attempts to propel carriages on common roads by means of steam.

With the application of steam power to the purposes of our great manufactories, of railway locomotion, and of the propulsion of ships, all are so familiar that it is not needful to dwell upon it. The peculiar work which has to be done by locomotive engines requires that they should occupy little space, and much skill has been shown in their construction and adaptation. Many of these are very powerful and very beautiful. Railways and locomotive engines on them were first brought into use in this country in the northern coal districts. The first steamer on the water was the *Comet*, which was constructed by Mr. Robert Bell, of Helensburgh, for passenger traffic, on the Clyde. What an amazing advance has been made since those times—from the coal-waggon and its poor engine to the saloon carriage of her Majesty, and the "flying" engine which suitably accompanies it; from the *Comet* to the *Great Eastern*, the Cunard liners, and the great war-ships! It is as if the nation would now stand still and be paralysed without steam and the steam-engine.

One of the most remarkable instances of the application of this power is to be seen in the steam-hammer, which can be so adjusted and worked that a watch may be placed under it, and the ponderous machine brought down upon it with such nicety that it is simply touched without having even its glass broken, while at the same time there is possessed by it a power which can weld tons of red-hot iron into a solid mass with the utmost ease and facility. Probably the largest steam-hammer in the world is that at Woolwich, which is used in the manufacture of large cast-steel guns. Of similar dimensions and power is the hammer at Perm, a town in the north-eastern part of Russia. The metal to be operated upon by these leviathan hammers is melted in vast furnaces, sometimes in several; that in Russia has connected with it no fewer than fourteen. Its anvil block weighs considerably more than 500 tons, and it was several months before the casting was cool enough to be uncovered and turned over.

But it would take volumes to tell of the wonders of steam and of the uses to which it is applied. Steam and coal and iron are the principal material means which have made this country great, and no country in the world is equal to this in the skilful manner in which it employs them.

## OUR IRONCLADS.

THE time was when people used to speak with confidence of the "wooden walls of old England," meaning thereby the great oak-built war-ships. But that time has now passed and gone. We now trust to ironclads.

The idea of strengthening the sides of ships, so as to render them capable of resisting attack, is, however, nearly as old as the art of navigation itself. From the time of the Norman freebooters, who protected themselves by ranging their bucklers along the sides of their vessels, down to the present, all nations have sought by means more or less perfect to make their ships impenetrable, and to render them invincible in battle. The first attempts at making ironclad vessels were made by the Normans in the twelfth century, when they put an armature or belt of iron around their vessels, just above the water-line. This belt was terminated by a spur, which answered the purpose of a ram. In some instances this armature was converted into a curtain of iron or brass, reaching above the bulwarks, for the protection of the combatants. The crusaders of the twelfth and thirteenth centuries protected their ships in a similar manner. Pedro of Aragon, in 1354, ordered the sides of his ships to be covered with leather or raw hide, to protect them against what were called incendiary compounds. Andrea Doria, who commanded in the expedition against Tunis in 1535, had one vessel in his fleet plated with lead, which was furnished by the knights of St. John of Jerusalem; and at the battle of Lepanto—1571—many of the Genoese ships were strengthened by blindages, or bulwarks, composed of heavy beams, old sails, cordage, and other material. In 1782 the Chevalier d'Arçon, on the suggestion of M. de Verdun, at the unsuccessful siege of Gibraltar, constructed and used ten floating batteries having their tops bomb-proof, and their sides protected by parapets six feet thick, composed of hard wood, reinforced by cork, wood, leather, and bars of iron. These floating batteries carried 214 guns of large calibre, of which seventy-two were reserves, and for several hours, at close range, they withstood the heavy fire of artillery concentrated upon them. They yielded finally only to red-hot shot, and all but one were burnt or blown up.

Some experiments were made at Metz and Gavres for the purpose of determining the power of different materials to resist penetration, in consequence of which General Paixhans recommended that the French vessels of war should be strengthened by plating them with iron; but this recommendation was rejected by the Board of Naval Construction in 1841. In 1835 Mr. John Podd Drake had proposed to the English Naval Department the protection of the machinery of steam men-of-war by iron plating $4\frac{1}{2}$ inches thick, and in 1841 he promulgated the idea of ironclad blockading-ships. The successful employment of ironclad floating batteries, by the English and French, against the forts in the Baltic,

at the time of the Crimean War, called special attention to the importance of vessels so protected; and at Toulon, on March 4th, 1858, the French Government began the construction of the *Gloire*, an iron-plated screw frigate of the first class. From that day wooden ships gave way to a new class practically impenetrable to the projectiles of artillery, and endowed with speed and sea-going qualities equal to those which they were destined to replace. The science of ordnance and gunnery also received a new impulse, inasmuch as guns of larger calibre and greater penetrating power became necessary. Every subsequent improvement in the construction of ironclad ships has been followed if not preceded by a corresponding improvement in artillery, and in no branch of human industry have greater ingenuity or more persistent effort been displayed.

Shortly after the *Gloire* was begun, the French laid the keels of several other vessels of a similar description. The construction of these ships was looked upon by all maritime nations as betraying an intention on the part of Napoleon III. to make France the principal naval power of the world, and was regarded by England as a direct challenge, which could only be properly met by the construction of a fleet of still more formidable vessels. Accordingly the Admiralty ordered the building of the *Warrior*, and shortly afterwards of the *Black Prince*, *Defence*, and *Queen*. These ships were the forerunners of a new fleet entirely of ironclads, built at an enormous cost, but making good the position of England as the first naval power of the world. The example of France and England was soon followed by other European powers. Austria undertook the construction of two frigates, and Italy of two corvettes in 1860.

Up to this time all the efforts of constructors had been directed to the building of vessels after the old patterns, simply using iron instead of wood, or to strengthening the wooden walls of old ships, without any essential modification of form or change of model. But the outbreak of the civil war in the United States gave a fresh impulse to invention in this direction. The seizure of the important points on the Mississippi iver, below Cairo, enabled the Southerners to erect batteries, and to stop navigation, and rendered it necessary for the Northern fleets to be accompanied by ironclads. Such vessels were, therefore, constructed and employed, several of them having turrets with guns.

The Southerners, shortly after the commencement of hostilities, seized the navy-yard at Norfolk, in the harbour of which the wooden frigate *Merrimac* had been scuttled and sunk. They raised her, cut down her sides, and converted her into an ironclad ram, which they called the

*Virginia*. She was covered with railroad iron, laid on an oak backing, and was well armed with powerful guns. This great and formidable ship, after having engaged and sunk several war-vessels of the enemy, was encountered on the 9th of March, 1862, by the *Monitor*, a novel ship, constructed by John Ericssen, of New York, which, after a brief but remarkable combat, disabled her and drove her back to Norfolk. This combat marks one of the most notable epochs in naval warfare, and changed the character of naval construction throughout the world. The essential feature of this vessel was a revolving turret, composed of wrought-iron plates, an inch thick, bolted together till a thickness of 8 inches had been obtained.

The results of this combat were far-reaching in effect. All maritime nations addressed themselves actively to the transformation of their old wooden steamships, wherever they were sound, by cutting down and plating their sides, and to the building of monitors, as well as ironclad frigates. Mr. E. J. Reed, Secretary of the Society of Naval Architects, was called, in England, to the post of chief constructor, and began at once a radical modification of the naval marine. In 1863, the *Bellerophon*, representing the ideas which Mr. Reed had carried into the English Admiralty, was put upon the stocks, and the *Warrior* and others soon followed. Ironclad ships are substantially of two forms or types: those in which the batteries are protected by armour laid upon the walls of the ships, such as the *Warrior*, *Hercules*, and *Bellerophon;* and those carrying their batteries in turrets, such as the *Glatton*, *Thunderer*, *Devastation*, and *Alexandra*. These ships are also divided into classes, according to their fitness for cruising, defending harbours, guarding coasts, or operating upon rivers and lakes. While there is a certain amount of similarity in all the vessels of each class, there are also many differences in details according to the intended use.

The *Warrior* is armed only at the middle with $4\frac{1}{2}$-inch plates, while both bow and stern, including the steering-gear, are exposed to shot and shell. In all the more recent English ships this central battery or "box" has been enlarged by a continuous belt of armour, extending from stem to stern, and protecting the region of the water-line and steering-gear.

The *Warrior's* armour is of uniform thickness; but in recent ships the most vital parts, such as the region of the water-line and over the machinery, have been further protected by thicker armour, additional backing, and iron bulkheads fitted inside. The *Warrior* possesses only broadside fire: all the latter vessels have their fighting capacity increased by bow and stern fire of greater or less extent. The *Warrior* has only her main-deck battery armour-plated: recent ships have a protected upper-deck battery. The *Warrior* has her guns well spread out: later ships carry their battery concentrated, and composed of much heavier guns. The *Warrior* was made extremely long, with a view to speed: recent ships are much shorter in proportion, and are handled more easily. The *Warrior* has a single-skinned hull and comparatively light framing: later ships are double-skinned, with deep strong framing, and water-tight compartments. The armour of the *Warrior*, as already stated, is only $4\frac{1}{2}$ inches thick: that of the *Bellerophon* is 6 inches, of the *Hercules* 9 inches, of the *Hotspur* 11 inches, and that on the sides of the monitors, *Glatton*, *Thunderer*, and *Devastation*, is 12 inches, while their turrets are 14 inches. Presuming that the resistance offered by armour-plates to penetration varies as the square of the thickness, which is approximately correct, the armour of the *Bellerophon* is nearly twice as strong as that of the *Warrior*, of the *Hercules* about seven times, of the *Hotspur* six times, of the *Glatton* seven times, and of the turrets of the latter nearly ten times. The guns, rifled, used by the *Warrior*, weigh $4\frac{3}{4}$ tons, those of the *Bellerophon* 12 tons, of the *Hercules* 18 tons, of the *Glatton* 25 tons, while those of the *Thunderer* and *Devastation* weigh 30 tons.

The necessity of carrying such armour and guns, and of firing ahead and astern, as well as from the broadside, has rendered essential many changes in the sizes, forms, and arrangements of the sides, decks, and batteries of armoured ships. The introduction of twin screws, and the necessity of having light-draught vessels for coast and harbour defence, have also led to further differences. When the first English ironclads were constructed, the most powerful guns used were 68-pounders, or 8-inch smooth-bore guns. The Americans then used 9 and 10-inch guns, and $4\frac{1}{2}$-inch armour-plating was deemed sufficient when properly backed and supported. This thickness of armour, backed in various ways, forms the protection of a large number of the English and French ironclads. In the first iron ships, the *Warrior*, *Black Prince*, *Achilles*, *Defence*, *Resistance*, *Hector*, and *Valiant*, the $4\frac{1}{2}$-inch armour was backed by 18 inches of teak, fitted outside the hulls; and in the wooden ships the armour was bolted on the outside of the planking.

In the *Minotaur* class the plating was increased to $5\frac{1}{2}$ inches, but the backing was reduced to 9 inches, so that practically the sides of the latter class are of the same strength as those of the *Warrior* class. In the *Bellerophon* the armour-plating is 6 inches, and the backing 10 inches, but it is still further strengthened by having the skin-plating $1\frac{1}{4}$ inch thick, or nearly an inch thicker than in the older iron-built vessels. The armour of sea-going broadside ships has, according to some of our best authorities, reached its greatest thickness in the *Hercules*, which has 9-inch armour at the water-line, 8-inch on the most important parts of the broadside, and 6-inch on the remainder, with teak backing 10 and 12 inches thick outside the $1\frac{1}{2}$-inch skin-plating. Below the lower deck, and down to the lower edge of the armour, the spaces known as the "wing passages" are filled in with solid teak backing, inside of which there is an iron skin $\frac{3}{4}$ inch thick, supported by vertical frames 7 inches deep. The total protection in the region of the water-line, therefore, consists of $11\frac{1}{4}$ inches of iron, of which 9 inches are in one thickness, and 40 inches of teak backing. The trial of a target at Shoeburyness, constructed to represent this part of the ship's side, proved that it was impenetrable to the 600-pounder rifle gun. But the maximum thickness of armour carried must not be considered to have been yet attained.

Coast-defence vessels and rams have been built to carry 11 and 12-inch armour, and ships have recently been constructed for sea-going purposes to carry 15, 18, and 20 inches of armour, either in turrets or broadsides. There can be little doubt that as improvements are made in the manufacture and working of heavy guns, corresponding additions will be made in the thickness of armour. It is hardly possible, indeed, to foresee in what way the competition between guns and ships will terminate. The armour of iron-built ironclads in the British navy ranges from $4\frac{1}{2}$ inches in thickness to 12 inches; the backing from 9 inches to 18 inches; and the skin-plating from $\frac{1}{2}$ inch to $1\frac{3}{4}$ inch. The wood-built ironclads have armour of from $4\frac{1}{2}$ inches to 6 inches, with a thickness of side of from $29\frac{1}{2}$ inches to $31\frac{1}{2}$.

In the earlier English ironclads the armour extended over a portion of the broadside only, as in the case of the *Warrior*, whose length is 380 feet, and the armoured portion only 213 feet, leaving the extremities of the ship entirely unprotected. At the ends of the armoured portion, iron-plated bulkheads are built across the ship, making with the side armour a central or "box" battery extending to a little more than 6 feet below the water-line. This box battery, or partial protection, is also adopted on the *Black Prince*, *Defence*, and *Resistance*, but has been modified by the addition of a belt of plating extending from the upper to the main decks, before and abaft the broadside armour, on the *Hector* and

*Valiant.* The main deck on which the guns are worked is thus protected throughout its entire length, but the extremities between wind and water are exposed. Both these plans of disposing the armour were considered unsatisfactory, and in the *Minotaur* and converted ships of what are called the *Caledonian* class, the "complete protection system," in which the armour extends from stem to stern, and 6 feet below the water-line, was adopted. This system is now followed in nearly all the monitors and turret-ships in all the great navies of the world.

The extraordinary development in the power of ordnance has led not only to increased thickness of armour, but to different modes of placing it. In the *Bellerophon* and *Hercules*, and in other large ships of the British navy, an arrangement of the armour consisting of a middle course between the *Warrior* and *Minotaur* has been adopted. The *Achilles*, a ship of the *Warrior* class, has had the water-line-belt added. This plan of plating is known as the central battery and armour-belt system. In this arrangement the great weight of the armour and battery is amidships, and the ends of the ship are not overloaded, as in the complete protection system.

The offensive powers of ironclads, in common with those of other ships of war, are measured by the number and power of their guns, the rapidity with which they may be loaded and fired, and the facility with which they may command all points within range. The wooden frigates of the British navy in use before the construction of ironclads carried 32-pounders and 68-pounders. But now, in the new style of vessel, ordnance of a much heavier description is employed. The *Thunderer*, the *Devastation*, and other ships, have 30-ton guns; while the great ironclads and turret-ships more recently built are armed with 81-ton guns, and even this last is not by any means likely to be the largest which may soon be employed.

A few words ought here to be said about iron-clad rams. The introduction of steam men-of war gave rise to numerous proposals for reviving the ancient method of naval warfare, that of disabling or sinking an enemy by ramming; and when the *Gloire* and the *Warrior* were built, their bows were designed, strengthened, and projected with this object in view. In all succeeding ironclads, more or less efficient provisions have been made to adapt the bows to the same purpose; and it is now the decided opinion of naval officers that every ironclad should be an unexceptionable ram. The victory of the Austrian over the Italian fleet at Lissa, in 1866, was in a great measure due to the services of the

Austrian ship *Ferdinand Max*, which rammed and sunk the frigate *Re d'Italia*, and damaged other ships severely. In order that a ship may be efficient as a ram, it is essentially necessary that she should be swift and handy under steam, so as to enable her not only to overtake her enemy, but to hit her directly and squarely in the side. These qualities are incompatible with either great size or great length. Hence the ram should have moderate dimensions and proportions, in combination with powerful machinery, twin screws, and improved means of steering. A new description of torpedo ram has been proposed, which will be very destructive. This vessel has been designed and is in course of preparation for the British navy. It will be heavily armoured, but will carry no guns, and thus will be unlike any war-ship now in existence.

The number is constantly receiving additions, but the following table shows the latest ascertained strength of the ironclad navies of the world :—

| Countries. | Number of ships. | Aggregate number of guns. |
|---|---|---|
| Great Britain | 58 | 761 |
| France | 44 | 357 |
| Germany | 6 | 79 |
| Austria | 11 | 166 |
| Italy | 22 | 207 |
| Spain | 7 | 145 |
| Holland | 20 | 61 |
| Denmark | 6 | 69 |
| Russia | 25 | 180 |
| Sweden and Norway | 13 | 23 |
| Turkey | 22 | 127 |
| Greece | 2 | — |
| United States | 48 | 121 |
| Brazil | 18 | 64 |
| Chili | 2 | — |
| Peru | 6 | 24 |
| General total | 310 | 2,384 |

At present there are several great turret ships, ironclads, in course of being added to the British navy—the *Inflexible*, the *Alexandra*, the *Nelson*, the *Temeraire*, and others. The cost of each such ship is between £400,000 and £500,000, and in some instances it is more. These later vessels have usually their turrets covered with 18 inches of iron, the sides of their citadels being coated with 24 inches of metal, while under the water-line they are protected with 16 inches of armour-plate. They carry, each, several 81-ton guns, which are so placed in the turrets that they can be turned in any direction. The engines are from 8,000 to 10,000 horse power, and the whole of this can be brought to bear upon the ram.

Many of our readers will recollect an occurrence which happened some years ago, which proves the destructive force of the ram even when brought to bear upon a powerful enemy. In this instance, however, the vessel which was sunk was a friend and not a foe. As the Channel fleet, consisting of five ironclads and a yacht, was cruising on the 1st of September, 1875, off the coast of Wicklow, in Ireland, the *Iron Duke*, in a fog, drove her ram into the *Vanguard*, accidentally, and sunk her. The two ships were twins, of the same dimensions, and armed in a precisely similar manner. The *Vanguard* was a fine ironclad vessel of 6,034 tons, carrying fourteen guns. She was launched in 1870, and cost £350,000; but as she sank, with armament and fittings on board, her value was over a half a million of money. Her armour is 6 inches thick, the backing 10 inches, and the skin-plating an inch and three-quarters. The hole made in her side by the collision was 15 feet by 4. She was built in water-tight compartments, but the doors were not closed, and the water rushed in by tons; and had it not been for the bravery and presence of mind of one of the crew, who rushed down and turned off the steam, there must have been an explosion and great loss of life. All on board were quickly transferred to the depredator—the *Iron Duke*—about twenty minutes only being required for the operation, when the *Vanguard* whirled round two or three times and then suddenly sank in deep water. And there she still lies. This incident shows what mighty machines these ironclad rams are, when they can thus injure even one another.

## BULBS.

AMONGST what may be called the curiosities of commerce, the trade in Dutch flower-roots may certainly have a place. About Autumn the advertising columns of our gardening periodicals are crammed with notices of the arrival of these bulbs from Holland, and the shop windows of our nurserymen and seedsmen are filled with the well-known, brown, dry-looking roots.

Of late years this branch of the business of our nurserymen has very much increased; the spread of a taste for gardening amongst all classes, together with the prevailing system of summer bedding,—which has led to a similar plan of treatment of spring flowers, so far, at least, as this is capable of being done,—has been the means of causing a greater demand for these bulbs. Under the general term of Dutch bulbs the nurseryman includes such spring flowering plants as the snowdrop, crocus, tulip, hyacinth, jonquil, narcissus, polyanthus, gladiolus, &c., immense quantities of which are annually brought to this country from Holland, where they are extensively grown on large farms or plantations for the special purpose of exportation. We propose here to give a few remarks on the history and cultivation of the four first-named plants, which form collectively and individually the large proportion of the Dutch bulb trade. The Snowdrop (*Galanthus nivalis*) belongs botanically to the same natural Order as the narcissus or daffodil, namely, the *Amaryllideæ*. It is one of the earliest harbingers of spring, opening its pure white blossoms sometimes as early as January, but mostly in February and March. It has a wide distribution in Central and Southern Europe, and was considered by Gerarde to be a native of Italy. It has been known in England, however, from an early period, and was cultivated before Gerarde's time; so long, indeed, has it been in cultivation, that it is by some considered as a native plant. The name, Snowdrop, is said to be derived from the similarity of the drooping flower to the eardrops worn by ladies in the sixteenth and seventeenth centuries. In some parts of Gloucestershire the flowers are still known as Candlemas bells, on account of their opening about Candlemas Day (Feb. 2).

The Crocus, of which several species are cultivated, belongs to an allied natural Order—the *Irideæ*, to which the gladiolus and iris or flag belong. The only British species is *Crocus nudiflorus*, the flowers of which are bright purple, opening in the autumn, and not in the spring. *Crocus vernus*, the flowers of which are either purple or white, is a spring-flowering species, but it is only naturalized with us, and is rare, occurring only in certain parts of the kingdom. From this latter species many of the cultivated garden sorts have been obtained; several other species, however, are much used: *Crocus sativus* is the most important in an economic point of view, as it is the plant from which the saffron of commerce is obtained, and which is composed of the yellow stigmas of the flowers.

It is not a native of Britain, but its cultivation was early introduced at Walden in Essex, where in the fifteenth and sixteenth centuries it was so largely grown that the place was afterwards called Saffron-Walden. By the early Greeks and Romans saffron was used not only in medicine but in cookery, and by the ladies as a cosmetic and for dying their hair. An old writer remarks that "saffron hath power to quicken the spirits, and the virtue thereof pierceth by-and-by to the heart, provoking laughter and merriness; and they say that those properties come by the influence of the sun, unto whome it is subject; from whome she is ayded by his subtill nature, bright and sweet-smelling." At the present time saffron is used chiefly as a colouring and flavouring ingredient, and is imported principally from Spain, France, and Italy. The Tulip is a well-known and very important spring plant, as it is suitable both for the open beds and for window gardening. One species only (*Tulipa sylvestris*) is found in this country, and this is a doubtful native, being considered by some botanists to have originated from discarded roots or bulbs of *Tulipa Gesneriana*, the species from which the numerous varieties of the florist have emanated. Norfolk and Suffolk are the only two counties where it possibly may exist in a wild state. It belongs to the natural Order *Liliaceæ*, familiar examples of which are to be found in the onion, asparagus, and the lily itself. The history of the Tulip is very remarkable : it was introduced to this country in 1577, having been first made known by Conrad Gesner, a native of Zurich, after whom the species is named, in 1559. The plant is said to have been originally brought from the Levant.

It was not till about the middle of the seventeenth century that Tulips became known as important articles of trade, after which they became the centre of much speculation, as much as 2,500 florins, and in some cases even 4,600 florins, are recorded as having represented the estimated value of a single root, the money changing hands without the bulbs being actually seen, or perhaps even in existence. They were objects merely of a speculative mania similar to that of railway scrip. They have always had much care bestowed upon them by florists, and at the present time there are hundreds of named varieties.

The most generally admired, perhaps, of all the spring bulbs is the Hyacinth, known to us now chiefly for window and pot culture, and admired for its massive spikes of flowers. Numerous varieties, of all shades of colour, and with single and double flowers, are brought to us yearly from Holland, where, like the tulip, crocus, snowdrop, and other bulbous plants, they are cultivated to a very large extent. The origin of the numerous forms we now possess is supposed to be *Hyacinthus orientalis* —a plant growing about Aleppo and Bagdad. A near ally to it is the Bluebell, which forms, as it were, a blue carpet to our woods in the spring; this is the *Hyacinthus nonscriptus* of Linnæus, though it is now known to us as *Scilla nutans*. The trade with Holland in spring-flowering bulbs, as we have before said, is of very great importance, immense quantities being sent yearly at the close of the summer season to our nurserymen and florists; and at the end of the year bulb sales by auction regularly take place in London.

That the cultivation of bulbs in Holland for exportation is a profitable business is evidenced from the fact that land near Haarlem suitable for Hyacinth-growing is worth about £200 per acre, and it is in the neighbourhood of Haarlem that most of the bulb farms are situated. It is estimated that the Hyacinth culture alone occupies 100 acres of land, and that from 400 to 500 acres more are planted with various other kinds of bulbs. It would seem that the low-lying Dutch lands, together with the moist atmosphere, are peculiarly suitable to this branch of culture. And in the preparation of a new bulb-ground care is taken to raise it just above the level by which it might be inundated, as a great deal of the land is flooded at certain seasons. The soil is composed of a large proportion of sand. A good deal of animal manure is mixed with it, but the bulbs are never put in the ground the same season that the manuring takes place. Potatoes are usually planted, which are said to grow to an enormous size, but of very poor quality, so that they are usually sold for feeding cattle. The ground occupied by Hyacinths one year is the next year planted with tulips and crocuses, and the third year either with narcissi, or, perhaps, again with potatoes.

On a large bulb farm the fields are formed into squares by closely-clipped hedges, which not only break the force of the winds, but prevent the sand from driving on to the plants, and so injuring them. The Hyacinth-beds are between three and four feet wide, and the bulbs are planted in rows about six inches apart, and covered with about three or four inches of soil. The different kinds of bulbs are kept together in masses, the colours of the flowers forming a most beautiful picture in the landscape. The bulbs are planted in September and October, and the beds are covered in winter with a layer of reeds to protect them from the frost. April is the period for flowering, and the flower-spikes are mostly taken off just at the time they are fully blown, and before they begin to wither, which practice is said to strengthen the bulbs and to make them more marketable. The flower-spikes, after being cut off, are of no value, but are thrown into heaps in the corners of the field. The bulbs ripen early in June, at which time they are taken out of the ground, and laid in the sun to dry, with their roots uppermost; they are next covered with dry sand for ten or twelve days, and the rootlets and leaves taken off; after which they are finally dried in the drying-house—a building fitted up with shelves, and through which air is allowed to enter and circulate freely. The bulbs are afterwards sorted, packed in the dry husk of buckwheat (*Fagopyrum esculentum*), an excellent material for packing where absolute dryness is necessary, and sent to England early in August.

Bulbs naturally increase by small bulbs being formed around their base, which production is of course taken advantage of by the growers in Holland, and when the young bulbs are of a sufficient size they are taken off and planted separately; but this natural process does not produce sufficient bulbs to meet the requirements of the growers; they therefore have recourse to the following methods of propagation:—The lower part of an old bulb is partially scooped out in a concave form; this cuts through, and exposes the edges of the imbricated scales of

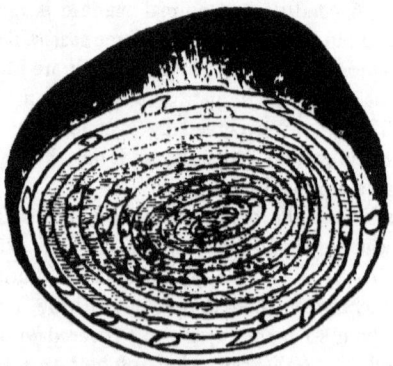

Hyacinth bulb with the base scooped out, showing the production of young bulbs.

which the bulbs are composed; they are then laid on the beds and covered with dry sand, and, after exposure for several days, are taken to the drying-house, where they are kept till autumn, and then planted in the ground. From the many edges of the scales young bulbs are formed in far greater numbers than if the old bulb had been left to itself.

Another plan, and one by which larger bulbs are produced, is to cut the old bulb across in four or five places, treat it as just described, and from the edges of the cuts young bulbs will appear; these young bulbs, after removal from the parent, take three years or more to bring to perfection —that is, before the period of blooming. New varieties are raised from

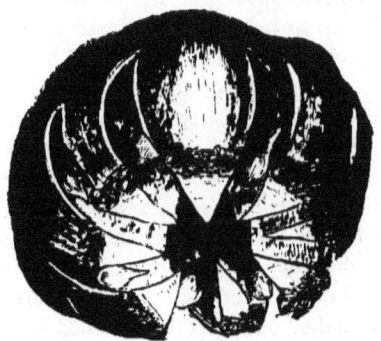

Hyacinth bulb cut across the base, for the purpose of increasing the production of young bulbs.

seed, which is a much more tedious operation, and, as it is similar to that of other plants, need not be recounted here. Sufficient has been said to show that spring flowers have a practical as well as a poetical aspect, and that, in these days of extended trade and commerce, those things which are apparently of little value are often of great importance.

## CATS, DOGS, AND OTHER HOUSEHOLD PETS.

ONE of the great advantages of having your dog extremely young is that he will grow up among the other pets in the house, and treat them almost as if they were other dogs. There is no necessary antagonism between animals in an artificial state. A dog which will hunt wild rabbits can be taught to take no notice of tame ones, and a most singular fact is recorded in Daniel's "Rural Sports" of a pack of foxhounds which used to pass utterly unnoticed a tame fox which was chained in the yard. It is often, however, difficult to make old cats and dogs agree, but the cat is generally the aggressor.

Puppies and kittens are naturally playfellows, and will acquire a very strong affection for one another; and it is wonderful how much boisterous play a kitten will take from a young dog. Occasionally a yell will be forced out of it at some extra rudeness, yet it submits to very much which must be extremely disagreeable from pure affection and complacency. I have seen a large retriever whose great delight was to carry about a kitten in his mouth. He was so gentle, however, that the kitten liked to have it done, and would come to him and ask for it. I have known a dog also fight furiously in defence of his own cat. In short, the old proverb of people "living like cat and dog" may be rendered utterly nonsensical by the commonest care and sense. I repeat that it is sometimes difficult to make an old cat unused to dogs an agreeable companion, but there are noticeable exceptions. I left my dog in the care of a lady once whose old cat had scarcely ever seen a dog, yet the friendship struck between them was so great that the cat would leave her kittens in charge of the dog; and when the dog was brought into the house run over, the cat licked its face until it died, and after it was buried in the garden tried to dig it up again.

We now come to a more difficult subject, that of cats and other domestic pets. Some amiable and excellent cats, those with peculiarly sharp heads, cannot be kept from birds by anything short of cruelty. In "Sandford and Merton," I think, the amiable Mr. Barlow cured the cat of going at the birds by placing a heated gridiron in front of the cage and letting the cat burn itself. We hope that he burnt his own fingers in this cruel experiment, but that is not recorded. Such remedies are not justifiable on any grounds. If you have a "bird" cat and birds, you must do one of two things: give the cat away to some one who does not keep birds, or hang your birds up out of reach. The latter remedy is extremely objectionable, because your birds do not get so tame as if they stood on a table. They are frightened if you put your face close to them, and you miss half the pleasure of having them by not seeing their innumerable pretty and characteristic ways. Tame canaries, for example, when habituated to being close to you, are among the most amusing and clever of animal companions. Some are occasionally exhibited on the Thames Embankment, at Charing Cross, and in Air Street, Regent Street, which seem to do almost everything but talk. They are perfectly free during performance, and are accompanied by a large and sleepy cat, not, however, of the bird sort.

Again, take those wonderful birds, the manikins; you might as well not

have them at all as hang them up to the ceiling. What makes them so valuable is not their song, for that is nothing, but the unutterable absurdity of their behaviour, which is as fantastic as any monkey. The pair will be sitting quietly, when the cock bird makes a curious little "wheetle" in his throat, which is all the song he knows. At once the hen bird pretends that she has never heard it before; she looks down and up, and in every direction, to see where this lovely sound comes from. He makes the noise again, and she discovers where it comes from, and gets into a state of admiration and delight. The last act in the farce is her looking down his throat while his bill is open to see how it is done. As regards white mice, dormice, and jerboas, you must mind the cat very carefully; and as to jerboas, with all their beauty I do not think that they are worth keeping. They are an expensive and delicate little beast, and give much more pleasure to curious friends than to yourself. They have no character and no brains; they are quite as great fools as the kangaroo, *or any other marsupial*, but they have not got the lumbering, stupid affection which the kangaroo has for his master. Of marsupial pets the koala is the most ridiculous and the oppossum the most intelligent, but of course they have never been brought alive into this country, as their food is unprocurable.

Dormice are very poor fun, and white mice very little better; still, they are very pretty and harmless, while the white mouse shows very considerable intelligence. The white mice will do things for the Italian boys which they will not do for us, but then they are handling them all day and half the night. These boys procure them in England, I am told, and they are very common in some warehouses. The other friend of the Savoyard boy, the marmot, is, I believe, a most entertaining and excellent companion, but I never had one, and so cannot speak of it.

The squirrel is really a charming little fellow, but you must get him young or he will be too wild for you. The best way to get him is to buy him in the streets off a man's arm, then you are sure that he is tame; if you buy one out of a cage, the chances are that you will never be able to let him loose with any comfort, and you must always remember that if you catch a recalcitrant squirrel without thick gloves on, he, with those little front teeth of his, will bite you pretty hard. It is difficult, however, to make a squirrel love you with any sort of love save "cupboard" love. He has very few brains, and those of no very high quality, behind that receding forehead of his: the most he ever thinks of you is that you are a movable tree to be climbed up, and to furnish nuts without any effort on his part. Still, his ways are so charming, nimble, and graceful, that he is worth having until you are tired of him. It is a very good and well-trained dog which will not hunt a squirrel; their rapid nervous motion brings out every hunting instinct in the dog. But then, on the other hand, it must be an extremely imbecile squirrel which allows a dog to catch it. The question is not between the dog and the squirrel, for the squirrel can take care of himself. The question is between you and your parents or guardians about the breakage of furniture. Looking at the matter as a whole, we do not entirely recommend squirrels. They climb up too much, and your sister's neck and a bedpost are alike to them; they require to get to the top of everything, and put their claws in pretty sharply in doing so. Cats never seem to take any notice of them, save that of making some excuse for removing from their company, with dignity combined with scorn.

I now pass to another very popular pet, "the restless cavy." That is the scientific name of the animal, as I discovered once in the Zoological Gardens at Clifton, where the name was written up outside a cage which seemed to contain nothing, but which certainly contained something alive, because a noise came from the interior, as of somebody trying to sing and making a mess of it. With the courage of my nation, I borrowed a parasol, and got over the enclosure, determined to face this strange wild beast or die. After poking into the den of this unknown animal for some time, out came three guinea-pigs. My science had been at fault, but my courage was beyond question.

Guinea-pigs are so called because they do not come from Guinea, but from Malacca, and because they have no possible connection with any kind of pig. They are pretty, comical little things, and show a trifle of affection and intelligence. They are very cleanly, and will live on anything edible, even refuse which the more stupid rabbit will cast away. The most amusing thing about them is the noise they make. Possibly, the nearest description of it is that of "outgribing," given by Humpty Dumpty, in "Alice through the Looking Glass:" "Something between bellowing and whistling, with a kind of sneeze in the middle." However, we cannot give a better simile for their noise than that. I had a very small one, who was extremely musical, but he went the way of all *my* guinea-pigs: he was killed by the rats.

The most extraordinary thing is that a great number of people believe that guinea-pigs will prevent rats coming near the place. My experience is that, next to rabbits, no animals are so likely to be attacked by them. Neither dogs nor cats will, I think, touch them; at least I speak under correction. You cannot make *friends* with a guinea-pig, as I believe you can with a rat, but he is an inoffensive little fellow, who really does his best; and which of us can do more?

I have alluded to the rat, the most intelligent, the most cunning, and the most disgusting of rodents. Tradition says that they can be tamed, but I should think it impossible. There seems to be an instinctive horror between the rat and the human race; we cannot set down the dreadful reason here, though it is undoubtedly a sound one. Our young friends must not for a moment confound the house-rat with the innocent water rat, or vole, an animal by no means so common as it was a few years ago. In rowing along the upper Thames, the most of the rats which you see on the banks are house-rats, which take to the water-side in summer-time. The vole, the "ditch-dog" of Shakespeare, you seldom see. The "old grey rat" of the same passage in "Lear" is now an extinct animal, at least I have never seen one. The present, or "Hanover" rat, has destroyed him. Household rats must, in the most humane manner possible, be destroyed; a dog which would not go at one is worthless.

Now we get naturally from those clever and horrible animals, the rats, to their victims,—that is to say, to those most foolish of created beings, rabbits. The money that is spent on these animals, both wild and tame, is very considerable. The wild ones show some instinct of self-preservation, the tame ones none at all. We have nothing to do with wild rabbits; that is a political question between farmers and landowners. We have only to do with tame ones. I have, at intervals, kept tame rabbits for about thirty years, and I can safely say, with a clear conscience, that I never saw the remotest glimmer of understanding in any one of them. They seem to exist only by instinct, while their very closely-related cousin, the hare, develops qualities, both in the wild and tame state, almost equal to those of the fox. Our forefathers taught hares to play on the tabor (see Harleian Collection of MSS., about A.D. 1400; also Ben Jonson's "Bartholomew Fair," A.D. 1610; again, the hare exhibited at Sadlers Wells about 1780). Cowper, the poet, and his hares are well known, but no one ever made a rabbit do anything except run away. No one ever dreams of making a pet of the more intelligent hare, and yet no one will hesitate to keep the foolish rabbit. Why is this? The answer is simple. One hare is like another, but the varieties of rabbits procured by selection and inter-breeding get in fantastic beauty and quaintness as near the limits of human imagination as do those of the dog or the pigeon. The smut-nosed Himalayan rabbit is the same animal as the great black lop-ear, or the wild rabbit of the fields; so the fantail pigeon is nothing but a developed blue rock, and the bloodhound is nothing more than a larger and differently coloured black-and-tan terrier. The rabbit has the power of developing into extraordinary forms, and so has become, what is called in the language of the day, a "fancy" animal; the hare, apparently, has not this faculty at all.

Of course I wish to say nothing against the keeping of rabbits; it is a good thing to have animals about you, for it is good to think that there are creatures weaker, and yet, in their way, stronger than yourself. Yet as you get on in age, rabbits get more and more unsatisfactory. The eternal contemplation of utterly selfish noodles is not always pleasant, and you may rest assured that if you were to die to-morrow, your pet rabbit would never miss you, so long as he was fed, any more than your horse would.

The habits of these animals are by no means agreeable in detail. A few hints may be necessary. Never, for instance, attempt to look at or touch the young ones until they begin to show for themselves. Give them mostly dry food, but always let them have water besides greenstuff. Take them up always behind the ears, and do not squeeze their bodies; see that they are continually fed, and judiciously; and lastly, when you are tired of them, sell them to some one who wants them, and spend the money on something more satisfactory.

The best of all rabbits in my opinion are the Himalayan: they seem to show some grains of sense (were such a thing possible), and they are certainly the prettiest of all rabbits. But when you have said the best about these animals as pets, what are they? Mere stupid crawling creatures. The meanest fowl or pigeon which ever strutted or flew about your house is worth fifty of such things. Dogs, cats, fowls, and pigeons are worth having; rabbits I hardly think are.

Before I tell you about fowls and pigeons, I will finish the "residuum" of pets. Lizards, especially the green ones, which you can get at Kennedy's in Covent Garden, smell; and they get into your bed and your shoes, and frighten your sisters. Snakes are an utter mistake: they get away and are reported lost at head-quarters, and then the maids refuse to go to bed and give warning. Tortoises are always missing, and getting found in strange places late at night, giving alarms scarcely inferior to that of fire; while monkeys are, on the whole, a terrible nuisance, from their habit of getting up early in the morning and waking every one. We are all happier, not to mention our friends, without those pets which I have mentioned in the "residuum."

## SUMMER GRASS.

THE sharpening of the scythe—very much like the guinea-fowl's cry—is not a musical sound in itself, and yet from its suggestiveness it blends into harmony with the natural melodies of its season. It tells us that summer is really come, and the news is confirmed by the luscious aroma that comes wandering from the house on the dew-cooled air in the early morning, when the little birds are beginning to chirp and twitter in the rippling leaves, and dropping to the ground in search of their early breakfasts, and distant barnyard cocks are giving and answering challenges.

The very names of grasses seem to tell of green seas of cool juiciness, becalmed, or billowing softly, in alternate sunshine and shadow, against hedges laden with blossom: rye-grass, oat-grass, yarrow, timothy, evergreen meadow, wood-meadow, trefoil, white clover, red clover, lucerne, bent, dogstail, foxtail, cockstoot, sheep's-fescue. And then the flowers that gleam in the grass and, under the dog-roses, on the hedge-banks, and the lush herbage in the ditches in which jars, barrels, baskets, and babies are laid to be out of the sun! Approaching bareness gives a touch of sadness to corn-harvest, autumn fruit-picking, and hopping; but the cut grass will spring again and yield another crop, while all around there is a rich abundance of nature not yet mature, but on the very brink of passing into ripeness, like a lovely maid on the eve of her bridal. "Pastoral farms" are "still green to the very doors." Yes, haysel and contemporary early fruit-picking seem to me the pleasantest of the year's farming festivals. One can feel towards them as a schoolboy does towards the first days of his summer holidays,—delightful in themselves, and with other delights beyond.

The grasses form one of the most interesting of Orders. What is the bamboo but a grass, a grass that grows a foot and a half a day, and towers into a tree, which supplies those who seek its shade with food and drink, buckets and ear-rings, bowls and baskets, blinds and brushes, bows and arrows and umbrellas, blow-tubes, flutes, guitars, cloth, paper, combs, and chairs?

What is the sugar-cane but another great grass, which was taken into cultivation in China and the South Seas before authentic history began? The almost as graceful maize, yielding both sugar and grain, is another. Sorghum, and rice, and millet, our English cereals, lemon-grass, pampas-grass, the common reed, used instead of laths by West of England plasterers, the plant that yields canary-seed, are all grasses.

But the grass that I like best is the English pasture-grass which I have named, or its equivalents abroad, waving or being cut down in its full summer beauty. Grahame, to whom few will now allow the name of poet, has, at any rate, left two picturesque lines—

> "The scythe lies glittering in the dewy wreath
> Of tedded grass, mingled with faded flowers."

Grass occurs in some of the most poetical passages in the Bible. Moab said to Midian that they would be licked up by Israel as the ox licketh up the grass on the field. "My doctrine shall drop as the rain, my speech shall distil as the dew, as the small rain upon the tender herb, and as the showers upon the grass." "He shall be as the light of the morning, when the sun riseth, even a morning without clouds; as the tender grass springing out of the earth by clear shining after rain."

"As the grass on the housetops" is Isaiah's graphic simile for a withered people. "He shall come down like rain upon the mown grass," sings the Psalm of Solomon; and in another key, the 103rd, "As for man, his days are as grass; as a flower of the field, so he flourisheth. For the wind passeth over it, and it is gone, and the place thereof shall know it no more."

Wordsworth, addressing a butterfly, says—

> "We'll talk of sunshine and of song,
> And summer days when we were young,
> Sweet childish days that were as long
> As twenty days are now."

Most of us have such "historians of our infancy." One of mine is sorrel waving its rusty spikes above summer grass. Having mentioned Wordsworth, let us try to call to mind as many places as we can in his poems in which summer grass is growing. We might spend a summer hour in a way less profitable as well as pleasing. From the ending of May to the beginning of July is our season. It is in summer, or at any rate, late spring grass, that little Sister Anne is doing her work of waste and ruin on the strawberry-blossoms. Summer grass is just stirring on the green graves to which the little cottage girl takes her little porringer, and sits down and eats her supper there. The grass of Barbara Lewthwaite's lamb is summer grass, for the green corn all day is rustling in its ears. It is on May grass that the two idle shepherd boys sit, trimming

their rusty hats with staghorn, whilst the fallen lamb, bleating back to its mother, is swimming round and round the black pool at the foot of the Force of Dungeon Ghyll. It is on a July evening that the homely priest of Ennerdale rises from the long stone seat beneath his cottage eaves, and goes to speak with Leonard of the dead who lie buried beneath the stoneless turf of his little churchyard. The noticeable man with large grey eyes plucks long blades of summer grass to make his pipes. It is over summer grass that ill-fated Ruth wanders, playing on her hemlock-stalk, whose music at evening in his homeward walk the Quantock wood-man hears; and all the summer long it is of grass she makes her couch. It is summer grass whose rustling afflicted Margaret dreads to hear. Her apprehensions come in crowds; the very shadows of the clouds have power to shake her as they pass. The grass waves as sadly in summer as in winter about the straggling heap of unhewn stones in Greenhead Ghyll, about which a few sheep stray, while kites are sailing overhead—all that remains to tell—that, and the oak, beneath which the sheep used to be clipped—of the family of three who once lived in the "Evening Star," which, through the silent hours, was made by the mother's industry to murmur as with the sound of summer flies. Over summer grass the robin chases the butterfly, whose wings in crimson are dressed, as bright a crimson as his own; when he is adjured, as the pious bird whom man loves best, to love him or leave him alone. Summer grass makes green the seven little islands where the seven lovely Campbells sleep; and the stream that flows out of the lake sings mournfully—oh! mournfully—the solitude of Binnorie.*

Summer grass and summer flowers grow in the lovely dell where no bird builds, no bee sucks, for there the Danish boy plays his harp and

---

\* "The stream that flows out of the lake,
    As through the glen it rambles,
    Repeats a moan o'er moss and stone
    For those seven lovely Campbells."

Compare, in Hawthorne's "Scarlet Letter":—

"But the little stream would not be comforted, and still kept telling its unintelligible secret of some very mournful mystery that had happened—or making a prophetic lamentation about something that was yet to happen—within the verge of the dismal forest. . . . The dell was to be left a solitude among its dark old trees, which, with their multitudinous tongues, would whisper long of what had passed there, and no mortal be the wiser. And the melancholy brook would add this other tale to the mystery with which its little heart was already overburdened, and whereof it still kept up a murmuring babble, with not a whit more cheerfulness of tone than for ages heretofore."

warbles the songs that make the far-off flocks listen, and the mountain ponies prick their ears.

> "Calm and gentle is his mien,
> Like a dead boy he is serene."

Poor Susan hears the thrush sing at the corner of Wood Street, and a river flows on through the vale of Cheapside; and in its midst she sees green pastures in the summer glory. Green with snmmer grass are Yarrow's holms, and sweet is Yarrow flowing. Home-bred kine are grazing in Burn-mill meadow, and the swan on still Saint Mary's Lake floats double, swan and shadow. About the baby's mossy grave, beside which, in her scarlet cloak, sits in all weathers the lonely woman, moaning, as she remembers the long ago time when summer leaves were green—

> "Oh, misery! Oh, misery!
> Oh, woe is me! Oh, misery!"——

there for full fifty yards around the grass still shakes upon the ground. The summer grass has vanished from the well, beside which the cruelly hunted hart had slept in summertide, whose waters he stirred with the last deep groan he breathed; but on grass fall the shadows of the Furness foxglove-bells, in which bees murmur by the hour, far more contented than hermits with their cells.

Old Adam whistles his way up the Haymarket, thrusting his hands into the waggons and smelling at the hay, and thinking of the fields he so often has mown, but must never mow again, down in pleasant Tilsbury Vale. Right across Bolton's verdant sod, towards the very house of God, comes gliding in with lovely gleam, serene and slow, soft and silent as a dream, the solitary White Doe of Rylston. The Wanderer tells of Margaret's husband seated at his loom ere the mower was abroad among the dewy grass, and of Margaret pacing to and fro through many a day of the warm summer, from a belt of hemp that girt her waist spinning the long-drawn thread, upon

> "—— that path,
> Now faint,—the grass has crept o'er its grey line."

Over the rough mountain summer grass comes the Solitary, dealing his words of comfort, and strings of ripe currants from a cabbage-leaf, to the sobbing little mourner.

> ——"They to the grave
> Are bearing him, my little one," he said,
> "To the dark pit, but he will feel no pain.
> His body is at rest, his soul in heaven."

If the "Excursion" is sometimes prosy, nevertheless its

> "fragrant air its coolness still retains,
> The herds and flocks are yet abroad to crop
> The dewy grass."

And elsewhere, although he does talk as if, for him, a glory had passed away from earth, to the grass, Wordsworth, more than any other poet, has given an enduring splendour.

## ABOUT A CATERPILLAR.

"I'D be a butterfly, happy and gay!" But who, if he could help it, would be a caterpillar? And yet, as an old saying hath it, "We must creep before we can go. We must walk before we can fly;" a decree significant of the beginnings and endings of more lives than are dreamed of in the philosophy of creeping things in general.

Look at the crawling, munching creature, contented and happy enough, satisfied with its lot, while that lot is cast on a good large cabbage-leaf, or, with hundreds of its fellows, swarming on the leaves of a young gooseberry-tree or a field of turnips. Repulsive being! its very contentment is revolting. We would fain inspire it with a few sentiments of a more exalted nature, and give it something to exist for, some object to sigh, to struggle for, and in vain to grasp. Yet what longings, what vain ambitions can ever equal the real future that lies before this despised being? whose life yet is unvaried by dreams of airy flight, or by any anticipations of his future; when he shall leave his present lowly condition to soar far beyond his present ken, mounting aloft on rapid wing, or balanced for a moment on the fair cup of the flower whose nectar he sips in passing, in place of now slowly munching the leaves, or sleeping on the remains of his heavy repast.

For the present to eat and to sleep seems to be the lot of these poor crawlers. But look within. *There* is far more than meets the eye. Beneath that mean form, its gaudy exterior and strange appendages of legs, of scales, and of teeth, a process is being carried on, a formation completing, a perfection advancing, contrasted marvellously with its exterior existence, and yet growing out of it, sustained by and assimilated from diet of the most unlikely unkind—the cabbage-leaf, disintegrated by a course of equal marches round its narrowing edge by the creature, whose rapacious tooth devours every inch that its feet can tread, the potato-field ravaged by the invading myriads, or the leaf and root of the forest tree. Yes, the future *imago* is forming now. Days of monotonous toil, of diligent accretion, or patient preparation, and of tedious torpor in the antechamber of mortality, shall result in that lovely winged thing, that shall float on the zephyr, and glitter in the noonday light, the wings, the antennæ, the exquisite plumage of various hues, the inconceivable lightness of the freight they bear, all wondrously contrasting with the form they left behind; and surely if colour, like sound, have its various waves and notes, that thing of beauty shall waft a song of praise to Heaven with every movement of its wings. Ah, yes! like that, and something more—not alone happy and gay, but blest for ever, "I'd be a butterfly," and gladly pass through the ordeal of all the strange, painful, and distressful vicissitudes that may prepare and form my fortune, for not to flutter for a day and perish in a night shall we arise from our imprisoning cell.

"The grovelling worm shall find his wings, and soar as fast and free
As the transfigured one, with lightning form:"

no ephemeral moth born but to die, rather to know no end, and leave mortality behind.

Yet, apart from parables, which kindle our hopes and enthusiasms, an inexorable philosophy still ask the question, which, so far unanswered, we may fairly leave to wiser heads, as to the uses of the caterpillar race in the economy of nature. Born to devour, and to be devoured in large proportion, it may satisfy the curiosity of some, that caterpillars furnish a savoury food for robins, and that the use of the robin is to devour the caterpillar, which he does right manfully, at the rate of three hundred for his breakfast; but it answers nought to the inquiries of those who seek a final cause in each atom of creation. The only end it seems to serve is what some call by the hard name of a "transposed end," an end cropping up in the path of its destiny and interrupting it—much in the way that a child gathers daisies and fulfils their transposed end of his amusement by hanging them in a chain round his neck. The caterpillar fulfils the transposed end of its existence, in the way of animal nutrition (though it never live to be a butterfly), albeit the good of man be never apparently reached, for he neither eats the caterpillar that devours the cabbage, nor does he even eat the robin that swallows the caterpillar that devours the cabbage; nay, further, the "transposed end" of the caterpillar affects man in the shape of a blight; for when the caterpillar eats the leaf, the fruit is rendered worthless by the absorption of the juices that should have fed the leaves. Then, what is the use? Could the caterpillar speak as well as eat—and why should it not? only it was sent into the world not to talk, but to do its duty—perhaps it might retort the query: Have you, my caustic friend, my utilitarian investigator, made the important discovery what you were sent into the world to do, and are you doing it? The use you'll find out by-and-bye; and, meanwhile, accept a suggestion very practically exemplified by our company of crawlers, as to a complete disentanglement from an old skin or a bad habit, for which an effort is required, that might have seemed in anticipation impossible. It is thus described, and is too interesting not to record at length:

"There is a phenomenon in the life of caterpillars which we ought to point out, and which has attracted the attention of the most illustrious observers. All caterpillars change their skins many times during their life. It is not indeed enough to say that they change their skins. The skins or cases they cast are so complete that they might be taken for entire caterpillars. The hairs, the cases of the legs, the nails with which the legs are provided, the hard and solid parts which cover the head, the teeth—all these are found in the skin which the insect abandons. What an operation for the poor little animal! The work is so enormous, so troublesome, that one cannot form a just idea of it. One or two days before this grand crisis the caterpillar leaves off eating, loses its usual activity, and becomes motionless and languid. Their colour fades, their skin dries little by little, they bow their backs, swell out their segments. At last this dried-up skin splits below the back, on the second or third ring, and lets us have a glimpse of a small portion of the new skin, easily to be recognized by the freshness and brightness of its colours.

"'When once the split has begun,' says Réaumur, 'it is easy for the insect to extend it; it continues to swell out that part of its body which is opposite the split. Very soon this part raises itself above the sides of the split; it does the work of a wedge, which elongates it: thus the split soon extends from the end or the commencement of the first ring, as far as the other side of the end of the fourth. The upper portion of the body which corresponds to these four rings is then laid bare, and the caterpillar has an opening sufficiently large to serve it as an egress, through which it can entirely leave its old skin. It curves the fore part, and draws it backward; by this movement it disengages its head from under its old envelope, and brings it up to the beginning of the crack; immediately upon this it raises it, and causes it to go out through this crack. The moment afterwards it stretches out its fore part, and lowers its head. There now remains for the caterpillar nothing but to draw its hinder part from the old case.

"This excessively laborious operation is finished in less than a minute The new lining which the caterpillar has just put on is fresh and bright in colour. But the animal is exhausted by its fast, and the efforts it has made, and requires a few hours in which to regain its equilibrium."

Apparently the caterpillar is an adventurous being, much addicted to attempting and never failing to accomplish the most difficult feats of the acrobat. Having, as we have just seen, succeeded in turning himself inside out, there are species which attain the yet more difficult art of suspending themselves head downwards or by the middle of their bodies before commencing the operation of forming the cocoon. The operation is attended with considerable difficulty, and is one of which a mathematician might be proud. It had escaped the observation of many naturalists, although the little creature which so successfully performs it is one of the most common of our English caterpillars, the little *Vanessa urticæ*, common on the stinging-nettle, and distinguished by numerous black specks on its dusky body. The plant on which they feed seems to afford too insecure a support for the intended chrysalis, and the insect, on the approach of its transformation, quits its usual resort and seeks some more convenient point of suspension, where in the following manner it commences operations. Threads are laid, in most admired disorder, as a covering to the surface of the body from which it desires to hang. To this earliest layer a fresh labyrinth of silken threads is added, covering a smaller surface, and so on, ever contracting the extent whilst thickening the central mass, and thus forming a little mound of loosely-woven fibre just firm enough to bear the weight about to be imposed upon it. If we had contrived such a mechanical device, should we not have cast about rather for a hook or a thorn to hang from, and woven the loop on the body to be suspended?

But the hook is there before, and has already answered many useful purposes, before this last, to the body of the caterpillar.

The membranous feet of the little creature are armed with tiny hooks of various lengths, with the aid of which it suspends itself. By wriggling contractions and elongations of its body, it pushes the hindermost legs against the hillock of silk so firmly as to entangle them in its meshes; it is then seen to "let go" and fall securely into a vertical position. It hangs there, but not idly, sometimes as long as twenty-four hours, engaged in the sober, staid operation of "splitting its sides" with labour — not with laughter—and when split, in folding downwards like a cast-off garment the striped and dusky skin, bristling with ebony spines, in which it crawled so long. No longer useful to its possessor, this garment must be not only folded into the smallest possible space, and gradually, by means of continued contortions, pushed upwards till it covers only the narrowest end of the chrysalis remains *in statu quo*. And how shall this be done? Let Blondin live and learn. The creature has neither legs nor arms, and must yet set itself free from the skin and reach the threads from which it is suspended. Its supple body has a contractile power which supplies the office of the limbs. Between two of its segments the insect seizes a portion of the folded skin so firmly as to support the entire body. It now curves slightly the hinder parts, and draws the tail entirely out of the sheath in which it was enclosed, and for an instant reposes before freeing itself entirely from the encumbrance. Curving the part below its tail, so that it can seize the thread to which it holds on, it gives its body a violent shock, which makes it spin round many times on its tail with great rapidity. During these pirouettes the chrysalis is acting against the skin, and the hooks of its legs fray the threads and break them or disentangle themselves. If unsuccessful in this effort, it begins to twirl itself in the opposite direction, and rarely fails the second time. It is from the golden hue of this chrysalis, which is sometimes brown with golden spots, and sometimes entirely golden, that the term chrysalis (from χρύσιος, golden) was suggested to the ancient naturalists. From this chrysalis emerges in due time—and that very short—the common, but most beautiful, tortoiseshell butterfly.

The yet more common and less richly tinted *Pieris brassicæ* is in its transformations a still more accomplished acrobat. It forms, like its neighbour of the nettle, a labyrinth of silk to hang from, but seems to prefer a horizontal to a perpendicular position, and acts accordingly, after having hooked itself firmly by the nails of the hinder feet to the point of suspension.

This caterpillar possesses the power of turning back its head on to its back after having lengthened its body to a certain point, and, with its six legs in the air, of reaching to its fifth ring. It can also, by bending sideways, bring its head, with the thread-spinning apparatus which is below, opposite and near to one of the membranous legs. The caterpillar begins operations by fixing on this point a single thread, the first of those that are intended to tie it up securely.

But how can it throw the thread over its head? The problem is almost as difficult as a boy's first essay when he has mounted his knickerbockers and must get his braces over his shoulders. It contrives to catch hold of the thread with its head, and, drawing it to the other side, it forms a loose loop over its doubled body. Having seen that this loop is firmly attached on both sides, it wriggles its head a little farther back, spins another thread from its tail, which it firmly attaches on the opposite side, and then, by a jerk, contrives to pass the thread over the crease between its head and neck. Again and again it repeats the same operation, until it has formed a loop strong enough to bear its weight, when it completes its somersault, and, in little more than a day, its transformation into the chrysalis is complete.

But many other caterpillars are not content with fastening their horny case to a branch or a rock. Before performing the feats we have described, they spin their houses of silk, in which they may undress and sleep, withdrawn from the vulgar gaze. Some work in communities, and make one large cocoon like a great silken bag supply the dressing-room for a large family. Others gum together a case of leaves; some take to masonry instead of carpentering or spinning, and gum together a shell of earth or mortar, kneaded with silk, and finely plastered within. If disturbed before they have completed their transformation, they will put out their head, and gather little grains of earth, which they entangle in silky threads, until the gap is completely closed.

Others again, especially in Australia, roof their abodes with shingle, after the fashion of our New Zealand colonists; little bits of bark being cut, and placed together with all the regularity of an experienced slater. In fact, there is no human mechanical art which may not find its prototype in insect architecture. Lake dwellings, cave men, bark huts, wigwams, woven tents, diving-bells, clay houses, existed long before man adapted the materials around him to the varied conditions in which he sought to make his home. In these varied dwellings, whether of the finest silk or he roughest masonry, the once grovelling caterpillar rests, sometimes only a few weeks, sometimes a year, till it emerges in due time to a new existence, in which, careless of food or clothing, it flits from flower to flower, its only care being now the reproduction of its species; and, having laid its eggs, a few showers or a windy day close the chapter of existence of the spangled butterfly.

## THE ACANTHUS.

THOUGH it is not every one who, on looking at a column in a building, is reminded of a tree, it is impossible, when once the suggestion is made, not to see that there is some connection between the two things. The simple pillar represents the trunk; the capital stands for the foliage. Take a roof: support it by trunks of trees; put a flat piece at the top, and another flat piece at the bottom, and you have a Tuscan column. We may conjecture that the idea of fluting a column came from grooves in the bark, and the idea of a more ornamental capital might be easily generated in the mind of some one who noticed, as it held up a roof, a tree-stem round the top of which leaves were sprouting. I saw such trees in a timber-yard the other day, and stood long in the street to admire them—there they were, column and capital, ready for the architect's hand to imitate in stone. We know as a matter of fact that the Greeks, at an early stage of their history, used to erect buildings entirely of wood, and that some of these were in no way distinguished except by that circumstance from certain other buildings in stone.

In the case of the architecture of Egypt the imitation of a tree is so plain—so "frank," to use the language of art criticism—that the palm or the lotus stands visibly before us in the supports of the temples. In the more refined architecture of Greece, the imitation is less frank, but the eye is more pleased. In the Ionic capital we see plainly the curl of the leaf, but we cannot fix the particular tree or plant. It so happens, however, that in the most ornate of the known orders of Grecian architecture we can unmistakably trace the acanthus—a well-known plant, for the most part growing as a weed, though some of its species are kept in hothouses, and the ordinary plant is grown in gardens for the beauty of its leaf, and out of regard for the use it has served in art.

It is said that the Corinthian capital—too familiar an object to require description—was suggested by seeing an acanthus growing under a slab of stone laid on the trunk of a tree; but we do not need the story, be it fancy or fact. When once it was understood that foliage was to form the capital of a column, almost any step towards the use of any particular leaf was easy. The acanthus is used in Gothic as well as in Grecian capitals. It is a noticeable fact that in Gothic, the capital is ornamented less in proportion as the rest of the building becomes what is called florid. In other words, there is usually much more leafage in the capital of Early Gothic than of Late Gothic.

There is no reason, considering the endless variety of beautiful forms in nature, why fresh types of capital should not be invented and used. The persistent use of a few fixed types, age after age, is an example of the tendency of things to keep on in particular lines of imitation till some change is violently made. In the general characteristics of the column (of whatever order) considered as the support of a building, no change can well be made. That a column should be broad at the base and taper towards the top (even if there be a slight bulge in the middle) is not only reasonable and natural imitation of a tree; it is obedience to the law of mechanics which tells us how to get the greatest amount of sustaining power out of a pillar.

## THE MICROSCOPE.

WE very much wish to do our young readers a good turn. Most of them, happily, have fathers and mothers, and bachelor uncles and maiden-lady aunts perhaps, and some have even big brothers. It is a most amiable weakness when such estimable relatives have a habit of giving presents to the younger ones. A microscope is a really capital thing for a present. Suppose that some of our elder friends try it. Most certainly they would find it yield a large return of gratitude, amusement, and instruction.

The instrument known as the microscope derives its name from two Greek words, μικρὸς, "small," and σκοπίω, "to view"—that is, to see or view such small objects as, without its aid, could not be seen. The honour of the invention is claimed by both the Italians and the Dutch; but the name of the inventor is unknown.

If we consider the microscope as an instrument of one lens only, it was probably known at a very early period; nay, to some extent, even the ancient Greeks and Romans must have been acquainted with it. At all events, spectacles were used as early as the thirteenth century; and, as the glasses of these were made of different convexities, and consequently of different magnifying powers, it is not unnatural to suppose that smaller convex lenses were made, and used for the examination of minute objects. " Burning-glasses " are spoken of at a very early date. These, of course, were magnifying glasses or microscopes of the simplest kind. There is in the French Cabinet of Medals a very ancient seal, beautifully executed, whose history can be traced very far back, and the engraving upon which is so minute that it is not all visible to the naked eye. The conclusion, therefore, is that both in the making of it and the reading of it, magnifying or microscopic glasses were employed and were necessary.

Sir David Brewster, at a meeting of the British Association in 1852, showed a plate of rock crystal worked into the form of a lens, which had been recently found among the ruins of Nineveh. Sir David was a high authority on such a subject, and he maintained that this lens had been intended for optical purposes, and that it never had been a mere personal ornament.

It is not difficult to fix the period when the microscope first began to be generally known, and to be used for the purpose of examining minute objects; for, although we do not know the name of the first inventor, we are acquainted with the names of those who first introduced it to public notice. Zacharias Jansens and his son are said to have made microscopes before the year 1590. About that time the ingenious Cornelius Drebell brought one made by them with him to England, and showed it to William Borrell and others. This particular instrument was possibly not strictly what is now called a microscope, but rather a kind of microscopic telescope. It was formed of a copper tube six feet long and one inch in diameter, supported by three brass pillars, which were fixed to a base of ebony, on which the objects to be viewed by the microscope were placed.

The single or simple microscope was invented and used long before any other. Even with that the beautiful forms of invisible nature were brought to view. This instrument consisted of a single lens of great magnifying power, which greatness of power made the field of view very small. Any of our young readers may, in a limited measure, realize for himself the sort of instrument which this was, by borrowing his grandmother's spectacles, and using one of the eyes as a magnifier, only remembering that the power of the single microscope lens was very high.

About the year 1665 small glass globules began to be occasionally applied to the single microscope, instead of convex lenses. Looking through these, instead of looking through the formerly used lens, an immense increase of magnifying power was obtained. In the Philosophical Transactions for 1696 Mr. Stephen Gray describes an experiment which carried him further than he had reached even with the globules. He tells us that he took on a pin a small portion of water which he knew contained some minute animalcules; this he laid on the end of a piece of brass wire, till there was formed somewhat more than a hemisphere of water; on then applying it to the eye, he found the animalcules enormously magnified, for those which were scarcely discernible with his glass globules with this appeared as large as ordinary-sized peas.

Dr. Hooke thus describes the method of using this water-microscope: "If," he says, "you are desirous of obtaining a microscope with one single refraction, and consequently capable of procuring the greatest clearness and brightness any one kind of microscope is capable of, spread a little of the fluid you intend to examine on a glass plate; bring this under one of your globules, then move it gently upwards till the fluid touches the globule, to which it will soon adhere, and that so firmly as to bear being moved a little backwards and forwards. By looking through the globule you will then have a perfect view of the animalcules in the drop."

The construction of the single microscope is so simple that it is susceptible of but little improvement, and has therefore undergone few alterations, and these have been mostly confined to the manner of mounting it, or to additions to its apparatus. The greatest improvement this instrument has received was made by Lieberkuhn, of Berlin, about the year 1740. It consists in placing the small lens in the centre of a highly polished speculum of silver, by which means a strong light is reflected upon the upper surface of an object, which is thus examined with great ease and pleasure. Before this it was almost impossible to examine small untransparent objects with any degree of exactness; for the dark side of the object being next the eye, and also overshadowed by the proximity of the instrument, its appearance was necessarily obscure and indistinct. Lieberkuhn's instrument was simply a piece of brass tube, about an inch long and an inch in diameter, which was provided with a cap at each extremity; the one end being fitted with a double-convex lens of half an inch in focal length, while the other carried a condensing lens three-quarters of an inch in diameter. With this instrument the inventor made many important anatomical and other observations. It is still much used for botanical purposes, in which case it is commonly fitted with a short handle.

Leeuwenhoeck's microscopes were rendered famous throughout all Europe by means of the numerous discoveries he made with them. His instruments were all single, and fitted up in a very convenient and simple manner. Each consisted of a very small double-convex lens, let into a socket between two plates riveted together, and pierced with a small hole. The object was placed on a silver point or needle, which, by means of screws adapted for that purpose, might be turned about, raised, or depressed at pleasure, and thus be brought nearer to, or removed farther from, the glass, as the eye of the observer, the nature of the object, and the convenient examination of its parts required.

It will be observed that the simple or single microscope is so called, not because it has necessarily only one glass,—it may have several, as we have seen,—but because it looks at objects *directly*, without the intervention of a reflector.

The three first compound microscopes that attract our notice are those of Dr. Hooke, Eustachio Divini, and Philip Bonnani. In the compound microscope it is the image which is contemplated instead of the object. A reflector is used, and the magnifying power is immensely increased. Hooke published an account of his instrument in 1667; a description of Divini's microscope was read at the Royal Society in 1668; and Bonnani issued a detail of the peculiarities of his in 1698. Sir Isaac Newton aided in the progress of discovery in regard to the microscope as well as the telescope; and onward to 1812 many distinguished names were added to the already lengthy roll of skilful men who had made improvements in the instrument. In 1812, with Dr. Wollaston, began a career of advancement which has been much more rapid than any that preceded it. Among the eminent men who have contributed to this result may be named Sir David Brewster, Frauenhofer, Selligues, Chevalier, Amici, Sir John Herschel, Professors Airy and Barlow, and others. Celebrated makers have aided much in this progress, and several English firms produce instruments which are unequalled in the world. Indeed, within the last forty or fifty years, by this means our sense of sight has been improved almost into a new faculty.

But it would be a great mistake to suppose that the possession of an efficient microscope is the sole requisite for the perception of the inexhaustible stores of beauty and of contrivance which this instrument is capable of disclosing to our view. The possession of a microscope scarcely constitutes a microscopist any more than does the acquisition of a musical instrument render its possessor a musician. Both the hand and the eye require to be instructed by habitual care and practice.

As seen through the microscope, the various departments of animal and vegetable life are full of beauty, and in their minutest details exhibit a completeness and a finish infinitely transcending the most exquisite and admired works of art. The scale of a sole, for example, is a marvellous pattern of regularity and delicacy. It is a kind of web, with a number of small points at one end, which fasten it to the body of the fish. Indeed, the scales of every fish are more beautifully woven than any texture which is found in the finest handiwork of man; and the different varieties have all peculiarities which are exclusively their own. Equal regularity and beauty are found in the structure of the feathers of birds, in the fibres of the flesh of animals, in the grain of the several kinds of wood, and in the forms of different salts and crystals. The dust on the wing of a moth or a butterfly, a single particle of which is so minute as to be invisible, is seen, when magnified, to be a beautifully-formed feather, and exhibits the most delicate and admirable arrangement in all its parts. In a moth there is a configuration entirely distinct from that of a butterfly, although, as seen by the naked eye, they seem to be very much alike; each has feathers different from those of the other, and every species has its own peculiarity.

To those who are possessed of microscopes, the following list of objects may be serviceable:—The scales of fishes; the dust on the wings of butterflies, moths, gnats, flies, and other insects; the flea, and mites in cheese; the eels, serpents, or little worm-like animals found in vinegar and paste; the animalcules existing in infusions of pepper, as well as of hay, grass, flowers, and other vegetable substances; the eye of the house-fly, the dragon-fly, and of various other insects; the legs of spiders; the claws of beetles; the wings of small flies; the eye of a lobster; slices of broom, lime-tree, dogwood, and oak; transverse sections or cross-cuttings of plants of various kinds; the leaves of trees, plants, and flowers; the fibres of a peacock's feather, and the feathers of other birds; the human hair, the hair of a mouse; the sting of a bee; the stings of a nettle; the small insects which infest flowers, fruits, and trees; seeds; mouldiness;

sponge; flakes of unmelted snow; the tails of fishes, and the webs between the toes of frogs, in which the circulation of the blood may be distinctly seen.

By means of the microscope we are brought into acquaintance with new living tribes in prodigious numbers, which, from their minuteness, would, without it, have escaped our observation altogether. How many of these invisible tribes there may be throughout the air, the waters, and the earth, is still unknown, but they doubtless far exceed the number of all other classes of living creatures combined.

The variety of animalcules is inconceivable. The forms of many of them are strikingly peculiar. They fight, they go in armies as if under the command of leaders. Ditches and pools also afford beautiful vegetable forms which are invisible to the naked eye, and these are perfect in their construction as the most gigantic trees. The unpopular little animal, the flea, appears a very beautiful and curious creature when examined by the microscope. It may be magnified with a very ordinary instrument to the extent of eight or ten inches in length, and a corresponding breadth. It is adorned with a curiously polished coat of armour of hard shelly scales, neatly jointed and folded over each other, and studded with long spikes somewhat like the quills of a porcupine. The general appearance of the insect is that of a beautiful piece of variegated tortoise-shell. Its head is furnished on each side with a beautiful and quick round black eye, behind which are cavities which are supposed to be its ears. Its apparatus for wounding and sucking, as well as its six many-jointed legs for leaping, are all distinctly visible. It can leap a hundred times its own length. Mites are crustaceous creatures, with a head which is small in proportion to the body. They have each two small eyes, and six and sometimes eight legs, each of which terminates in two hooked claws. It has a sharp snout, and a mouth like that of a mole. It would take 91,120,000 mites' eggs to equal in size one pigeon's egg.

In all creatures the eye is a striking object, but as seen through this instrument the eyes of insects are so peculiar as to excite our highest admiration. On the heads of beetles, bees, common flies, butterflies, and other insects, may be perceived two protuberances, which contain a prodigious number of small transparent hemispheres, placed with the utmost regularity, in lines crossing each other like latticework. These are a collection of eyes, which, like so many mirrors, reflect the images of surrounding objects. In some insects there are many thousands of them. The farina of flowers looks to the naked eye like simple dust, but when magnified it is seen to be finely constructed, and of great variety according to the character of the plant to which it belongs. Leaves are among the most delicate and gorgeous forms in nature. The leaf of the box is supposed to contain upon its two sides as many as 344,180 pores; and the back of a rose-tree-leaf looks as if diapered with silver. The cross-cut slices of plants or trees exhibit the most variegated and wonderful arrangement. A piece of human skin is marvellous in its mechanism. Ten thousand of the fine threads of a spider's web are not so thick as a human hair—a fineness this which is beyond our conception.

A small needle, highly polished, when viewed with a high magnifying power, appears neither round nor flat, but full of holes and scratches, and as broad and blunt at the point as the end of a poker. On the other hand, the sting of a bee, or the proboscis of a butterfly or a flea, appears, when examined by the microscope, to be formed with the most surprising beauty and regularity. The sting of a bee shows a polish without the least flaw, blemish, or inequality, and ends in a point too fine to be discerned. On submitting the microscope to the edge of a very keen razor, it appears as broad as the back of a very thick knife, and is rough and full of notches. The finest cambric or silk that human skill can produce resembles ill-made twine or ropes, and seems scarcely fit to be used as a door-mat, whereas the silkworm's web appears perfectly smooth and shining, and everywhere equal.

How wondrous are the facts with which this instrument makes us acquainted! When we look through this small tube, a new world of beings, before unseen and unsuspected, start into view, presenting the most unaccustomed forms, and performing, with monstrous organs, evolutions the most startling and grotesque, such as our imagination never pictured in its wildest moments. These beings are constantly around us; everything swarms with their tiny vitality; they dart through each pool of water; they bask in myriads upon a blade of herbage; their germs float constantly through the atmosphere; they even increase so rapidly, and have so abounded, that their remains form solid strata of the earth's surface. Yet, without the microscope, their existence must have been for ever imperceptible to our senses, and therefore unknown. The disclosures of the telescope, as it sweeps the heavens and shows to us an infinitude of worlds and systems reaching into incalculable space, scarcely more display the omnipotence of creative power than does the microscope directed to the narrow sphere of a drop of water from a pond, or a particle of the herbage which grows upon its bank—" a universe within the compass of a point."

## WATERY WASTES.

NOTWITHSTANDING the numerous geological changes which have been, at various periods of the earth's history, rung upon its surface, there appears to be no law in physical geography more persistent than that which bids "the waters be gathered into one place," so that the "dry land may appear." Nearly three-fourths of our planet's surface is covered by water, chiefly, of course, as seas and oceans. On the remainder—a little more than one-quarter of the entire area, and much of that inaccessible as mountains, or bound in realms of eternal ice—poor humanity is nursed as in a terrestrial cradle. This limited space is the grave-yard not only of our own race, but that of every species of animal and plant, terrestrial and marine, which has come into being since the Lord commanded "the waters to bring forth abundantly!" It is a little space contended for by untold millions of living creatures besides men. The rocks over which the soils are strewn, and whose produce feeds this various creation, are themselves huge sepulchres "written within and without" of the life that has been. Even with regard to our earth as a planet, there is a Plan and a Purpose silently but surely being worked out during the slowly rolling ages!

It is the fact that, as a rule, the water is separated from the land, which enables terrestrial life to exist so abundantly: wherever we have a "Serbonian bog" kind of mixture, there do we find that both animals and plants are few and undiversified. But the operations of physical geography are like the delicate mechanism of a well-constructed and highly sensitive watch. The slightest variation of the "regulator" causes it to run either too fast or too slow. So is it with the terrestrial portion of the earth's surface. The immense vapours lifted by solar heat from the surface of its vast waste of waters, are carried by the winds and drifted against mountain-sides or hill-summits. They there condense into rain, trickle down as mountain torrents, run together as rills and rivulets, join each other and form rivers, and anon they find themselves once again part of the huge volume of oceanic waters whence they were originally raised. Like the diurnal revolutions of the earth on its axis, which go on through a practical eternity of time, all things earthly revolve in cycles. The process just described has been going on after the same invariable order ever since God commanded the "dry land to appear."

In spite of the general law to which we have made allusion, that the water and the land should remain separated, the boundaries of the two are liable to a constant change. As a rule, it is the former which attacks the latter. Agencies far away removed may bring about an unsettlement of the mutual relationship, although there can be little doubt that man is a great interfering agent. His power for good and evil can be made manifest in the natural as in the moral world. As the area of dry land is only as one to three of that of water, it follows that the latter is always in a state representing that of an invader, and the former in that of defence. The fairest lands may be inundated and temporarily destroyed. Even in England we have had evidence repeatedly of the devastative power of water. Those terrible results of the bursting of the Holmfirth Reservoir in 1851, and of that at Sheffield in 1864, will have had too vividly impressed upon people's minds the destructive forces held in check wherever there are water-works. "Fire and water," says an old proverb, "are good servants, but bad masters!" Notwithstanding the wonderful manner with which the physical geography of the greater part of Europe has been subdued by man, his conquest has not been effectual. Every spring, news comes to us of the chief rivers overflowing their banks, and laying waste much of the fertile lands they traverse. It would almost seem as if the violence of waters poured down European rivers were greater than formerly. The question we have to ask is, whether man can really affect the arterial drainage of the country he inhabits? In many respects, it will be held that this is a most important subject for discussion. Apart from the actual destruction of property which takes place, we must not forget the actual physical and mental suffering which ensues. Those who have visited an inundated town, and seen the lowlying and poor parts of it so flooded that boats plied up the narrow and unsightly streets, and relieved the terrified and agonized inhabitants who had taken refuge in the upper rooms, will readily grant that, much as we have subordinated the forces of nature, we have not yet conquered them. At intervals of a few years we hear of floods in the south of France, especially in the neighbourhood of Toulouse, when scores of people are killed, cattle destroyed, property annihilated, and the hard earnings of a thrifty people vanish like vapour in a single night! Can nothing be done to stay this kind of thing? Are we to be year after year the victims of physical geographical circumstances of this kind? If the waters are not separated from the dry land, whose fault is it that the Divine fiat is not strictly carried out?

This brings us to note a few of the great floods which have become historical through the great destruction both of life and property which took place through their agency. It is to be doubted, however, whether the more insignificant occurrences of every winter and spring are not, in the long run, more devastatory. Almost every country watered by large rivers has a fluvial history of catastrophes associated with them. As the magnitude of a river is dependent upon the area of the watershed and the varying amount of the fall, it follows that the magnitude of river-overflows, or inundations, is usually proportionate to the size of the rivers themselves. Every year there are immense inundations in the valley of the Amazon and in the valleys of its tributary rivers. These are regarded as part of the Amazon river system, and a peculiar fauna and flora is more or less adapted to these conditions. The floods of our Indian rivers are also commensurate with the great magnitude of the streams, those of the Ganges being almost annual in their occurrence. The overflows of the Nile are historical, and occur as regularly as the return of summer. Such inundations as these can hardly be called cataclysmal. In the case of the Nile, the fertility of Egypt depends upon it; and an ancient system of agriculture has sprung up in that country thoroughly adapted to the physical effects produced annually by the Nile floods. As these inundations always occur where the land is lowest lying, and as the waters cover the low-lying land when they are discoloured with the muddy matters brought down by the rains from the higher grounds, it follows that in course of time such inundations create their own natural checks. Every year a thin layer of alluvium soil is strewn over the areas occupied by the inundations, and thus the low-lying or swampy land is eventually raised high enough to confine the swollen waters of the rivers to their natural channels. It was an ancient saying that "Egypt was the gift of the Nile," and geological investigation proves that the annually added loams brought down by that river are some hundreds of feet in thickness. In every river valley we have peculiar black soils known as *alluvium*, and these are known to be the accumulated sediments brought down in the course of ages by the adjacent rivers.

Rivers, therefore, may be recuperatory in their action, as well as destructive. But it is astonishing what even drops of rain-water can do when combined. Sir Charles Lyall mentions that when he was travelling in Georgia and Alabama, in 1846, he saw hundreds of valleys commenced where the ancient forests had been cut or burnt down. There can be little doubt that there is a distinct connection between the rain-fall and mean temperature of a country, and the number of trees the latter may possess. Trees are magnificent regulators of climature. They are to it what the pair of revolving "governor-balls" are to a stationary engine. When the engine is going too fast, the "governor-balls" distend, and "throttle" or compress the aperture whence the motive steam-power is issuing. When the engine is working slowly, the balls droop, and so open the valve as to allow more steam to issue. The same with the woods and forests of a country. When the rainy seasons are on, every tree and plant absorbs some of the moisture, and stores it away in its own tissues. It thus prevents great quantities from flowing off the surface, and gathering into rills and rivulets, and so swelling the main rivers as to cause them to overflow their lowest-lying banks. During periods of drought, the leaves of the same forest give out the moisture they consumed into the atmosphere, and so prevent its being as dry and parching as it otherwise would have been. During the hours of night, also, the surfaces of the leaves become colder than the air, and thus the moisture contained in the latter is condensed upon them as dews. In many parts of Arabia this is the only kind of waterfall with which the parched earth is visited.

The destruction of woods and forests, therefore, is always attended by an alteration of climate for the worse. It becomes more irregular. There are now alternate periods of rain-fall, and consequent floods, and of droughts. No country in the world has been so altered in this respect as the United States, for in no other has there been so much clearing of forest land within the last two centuries. In Italy great destruction of forests has taken place, and the summers are now more arid and hot in consequence. The reader will find in the Hon. G. Marsh's, "Physical Geography as influenced by Human Action," a long list of parallel cases, where men have unconsciously modified the climate of the country in which they dwelt. Sometimes this is for the better, as in the case of the fen districts of Cambridgeshire. These, which were watery wastes during the times of the early Norman kings, are now well drained; and rich crops smile and wave where once unbounded sheets of water stretched from horizon to horizon, only varied by islands of water-reeds. The temperature of the fen country is ten degrees higher than it was when undrained. The same fact is characteristic of the neighbourhood of Bolton, in Lancashire. The bogs and moors are drained and under cultivation, and the heat of the sun, once expended in lifting the waters from the surface by evaporation, is more usefully exercised in warming the land, and in thus promoting the growth of corn. Man, therefore, may "be a co-worker with God," even in his influence to modify for the better the climate of his native land! If it be a good thing to cause two blades of grass to grow where only one grew before, we hold that it is equally wise to plant trees where none grew before. Our English towns are often dreadfully unlovely, when they might be rendered bright and cheerful by such "boulevards" as we see in continental towns, as well as in those of North America. There is a tendency for us to recognize this fact, and tree-planting is taking place among us more than it ever has done. Let us thoroughly recognize its importance, in that trees not only beautify our towns, but they remove carbonic acid, and give out oxygen where that beneficent gas is most needed. Moreover, we have seen that the regularity of the order of climatal changes is much affected by the extent of arboreal growth; so that by planting more trees we may incidentally be the means of decreasing those floods and inundations we have been considering.

How the surface of the ground can be cut up by the heavy rains which take place after extensive forests have been cut down, is well shown in Lyell's "Principles." In Georgia, where the land was cleared of forest in 1826, the surface was quite level. Cracks began to occur in the ground when the latter was bared of timber. In these the sudden rains washed, so as to widen the walls, and wear them backward, until, in a short space of twenty years, a chasm had been formed nearly sixty feet deep, three hundred yards long, and from seven to sixty yards wide. In the neighbourhood of hills and mountains, in the northern hemisphere at least, the occurence of extensive peat-bogs may arrest the rapid flow of surface rain-water, and thus prevent floods; although the presence of these bogs must make the climate colder. In the north of Ireland we have most extensive peat-bogs, always soaked with water like a sponge. At the base of the highlands of Scotland, and of the Welsh and Cambrian mountains, we have extensive areas occupied by swamps. The rain-water trickling down the hard and imporous rocks would gather into rapid torrents, and be poured suddenly into the valleys, if it were not for the arresting power of these swamps. As it is, the sources of the rivers are usually from these bogs, where the excess of water is slowly and constantly given out. Nature works in many ways, and we have seen that much of the devastation of floods may be due to the thoughtlessness of man himself, in cutting down the forests which regulate the climate of a country, or in indolently neglecting to plant those trees which Providence has so plainly utilized as agents for good.

# ANIMAL DEFENCES.

ANY person who has had the opportunity of observing closely the habits and private life of animals must have often been astonished at the manner in which various creatures often combine, either for their mutual protection or mutual benefit.

We have, fortunately, had many opportunities of watching the conduct of various creatures in their native homes, and the delight that any lover of nature experiences in thus contemplating the wise acts of the animal creation far exceeds the savage joy of the mere slaughterer or sportsman.

Hours and days have been happily passed whilst watching the skilful golden oriels weaving their retort-shaped nests among the pendent branches which overhang an African stream. Often have we enjoyed a good laugh as we witnessed the futile attempts of an inexperienced grey monkey to grasp the nests of these birds and extract the eggs, attempts which almost always resulted in giving the adventurer a ducking. Whether we examine the skilled details of work shown in a beehive or ants' nest, the combined efforts of a pack of wild dogs to hunt down their prey, or the architectural skill of a village of beavers, we may invariably find traces of that same great Wisdom which holds a planet in its orbit and makes the world a sphere.

There are many creatures, however, to which we are not accustomed to attribute any special powers of skill or combination, and which we usually regard as stupid and almost unworthy of notice. Thus, who would be disposed to believe a rat a very clever fellow in his way, and able to plot and carry out a most formidable rebellion against a tyrant? Yet such a case happened within our own experience.

A friend of our own, a skilled naturalist, possessed a cat, which was rather old, though still strong and active. This cat was a terror to a colony of rats which inhabited the neighbouring pig-styes, banks, and hedgerows. Many a rat was brought by Pussy, and deposited with great pride at her master's feet.

During several weeks this sport continued, but one morning Pussy came in from her kennel looking dirty, rumpled, and scarcely able to crawl. The naturalist examined his pet, and his skilled eye soon saw the cause: Pussy was severely bitten by rats in twenty or thirty places. An examination of the scene of action plainly showed that there had been a battle royal. At least a dozen rats must have combined, and coming on Pussy in a body, had so punished her that she died a few days afterwards from the effects of their bites.

Not long since we were passing a poultry-yard in which were several turkeys and fowls. Whilst watching these creatures, a dispute occurred between a turkey and a hen relative to some food. The old hen cackled forth her displeasure loudly, when instantly the turkeys rushed to the scene of dispute and surrounded the disputing turkey. Forming a ring round him, they drooped their wings and lowered their heads, whilst they all uttered a low grumbling kind of sound. The turkey in their midst flapped its wings in a despairing manner, and "gobbled" loudly.

This strange scene lasted fully a minute, after which one of the largest turkeys jumped at the prisoner and pecked him severely; then another rushed at him, and so on, each turkey giving several pecks at the one that had evidently been tried by a jury of his fellows, and found guilty of trying to rob a hen of its food.

Among the larger animals such cases are by no means uncommon. Elephants, we have often heard, in their wild state signal to their fellows when danger is near; and we are convinced that these creatures have several calls, or trumpet-like sounds, which mean special things, such as "danger," "feed," "all right," etc.

A very curious case of a combination on the part of animals to rid themselves of a foe occurred near the Winterberg, a mountain to the north of the eastern frontier of the Cape of Good Hope.

In this locality there were several troops of baboons, young and old, which resided in the deep rocky ravines, and gambolled among the fearful precipices around. Very human were these creatures in their appearance and habits, especially when suddenly alarmed; the mammas were seen to catch up their young ones, who clung round their parents' necks, and were thus carried rapidly to the summit of the rocks, where they would grimace and cough out their defiance at the intruder who had ventured into their domain.

An enemy, however, once found his way into their stronghold, and this was an enemy hungry, cunning, and powerful. It was a Cape leopard. Crouching down among the long grass, or amidst the crevices of the rocks, the leopard would suddenly spring upon a young baboon, and actually devour it before the eyes of its screeching parents. Strong as is a baboon, the leopard is yet far stronger, and with its terrible claws could soon tear to pieces the largest male baboon.

During some days the leopard feasted on baboons, but at length these creatures combined, and jointly attacked the leopard. They did not really mean to risk a pitched battle with him, for these creatures evidently knew and respected his great powers. They had, too, as the result proved, determined on a safer and more crafty method of proceeding.

The leopard, fearing the combined strength of his adversaries, left their neighbourhood, and retreated across country, but he was followed by nearly all the large baboons.

On went the leopard; on followed the baboons. The day was hot, and the leopard disliked this perpetual tramping, and so tried to seek a retreat and lie down and rest. Then it was that the baboons closed round and worried him. Soon, too, he began to thirst, his tongue hanging out of his mouth, and the white foam covering his jaws.

Water was soon scented by the hunted brute, and to this it rapidly made its way; but now the baboons became frantic: they closed on to the leopard, some by their great activity actually tearing him with their sharp teeth, and the creature could not drink. The baboons could relieve one another, and some could eat and drink too whilst their companions continued worrying the leopard.

During two days and a night the country for several miles along the course of these creatures was startled by the cries of pursuer and pursued, and several farmers were witnesses from a distance of portions of the scene here described. They would not interfere, but watched the baboons' method of administering justice.

Worn-out with exhaustion and thirst, the leopard at length could totter on no farther, and sank to the ground a prey to the baboons, who, in spite of his claws and teeth, which were yet formidable, attacked him with their whole force and soon tore him to pieces, they themselves escaping with only a few severe scratches.

Assembling their forces, the baboons returned rapidly to their stronghold, where they were welcomed by their females and young with choruses of loud and triumphant barks, which were continued during the greater part of the night, whilst for several days the excitement did not seem to calm down, but was shown by the unusual noises which proceeded from this curious colony.

Such an incident as the preceding may seem strange and unlikely to those who have not seen animals in entire freedom and left to their own instinct or reason, but our personal experience on many other occasions has taught us that it is not uncommon, and we do not therefore hesitate to record it in these pages. Another singular incident was related to us by a credible witness.

Amidst the deeply wooded ravines of a range of mountains on the eastern frontier of the Cape of Good Hope, a large colony of the pig-faced baboons were located. These creatures had found there a safe resting-place for many generations; so steep and dangerous were the cliffs, that no creature except a baboon could dare wander among them, and so the animals were safe and happy.

The traveller in that wild region would find his arrival announced from hill-tops by a chorus of wild weird-like coughs or barks, whilst these semi-human animals could be seen on the side of naturally-formed walls, of a thousand feet deep, grimacing at and threatening the solitary traveller who had intruded into this domain; a domain of which a king might well be proud.

This part of Africa has been gifted with a lovely climate, and with an air that is inhaled with effects similar to those produced by quaffing champagne. No wonder that the chameleon is found in this neighbourhood—a creature fabulously said to feed on air—for it has a glorious feast if it feed on the scented air of the Amatola Mountains.

Here are steep rocky precipices; sheltered glens, each with bright flowery shrubs, whose purple and crimson blossoms give a distinct colouring to even the distant glens; whilst a sea of mighty hills roll one after the other, far as the eye can reach, boundless and desolate, yet lovely as a Paradise. It is amidst these regions that the grey vulture floats like a thistle-down high up in the heavens, where the eagle hisses through the air on his prey, and where the baboon scampers at will, the legitimate and hereditary possessor of the soil.

Human-like almost in form, the baboon seems nearly human in his passions, as the following anecdote will show.

Some miles from the Amatola, and separated from them by an intervening plain, was another rocky stronghold, in which another colony of baboons were located. These latter, to the inexperienced eye, showed no distinct peculiarities from their neighbours in the Amatola, yet there were men whose keen perceptions were able to discover distinct peculiarities between the two races, and to be able to tell which was an Amatola baboon and which a denizen of the Chumie.

The baboons themselves did not fraternize, and if by chance stray baboons from each colony met one another in their wanderings, a regular fight ensued.

When the sun sank beneath the Chumie hills, the baboons from that region would sit on the most giddy precipices, and bark forth a defiance to the distant mountains. In that clear atmosphere sound travels a long distance, and is heard during a still evening at almost fabulous distances. Thus the barks and the coughs of the baboons at one district were heard and replied to by the creatures some miles off, in the Amatola.

To the uninitiated these mere animal barks seemed to mean nothing, but to the keen ears and comprehensive senses of the baboons they conveyed the direst insults and most defiant challenges.

Human nature has its limits of endurance, and so has baboon nature; thus, after a particularly warm summer day, during which, probably, the creatures' blood became additionally heated, the evening challenges were given and answered with unusual vehemence. The moon rose bright and full, and the night was calm and lovely, and it seemed strange that all nature should not be at peace; but shortly after midnight the Chumie rocks and precipices resounded with screams, barks, and fiend-like sounds, as though a legion of demons had broken loose and were fighting among themselves.

For hours these fearful sounds were heard, and the few settlers then in that neighbourhood listened with astonishment, not knowing whether these noises indicated a coming storm of unusual force, or were the indicators of some convulsion of nature. Towards daybreak they gradually ceased, and the men whose rest had been disturbed armed themselves and cautiously approached the scene of the midnight tumult.

The cause of the alarming disturbance was then manifest. The baboons of the Amatola had long borne the challenges and insults of their neighbours of the Chumie; they had listened to their taunts, and had burned with a desire for vengeance. At length an attack was organized, and on the night in question the male baboons of the Amatola assaulted the colony of the Chumie, and a fearful fight ensued.

The baboon's method of attack is singular and formidable; his muscular power is enormous, whilst the crushing power of his jaws is inferior to that of many smaller animals: when once he grips with his jaws, however, he can hold on, and so he combines his powers, by seizing his antagonist with his teeth, grasping him at the same time with his powerful arms, and then pushing him from him, so that he tears out the piece which he has in his mouth. By this means we have seen a large dog so maimed in a few seconds by a baboon, that the former was obliged to be shot, as there was no hope of its recovery.

The result of the night attack which we have described was, that nearly one hundred baboons were found dead or dying by the hunters who visited the scene of action, whilst it was remarked that the coughs and barks which had previously disturbed the evenings almost entirely ceased, as though each party had gained a certain amount of respect for the other by the experience gained during the midnight battle.

In England we may often see a combination formed by small birds, who chase a hawk or kestrel which has invaded their domain: the result usually is that the bird of prey retreats, though it be powerful enough to crush with its talons any one of its pursuers.

Hence we see the practical result of a combination against a foe or a difficulty, when we note the habits of the animal kingdom, and thus we may learn how much may be accomplished by ourselves when we combine hearts and hands for the general good: families united for one purpose, men working as brothers with one aim, and a nation combined for a nation's good. Whilst unity gives strength and power, and defies a foe, the house divided against itself shall fall.

www.ingramcontent.com/pod-product-compliance
Lightning Source LLC
Chambersburg PA
CBHW030347230426
43664CB00007BB/557